Particle-Stabilized Emulsions and Colloids

Formation and Applications

RSC Soft Matter Series

Series Editors:
Hans-Jürgen Butt, *Max Planck Institute for Polymer Research, Germany*
Ian W. Hamley, *University of Reading, UK*
Howard A. Stone, *Princeton University, USA*
Chi Wu, *The Chinese University of Hong Kong, China*

Titles in this Series:
1: Functional Molecular Gels
2: Hydrogels in Cell-Based Therapies
3: Particle-Stabilized Emulsions and Colloids: Formation and Applications

How to obtain future titles on publication:
A standing order plan is available for this series. A standing order will bring delivery of each new volume immediately on publication.

For further information please contact:
Book Sales Department, Royal Society of Chemistry, Thomas Graham House, Science Park, Milton Road, Cambridge, CB4 0WF, UK
Telephone: +44 (0)1223 420066, Fax: +44 (0)1223 420247
Email: booksales@rsc.org
Visit our website at www.rsc.org/books

Particle-Stabilized Emulsions and Colloids
Formation and Applications

Edited by

To Ngai
The Chinese University of Hong Kong, Shatin, China
Email: tongai@cuhk.edu.hk

Stefan A. F. Bon
University of Warwick, Coventry, UK
Email: s.bon@warwick.ac.uk

THE QUEEN'S AWARDS
FOR ENTERPRISE:
INTERNATIONAL TRADE
2013

RSC Soft Matter No. 3

Print ISBN: 978-1-84973-881-1
PDF eISBN: 978-1-78262-014-3
ISSN: 2048-7681

A catalogue record for this book is available from the British Library

Published by The Royal Society of Chemistry,
Thomas Graham House, Science Park, Milton Road,
Cambridge CB4 0WF, UK

Registered Charity Number 207890

For further information see our web site at www.rsc.org

Printed and bound by CPI Group (UK) Ltd, Croydon, CR0 4YY

Preface

The stabilization of oil and water droplets by solid particles has been known for over a century. Today, these particle-stabilized emulsions are generally referred to as Pickering emulsions. They are encountered in personal care products, the food industry and have long been used in oil recovery and mineral processes. The generality of emulsion or foam stability against coalescence has been well demonstrated by the use of many different types of particles. In addition, many studies have been devoted to the elucidation of the mechanisms behind particle stabilization by focusing on the influence of particle size, hydrophobicity, surface roughness and shape. The current challenge entails translating knowledge about emulsion stabilization by colloidal particles into useful industrial applications. This book will discuss very recent studies on both fundamental properties of particles at fluid interfaces and their emerging applications.

After a brief introduction in Chapter 1 by Bon, subsequent chapters are devoted to various groups of target formation and application of Pickering emulsions. Park *et al.* (Chapter 2) give a comprehensive overview of the interactions and micromechanics of various colloids confined at fluid–fluid interfaces. Zhao and Tian (Chapter 3) discuss emulsions and colloidal particles stabilized by clay layers with different polymer brushes, amphiphilic gold nanoparticles and Janus structures. Bon (Chapter 4) describes Pickering suspension, mini-emulsion and emulsion polymerization using solid particles as the efficiency stabilizers, while Li and Ngai (Chapter 5) discuss using soft, stimuli-sensitive microgel particles in controlling emulsions. Clegg *et al.* (Chapter 6) describe the current state of bicontinuous interfacially jammed emulsion gel (bijel) research. Nonomura (Chapter 7) and Wang *et al.* (Chapter 8) discuss multiple, non-spherical emulsions and bicontinuous structures stabilized by solid particles as well applications of these complex systems in the area of functional materials. Biggs and Cayre

RSC Soft Matter No. 3
Particle-Stabilized Emulsions and Colloids: Formation and Applications
Edited by To Ngai and Stefan A. F. Bon
© The Royal Society of Chemistry 2015
Published by the Royal Society of Chemistry, www.rsc.org

(Chapter 9) describe recent academic work that has used Pickering emulsions as precursor structures for the preparation of hollow spheres or microcapsules. Pichot *et al.* (Chapter 10) discuss the recent advances in terms of both food-grade solid particle formation and structuring of emulsions using particles suitable for use in food. Xu *et al.* (Chapter 11) describe the use of Pickering emulsions in advanced oil sands extraction and mineral processes.

The colloidal particles trapped at liquid–liquid interfaces will remain a rich topic in the future as many aspects of their behaviour remain poorly understood. This book does not offer a complete description of particle-stabilized emulsions and their properties; instead, it provides a snapshot of some of the recent achievements in both fundamental properties of particles at fluid interfaces and their new applications. We are sure that this book will serve to provide a background for researchers and graduate students working in this important area and also to provide a source of inspiration for future work in this field.

We would like to sincerely acknowledge our colleagues and friends who have contributed with passion and expertise to this book. In addition, our thanks go to the editorial team from RSC Publishing for their assistance in preparing this book.

To Ngai
Stefan A. F. Bon

Contents

RSC Soft Matter No. 3
Particle-Stabilized Emulsions and Colloids: Formation and Applications
Edited by To Ngai and Stefan A. F. Bon
© The Royal Society of Chemistry 2015
Published by the Royal Society of Chemistry, www.rsc.org

Chapter 11 Particle-Stabilized Emulsions in Heavy Oil Processing 283
David Harbottle, Chen Liang, Nayef El-Thaher, Qingxia Liu, Jacob Masliyah and Zhenghe Xu

CHAPTER 1

The Phenomenon of Pickering Stabilization: A Basic Introduction

STEFAN A. F. BON

Department of Chemistry, University of Warwick, Coventry CV4 7AL, UK
Email: s.bon@warwick.ac.uk

1.1 A Brief Historic Perspective on Pickering Stabilization

The ability of solid particles to adhere to soft deformable interfaces, for example to the surface of emulsion droplets or bubbles, is currently the subject of renewed interest in material science. The phenomenon that solid particles can reside at the interface of droplets and bubbles, thereby providing them with resistance against coalescence or fusion, and (debatable) coarsening or Ostwald ripening, is known as Pickering stabilization and named after Spencer Umfreville Pickering.[1] Food science and flotation technology show a steady stream of research over the 20th century using Pickering stabilization in, for example, table spread/margarine formulations, where fat crystals sit on the surface of water droplets dispersed into the oil matrix. Interestingly, the origins of Pickering stabilization in the area of (froth) flotation lie further back than the cited works by Pickering (1907) and Ramsden (1903).[1,2] Patents by William Haynes[3] (1860) and the Bessel brothers[4] (1877) clearly reported the phenomenon, the latter patent interestingly illustrating the concept with graphite flakes attached to bubbles.

RSC Soft Matter No. 3
Particle-Stabilized Emulsions and Colloids: Formation and Applications
Edited by To Ngai and Stefan A. F. Bon
© The Royal Society of Chemistry 2015
Published by the Royal Society of Chemistry, www.rsc.org

In the area of polymer chemistry the idea of using solid particles as stabilizers for the fabrication of polymer beads by suspension polymerization was explored to some extent from the 1930s to the 1950s.[5-7] A revival of the concept of using solid particles as stabilizers in heterogeneous polymerizations did not emerge until 50 years later with the development of Pickering mini-emulsion polymerization[8-10] and Pickering emulsion polymerization.[11-13] The idea of using Pickering stabilization as a way of assembling colloidal particles into intricate supracolloidal structures drew attention from the soft matter physics crowd initiated by the works of Velev *et al.*[14-16] and Dinsmore and coworkers,[17] the latter coining the term 'colloidosomes' for the semi-permeable hollow structures made by assembly of particles onto droplets. Not only does the fabrication of supracolloidal structures receive great attention, but also the underlying physics is studied and discussed widely, for example looking at why particles adhere to a liquid–liquid interface, how strong the interaction energy is, and what the interplay between particles at the droplet surface is.

1.2   A Basic Physical Understanding of Pickering Stabilization

This short introductory chapter does not aim to provide a thorough literature review of the underlying physics of Pickering stabilization, but merely to give the reader a basic understanding. The question of why a particle would prefer to sit at the interface of an emulsion droplet instead of being dispersed in either the water or oil phase has already been raised and discussed by, for example, Hildebrand and coworkers in 1923.[18] They said that for solid particles to adhere to and be collected at the surface of emulsion droplets, the powder had to be wetted by both liquids. They stated that in general particles have a preference for one of the two liquids, which meant that the particles would reside for longer in that liquid. They described how the assembly of particles onto the oil–water interface will cause the interface to bend in the direction of the more poorly wetting liquid, thereby facilitating its emulsification into droplets. They concluded that the type of emulsion, *i.e.* oil-in-water or water-in-oil, could be predicted on the basis of this wettability, and thus on the basis of the contact angle of the interface with the solid.

To describe the behaviour of a single particle at the liquid–liquid interface we often see the following expression for the adhesion energy as a function of the contact angle:

$$\Delta E = \pi R^2 \sigma_{12} (1 \pm \cos\theta_{12})^2$$

Care must be taken to define the right contact angle and whether to use a plus or minus sign within the brackets to calculate the escape energy needed to remove the particle from the interface and place it into either phase 1 or 2. This can lead to confusion. An approach that circumvents this issue is to

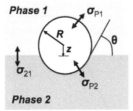

Figure 1.1   Schematic representation of a spherical particle at a liquid–liquid interface.

calculate the energy well completely, as reported by Pieranski in 1980.[19] In his work, he studied the adhesion of polystyrene spheres at the air–water interface.

Imagine a thermodynamic type of experiment as shown in Figure 1.1. We take a perfectly smooth spherical particle that we disperse in a liquid phase, which we call phase 1. We ignore all dynamics (kinetics) and external force fields, such as gravitational, electrical, optical and magnetic. We also do not consider any surface roughness of the particle, and we ignore any ionic (Coulombic) interactions, dielectric effects and thus van der Waals interactions. The question we pose is what would happen to the free energy if we move the particle, which is dispersed in phase 1, all the way to phase 2, and thus through the interface?

To answer this question we need to take into account the interfacial energies upon placing the particle at various heights, z. These are the energy between the particle and phase 1, E_{P1}, the energy between the particle and phase 2, E_{P2}, and the energy between phase 1 and phase 2, E_{12}. For this we need to know the interfacial tensions between the particle and phase 1, σ_{P1}, the particle and phase 2, σ_{P2}, and the interfacial tension between phase 1 and phase 2, σ_{12}, and multiply these by the respective contact areas.

$$z_0 = \frac{z}{R}; \; S_T = 4\pi R^2; \; A_T = \pi R^2$$

$$E_{P1} = \sigma_{P1} S_T \frac{(1 + z_0)}{2}$$

$$E_{P2} = \sigma_{P2} S_T \frac{(1 - z_0)}{2}$$

$$E_{12} = -\sigma_{12} A_T (1 - z_0^2)$$

The sum of the three energies and its division by $k_B T$ leads to the following quadratic expression for the energy well (see also Figure 1.2):

$$E_0 = \frac{E_{P1} + E_{P2} + E_{12}}{k_B T} = \left[\frac{\pi R^2 \sigma_{12}}{k_B T}\right] \left(z_0^2 + \frac{2(\sigma_{P1} - \sigma_{P2})}{\sigma_{12}} z_0 + \frac{2(\sigma_{P1} + \sigma_{P2})}{\sigma_{12}} - 1\right)$$

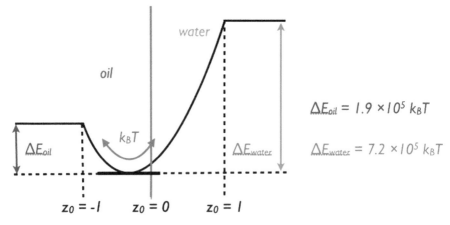

Figure 1.2   Schematic representation of the quadratic energy well for a polystyrene spherical particle (radius = 100 nm) at a water–hexadecane interface.

The equilibrium position of the particle can be easily found from $dE_0/dz_0 = 0$, which yields:

$$z_0^{min} = \frac{\sigma_{P2} - \sigma_{P1}}{\sigma_{12}}$$

For values of z_0^{min} between -1 and 1 the particle adheres to the liquid–liquid interface. The energy it will take to remove the particle from the interface into either the bulk of phase 1 or phase 2 can easily be obtained from:

$$\Delta E_1 = E_0(z_0 = 1) - E_0(z_0^{min})$$
$$\Delta E_2 = E_0(z_0 = -1) - E_0(z_0^{min})$$

An example for a polystyrene sphere of radius 100 nm at the water–hexadecane interface (water = phase 1) using $\sigma_{P1} = 32$ mN m^{-1}, $\sigma_{P2} = 14.6$ mN m^{-1} and $\sigma_{12} = 53.5$ mN m^{-1} yields values needed for the particle to escape the well and enter the water and oil phase of 7.2×10^5 $k_B T$ and 1.9×10^5 $k_B T$, respectively.

This continuous model gives a good first approximation for the magnitude of the energy well in which the particles are trapped at the interface, and an idea of what energy is needed for a particle to escape from the interface into one of the two liquid phases. It is, however, a rather crude estimation of reality, not least because of the number of assumptions made under which the model is valid (remember no gravity, no charges, no dynamics, *etc.*), and because we assume that the particle is a smooth perfect sphere.

In addition to these restrictions the careful observer can also see from Figure 1.1 that in this 2D picture we have overlooked two points at which three phases are in direct contact with each other. This three-phase inter-action in 3D is a contact line in the form of a circle between the two liquids and the particle. Gibbs already pointed out that qualitatively this three-phase contact line could be seen as a one-dimensional equivalent of the surface tension (for the latter multiplying with contact area gives interfacial energy), and is referred to as line tension, τ (multiplying by the circumference of the circle gives energy). Aveyard and Clint included line tension in the Pieranski model by providing this extra term:[20]

$$E^{\text{line}} = \frac{2\pi R \tau}{k_B T} \sqrt{(1 - z_0^2)}$$

1.2.1 Pickering Stabilizers of Nanoscale Dimensions

Line tension becomes important for smaller spherical particles as it scales linearly with the radius of the particle, whereas surface tension scales quadratically. The experimental difficulty that remains is to determine qualitative values for line tension. Questions that arise when we shrink our particle to nanoscale dimensions are: can we still assume that the liquid–liquid interface is flat and can we still assume continuous wetting? Cheung and Bon carried out molecular simulations and showed that these as-sumptions break down for small particles and that capillary waves on the liquid–liquid interface and discrete wetting of the particle by individual molecules needs to be taken into account.[21] The non-flat nature of the liquid–liquid interface widens the interaction distance and thus broadens the energy well in comparison the Pieranski model, whereas the combined effects of line tension with discrete wetting led to deeper energy wells, with deviations of up to 50% in adhesion energy, showing that nanoparticles stick considerably better to interfaces than can be predicted on the basis of the Pieranski model.

1.2.2 Pickering Stabilizers with a Rough Surface

When we add roughness to a spherical Pickering stabilizer we can see that the total contact area between the particle and the liquid phase(s) increases dramatically, whereas the 'circular' area that get taken away by placing the particle at the liquid–liquid interface remains relatively unchanged (see it as noise over a circle that cancels out). In systems where we ignore the effects of line tension we can see that the energy levels for dispersing the particle in phase 1 and 2 have considerably higher values. The net effect is that the energy needed for the particle to escape the interface drops considerably. This means that rough particles do not adhere strongly to the interface. This effect is nicely observed experimentally in work by Ballard and Bon for Lycopodium spores that were decorated in a mesh of interpenetrating poly-mer nanoparticles and used as patchy Pickering stabilizers.[22] Theoretical

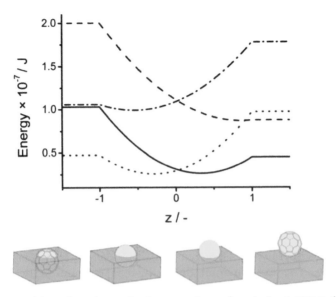

Figure 1.3   Position of a sphere of polystyrene (▬▬) and of polyHEMA (•••) and a buckyball type structure of polystyrene (— — —) and polyHEMA (—•—) at the oil–water interface and illustration of predicted contact angles for (from left to right) a buckyball type structure of polyHEMA, a polyHEMA sphere, a polystyrene sphere and a polystyrene buckyball structure. Reprinted with permission from reference 22. © 2011 Royal Society of Chemistry.

comparison between smooth and rough particles (modelled with a buckyball type mesh) indeed led to the behaviour explained above (see Figure 1.3).

1.2.3   Non-Spherical Pickering Stabilizers

The take-home message is that non-spherical particles have the potential to adopt multiple orientations once adhered to a liquid–liquid interface. In order to understand and verify experimentally observed orientations it is important to calculate the entire energy landscape because, in addition to thermodynamic minima, kinetically trapped states (unexpected orientations) can occur. In our work we showed that an isolated microscopic haematite particle of super-ellipsoidal shape adopts three orientations when adhered to a hexadecane–water interface.[23]

Two of the orientations, and estimates of their relative populations, can be assigned to two thermodynamic minima on the energy landscape. This was determined by both free-energy minimization and particle trajectory simulations. The third orientation was found to correspond to a kinetically trapped state, existing on certain particle trajectories in a region of a negligible gradient in free energy.

References

1. S. U. Pickering, *J. Chem. Soc. Trans.*, 1907, **91**, 2001–2021.
2. W. Ramsden, *Phil. Trans. R. Soc. London*, 1903, **72**, 156–164.
3. W. Haynes, 1860, Berlin Patent 488. February 23.
4. G. Bessel, 1877, Berlin Patent 42. July 2.
5. von E. Trommsdorff, *Die Makromol. Chemie*, 1954, **13**, 76–89.
6. O. Röhm and E. Trommsdorff, 1939, US Patent 2,171,765.
7. W. P. Hohenstein, 1950, US Patent 2,524,627.
8. S. Cauvin, P. J. Colver and S. A. F. Bon, *Macromolecules*, 2005, **38**, 7887–7889.
9. S. A. F. Bon and P. J. Colver, *Langmuir*, 2007, **23**, 8316–8322.
10. S. Fortuna, C. A. L. Colard, A. Troisi and S. A. F. Bon, *Langmuir*, 2009, **25**, 12399–12403.
11. P. J. Colver, C. A. L. Colard and S. A. F. Bon, *J. Am. Chem. Soc.*, 2008, **130**, 16850.
12. C. A. L. Colard, R. F. A. Teixeira and S. A. F. Bon, *Langmuir*, 2010, **26**, 7915–7921.
13. R. F. A. Teixeira, H. S. McKenzie, A. A. Boyd and S. A. F. Bon, *Macromolecules*, 2011, **44**, 7415–7422.
14. O. D. Velev, K. Furusawa and K. Nagayama, *Langmuir*, 1996, **12**, 2374–2384.
15. O. D. Velev, K. Furusawa and K. Nagayama, *Langmuir*, 1996, **12**, 2385–2391.
16. O. D. Velev and K. Nagayama, *Langmuir*, 1997, **13**, 1856–1859.
17. A. D. Dinsmore, M. F. Hsu, M. G. Nikolaides, M. Marquez, A. R. Bausch and D. A. Weitz, *Science*, 2002, **298**, 1006–1009.
18. P. Finkle, H. D. Draper and J. H. Hildebrand, *J. Am. Chem. Soc.*, 1923, **45**, 2780–2788.
19. P. Pieranski, *Phys. Rev. Lett.*, 1980, **45**, 569–572.
20. R. Aveyard and J. H. Clint, *J. Chem. Soc. Faraday Trans.*, 1996, **92**, 85.
21. D. Cheung and S. Bon, *Phys. Rev. Lett.*, 2009, **102**, 066103.
22. N. Ballard and S. A. F. Bon, *Polym. Chem.*, 2011, **2**, 823.
23. A. R. Morgan, N. Ballard, L. A. Rochford, G. Nurumbetov, T. S. Skelhon and S. A. F. Bon, *Soft Matter*, 2013, **9**, 487.

CHAPTER 2

Interactions and Conformations of Particles at Fluid-Fluid Interfaces

BUM JUN PARK,*[a] DAEYEON LEE[b] AND ERIC M. FURST[c]

[a] Department of Chemical Engineering, Kyung Hee University, Yongin-si, Gyeonggi-do, 446-701, South Korea; [b] Department of Chemical and Biomolecular Engineering, University of Pennsylvania, Philadelphia, PA 19104, USA; [c] Department of Chemical and Biomolecular Engineering, University of Delaware, Newark, DE 19716, USA
*Email: bjpark@khu.ac.kr

2.1 Introduction

Colloidal particles will typically adsorb strongly and irreversibly to fluid–fluid interfaces.[1,2] Adsorption enables colloids to stabilize interfaces, opening up new potential applications for conventional colloids as alternatives for expensive and potentially environmentally hazardous molecular amphiphiles (*i.e.* surfactants).[1,2] An emulsion system in which solid particles are used as surface active additives is called a Pickering–Ramsden emulsion.[3,4] The presence of solid particles leads to lowering of the interfacial tension between the immiscible fluid phases and imparts kinetic stability to the emulsion droplets, impeding their flocculation, coalescence and consequent creaming.

There are several advantages of studying two-dimensional (2D) colloidal systems. First, 2D colloids are excellent models for well-defined three-phase systems and enable one to evaluate particle behaviours at multi-phasic fluid interfaces.[2,5] Second, the strong adsorption of colloids to 2D interfaces

RSC Soft Matter No. 3
Particle-Stabilized Emulsions and Colloids: Formation and Applications
Edited by To Ngai and Stefan A. F. Bon
© The Royal Society of Chemistry 2015
Published by the Royal Society of Chemistry, www.rsc.org

suppresses their out-of-plane movements to allow the direct visualization of microstructure and rheology.[6–9] Third, the interactions between particles and the microstructures that follow (including two-dimensional crystals, gels and fluid-like phases) can be easily manipulated by the addition of additives, such as electrolytes and surfactants.[9–12] This in turn alters the particle wettability, the interfacial tension and the rheological properties of the interface. This chapter aims to describe and understand the interactions, configurations, microstructures and micromechanics of various types of colloidal particles confined at fluid–fluid interfaces. This chapter also connects small-scale measurements between individual particles to their macroscopic phase behaviour and rheological properties.

2.2 Theoretical Background

2.2.1 Particle Wettability and Attachment Energy

2.2.1.1 Homogeneous Particles

For a colloidal particle that is a few micrometres or less in diameter, its attachment energy to a fluid–fluid interface can be calculated by considering the surface area (S) exposed to each fluid phase, the volume (V) confined in each fluid, and the corresponding surface tension (γ). In this case, the interface remains flat because the Bond number ($Bo = \rho g a^2 / \gamma$ where g is the gravitational acceleration, a is the particle radius, ρ is the density of fluid), which is the ratio of the gravitational force and the surface tension, is sufficiently small. For a particle that adsorbs to the oil–water interface from the aqueous phase, the attachment energy is given by the difference in the energy of the system before and after the adsorption (Figure 2.1)[2]

$$\Delta E_{Iw} = E_I - E_w \tag{2.1}$$

The first term on the right-hand side (E_I) is the energy when the particle is located at the interface and can be expressed as:

$$E_I = S_w \gamma_{sw} + S_o \gamma_{so} + S_I^{(2)} \gamma_{ow} + V_w P_w + V_o P_o \tag{2.2}$$

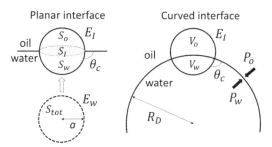

Figure 2.1 Schematics of the attachment energy of a homogeneous particle to an oil–water interface.

The second term (E_W) indicates the energy when the particle is completely immersed in the water phase:

$$E_W = S_{tot}\gamma_{sw} + S_I^{(1)}\gamma_{ow} + V_{tot}P_w \tag{2.3}$$

where the total surface area of the particle is $S_{tot} = S_w + S_o$. Note that $S_I^{(2)}$ and $S_I^{(1)}$ are the surface area of the oil–water interface with and without the particle. The displaced area, which is the cross-sectional area displaced by the presence of the particle at the interface, is defined as $S_I = S_I^{(1)} - S_I^{(2)}$. Therefore, by combining Equation (2.1) with the Young's equation, $\gamma_{ow}\cos\theta_c = \gamma_{so} - \gamma_{sw}$, where θ_c is the three-phase contact angle, the attachment energy can be expressed as:

$$\Delta E_{Iw} = S_o(-\gamma_{sw} + \gamma_{so}) - S_I\gamma_{ow} + V_o(P_o - P_w)$$
$$= \gamma_{ow}(S_o \cos\theta_c - S_I - 2V_o/R_D) \tag{2.4}$$

The Laplace pressure across the oil–water interface is given by $P_w - P_o = 2\gamma_{ow}/R_D$, where R_D is the radius of curvature (Figure 2.1). Similarly, the attachment energy from the oil phase to the oil–water interface is:

$$\Delta E_{Io} = -S_w(-\gamma_{sw} + \gamma_{so}) - S_I\gamma_{ow} - V_w(P_o - P_w)$$
$$= -\gamma_{ow}(S_w \cos\theta_c + S_I - 2V_w/R_D) \tag{2.5}$$

In a particular case when a spherical particle is attached to a planar oil–water interface, the substitution of the geometrical relations yields:[1]

$$\Delta E_{Iw} = -\pi a^2\gamma_{ow}(1 - \cos\theta_c)^2 \quad \text{from the water phase} \tag{2.6}$$

$$\Delta E_{Io} = -\pi a^2\gamma_{ow}(1 + \cos\theta_c)^2 \quad \text{from the oil phase} \tag{2.7}$$

The attachment energy of a spherical particle with radius of $a = 1$ µm at an oil–water interface with $\gamma_{ow} \approx 50$ mN m^{-1} is approximately 10^8 k_BT, which is significantly larger than the room temperature thermal energy, k_BT, suggesting that the adsorption of the particle to the interface is strong and essentially irreversible.

2.2.1.2 Amphiphilic Janus Particles

Janus particles with two different sides (*e.g.* one is polar and the other is apolar) can also be used as effective solid surfactants. The wettability of each side can be characterized by the three-phase contact angle (θ_A or θ_P) of a homogeneous sphere, as shown in Figure 2.2. The corresponding attachment energy is obtained using a similar approach to that in the

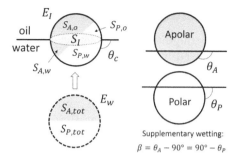

Figure 2.2 Schematics of the attachment energy for Janus particles.

previous section. In this case, the energies of a particle confined at the oil–water interface and immersed in the water phase are expressed as:[13–18]

$$E_{\mathrm{I}} = S_{\mathrm{Ao}}\gamma_{\mathrm{Ao}} + S_{\mathrm{Aw}}\gamma_{\mathrm{Aw}} + S_{\mathrm{Po}}\gamma_{\mathrm{Po}} + S_{\mathrm{Pw}}\gamma_{\mathrm{Pw}} + S_{\mathrm{I}}^{(2)}\gamma_{\mathrm{ow}} + V_{\mathrm{w}}P_{\mathrm{w}} + V_{\mathrm{o}}P_{\mathrm{o}} \qquad (2.8)$$

$$E_{\mathrm{w}} = S_{\mathrm{A,tot}}\gamma_{\mathrm{Aw}} + S_{\mathrm{P,tot}}\gamma_{\mathrm{Pw}} + S_{\mathrm{I}}^{(1)}\gamma_{\mathrm{ow}} + V_{\mathrm{tot}}P_{\mathrm{w}}. \qquad (2.9)$$

The subscripts A and P indicate the apolar (or non-polar) and polar surfaces of the particles, respectively (Figure 2.2). By combining the Laplace pressure and the Young's equations, $\gamma_{\mathrm{ow}}\cos\theta_{\mathrm{P}} = \gamma_{\mathrm{Po}} - \gamma_{\mathrm{Pw}}$ and $\gamma_{\mathrm{ow}}\cos\theta_{\mathrm{A}} = \gamma_{\mathrm{Ao}} - \gamma_{\mathrm{Aw}}$, for the two polar and apolar surfaces, respectively, the resulting attachment energy from each fluid phase is given by:

$$\Delta E_{\mathrm{Iw}} = \gamma_{\mathrm{ow}}\left(S_{\mathrm{Ao}}\cos\theta_{\mathrm{A}} + S_{\mathrm{Po}}\cos\theta_{\mathrm{P}} - S_{\mathrm{I}} - \frac{2V_{\mathrm{o}}}{R_{\mathrm{D}}} \right) \quad \text{from the water phase} \quad (2.10)$$

$$\Delta E_{\mathrm{Io}} = -\gamma_{\mathrm{ow}}\left(S_{\mathrm{Aw}}\cos\theta_{\mathrm{A}} + S_{\mathrm{Pw}}\cos\theta_{\mathrm{P}} + S_{\mathrm{I}} - \frac{2V_{\mathrm{w}}}{R_{\mathrm{D}}} \right) \quad \text{from the oil phase} \quad (2.11)$$

To determine the surface area of each side in contact with each fluid phase, a numerical method based on the hit-and-miss Monte Carlo method has been developed.[16,17] A sufficiently large number of points (N_{tot}) are homogeneously generated on the particle surface. The fraction of the points ($P_{\mathrm{ij}} = N_{\mathrm{ij}}/N_{\mathrm{tot}}$) on i-surface (apolar or polar) in contact with j-fluid is numerically calculated. The surface area is then given by $S_{\mathrm{ij}} = P_{\mathrm{ij}}S_{\mathrm{tot}}$. In this process, it is assumed that the contact line between the particle and the interface is smooth. The effect of line tension at the three-phase contact line is negligible for particles that are larger than a few nanometres in size. Gravity-induced interface deformation is also negligible in the small Bond number regime.[13,14] Interface deformation around non-spherical Janus particles could possibly occur. However, theoretical and experimental studies demonstrated that the flat interface assumption on the attachment energy calculation is valid to facilitate the determination of equilibrium configurations resulting from the energy minima in the energy profiles.[16,17,19,20]

2.2.2 Interactions Between Particles

The classical DLVO (Derjaguin–Landau–Verwey–Overbeek) description of colloidal interactions, based on an electrostatic double layer repulsion and van der Waals attraction, captures many of the behaviours of colloids dispersed in a single fluid medium.[21] In aqueous suspensions, these interactions persist over tens to hundreds of nanometres. In contrast, the interactions observed between colloids at fluid–fluid interfaces are abnormally strong over separations that are tens to hundreds of times larger compared to those in the single fluid phase. Clearly, DLVO models of the interactions do not capture the complete interaction potential between particles in 2D systems.[2] Two interactions at the 2D interface are of principal concern: electrostatic interactions and capillary interactions.

2.2.2.1 *Electrostatic Repulsive Interactions*

The abnormally long-ranged repulsive interactions between colloids at a fluid–fluid interface are due to charge dissociation in the polar solution (*e.g.* water) and the presence of this dissociated charge near the low permeability (non-polar) medium. This mechanism was first proposed by Pieranski and followed by an analysis by Hurd.[22,23] The asymmetric surface charge distribution across the interface leads to a dipole perpendicular to the interface which leads to the strong repulsive interactions between particles.

Here we summarize the key points of Hurd's derivation. First, the analytical expression of the Hurd model is obtained by solving the Poisson–Boltzmann (P–B) equation at an air–water interface.[22,24] There are several assumptions: (i) negligible ion–ion correlations in a dilute electrolyte concentration; (ii) over large separations ($r \gg a$), the charge can be approximated by point charges; (iii) the linearized Debye–Hückel approximation is valid for the low surface charge density of particles. Taking this latter approximation, the asymptotic form of the interaction between two point charges located at the interface with large separations is given by:

$$U(r) = \frac{2\mu^2}{r^3} \tag{2.12}$$

where the effective dipole is $\mu = q/(\varepsilon_w \kappa)$ and the inverse Debye screening length is $\kappa = \sqrt{(1000 e^2 N_A 2I)/(\varepsilon_w \varepsilon_0 k_B T)}$. N_A is Avogadro's number, ε_0 is the vacuum permittivity, ε_w is the dielectric constant of water, and I is the ionic strength of the solution. At an oil–water interface, Equation 2.12 can be rewritten:

$$F_{el}(r) = -\frac{dU_{el}(r)}{dr} = \frac{3\varepsilon_{oil} q^2 \kappa^{-2}}{2\pi \varepsilon_0 \varepsilon_w^2 r^4} \tag{2.13}$$

For particles with a high surface charge, the counterion concentration near the particle surface is not described by the Debye–Hückel approximation due to the significant non-linearity of the P–B equation.[21] In this case, the surface charge should be renormalized by considering an effective charge (q_{eff}) which replaces the original bare charge (q) in Equation 2.13:[25]

$$F_{renor}(r) = \frac{3\varepsilon_{oil}q_{eff}^2\kappa^{-2}}{2\pi\varepsilon_0\varepsilon_w^2 r^4} = F_0\left(\frac{a}{r}\right)^4 \tag{2.14}$$

The effective charge is given by:

$$q_{eff} \approx \frac{4\pi\varepsilon_0\varepsilon_w a^2\kappa(1+\cos\theta_c)k_B T}{e}\ln\left(\frac{\sigma_s^*}{\kappa^*}\right) \tag{2.15}$$

where the reduced inverse screening length is $\kappa^* = \kappa a$ and the reduced charge density is $\sigma_s^* = \sigma_s(ea)/(k_B T\varepsilon_0\varepsilon_w)$. Notably, the original expression of Hurd's model, Equation 2.13, shows a strong dependence of the interaction force on the Debye screening length, $F_{el} \sim \kappa^{-2}$, whereas the scaling based on the charge renormalization (Equations 2.14 and 2.15) predicts a much weaker salt dependence, $F_{renor} \sim \kappa^{-0.8...-0.4}$.

The interaction magnitude estimated by the charge renormalization, however, is significantly lower than those reported in a consensus of experimental results.[12,25] Masschaele *et al.*[26] measured the interaction force using several independent methods, such as the use of the Boltzmann distribution, $U(r) = -k_B T\ln g(r)$, with the pair correlation function $g(r)$ at an infinitesimal particle concentration ($\phi_s \rightarrow 0$), laser tweezer measurements between particle pairs, and interfacial shear modulus predicted using the Zwanzig–Mountain equation, $G'_{\infty,2D}(\phi_s) = \frac{8\phi_m^{2D}n}{9\pi^2}\frac{\partial^2 U}{\partial r^2}$, where $\phi_m^{2D} = 0.906$ is the maximum packing fraction at the 2D interface and $n = 6$ is the number of nearest neighbours. Depending on the surface coverage of the particles, the high-frequency shear modulus ($G'_{\infty,2D}$) can be obtained by either analysing the thermal strain fluctuations of a 2D colloidal crystal at the high surface coverage or measuring the interfacial shear rheology using a magnetic rod rheometer. These experimentally measured pair interaction forces are consistent with each other and the previous laser tweezer measurements performed by Park *et al.*[27] Based on this experimental study, the investigators proposed a role for the finite size of counterions in the Stern layer. In this case, the dipole can be generated in the condensed ion layer, which is normally ignored by the point-charge approximation of the Gouy–Chapman theory of the double layer used in the P–B model. The magnitude is characterized by the thickness of the dense layer and the colloidal bare charge, $\mu \sim d_{st}q_{bare}$. By using the scaling analysis, this dipole accounts for the strong repulsive interactions and the discrepancy of the interaction magnitude between the charge renormalization theory and the experimental measurements.

The above models are based entirely on charge dissociation in the polar phase. Another model of the long-range repulsion was suggested by Aveyard and coworkers in 2000, in which a small amount of surface charges were proposed to dissociate by residual water in the non-polar solution.[28,29] Hypothetically, this residual water could be captured in the rough particle surface. Due to the absence of the charge screening effect through the non-polar fluid, the resulting interactions should be extremely strong. They postulated that the repulsion is dominated by the presence of a small amount of surface residual charge in the non-polar solution (*i.e.* oil or air). For example, a point charge that can be dissociated in the oil phase is proportional to the surface charge density (σ), the three-phase contact angle (θ_c), the particle size (a) and the frictional degree of charge dissociation (α_{oil}):

$$q_{oil} = 2\pi a^2 \sigma (1 - \cos\theta_c)\alpha_{oil} \tag{2.16}$$

The asymptotic form of the interaction force in large separations is expressed as:

$$F_{el,2} = \frac{6q_{oil}^2 \zeta^2}{4\pi\varepsilon_{oil}\varepsilon_0 r^4} \tag{2.17}$$

where $\zeta = a(3 + \cos\theta_c)/2$ is a vertical distance of the point charge above the interface. This equation is equivalent to the force between a point charge on one particle, and an effective dipole ($\mu_{eff} = 2\zeta q_{oil}$) that results from a point charge on the other particle and the corresponding image charge in the aqueous phase. The proposed role of charge dissociation in the non-polar phase was based on experiments that could not discern any dependence on the ionic strength of the aqueous phase. Indeed, subsequent experiments, averaged over many particle pair interactions, did demonstrate a salt dependence, although it was much weaker than Hurd originally proposed.[12] Attempts to resolve this discrepancy using a charge renormalization model appear to capture this weaker dependence,[25] but have not to date been analysed further in the context of the role of the Stern layer ions, as discussed earlier, which would bring the interaction magnitude in line with experiments. These issues, along with the fact that the existence of dissociated charge in the non-polar phase has not been decisively ruled out, remain interesting and open questions.

2.2.2.2 Attractive Interactions

Colloidal particles at fluid–fluid interfaces also experience attractive interactions. There are several phenomena that could cause these attractions, including deformation of the interface driven by gravity, thermal fluctuations, dipolar electric fields, and undulating menisci.[30–36] The first two cases are limited to sub-millimetre and sub-nanometre sized particles, respectively. For typical colloidal particles with micrometre dimensions, these two factors are negligible. Alternatively, Stamou and coworkers suggested that the interface undulation around a particle leads to an interface deformation that subsequently results in induced capillary forces between particles.[35] The shape of the interface can be determined by the

Young–Laplace equation, $\nabla^2 h(r,\varphi) = 0$, in cylindrical coordinates. Using multipole expansion, the general solution of this equation is given by:

$$h(r, \varphi) = \sum_{m=0}^{\infty} H_m \left(\frac{r_c}{r}\right)^m \left[\cos\left(m(\varphi - \varphi_{m,0})\right) + \sin(m(\varphi - \varphi_{m,0}))\right] \qquad (2.18)$$

where H_m is the expansion coefficient and $\varphi_{m,0}$ is the phase angle. In the absence of external force (*i.e.* gravity, external torque, *etc.*), the monopole ($m = 0$) and dipole ($m = 1$) terms are unstable. The lowest order solution that is stable is therefore the quadrupole term ($m = 2$), which is the dominant term in a far field. The surface profile of the quadrupolar interface deformation is:

$$h^{(2)}(r, \varphi) = H_2 \cos(2(\varphi - \varphi_1)) \left(\frac{r_c}{r}\right)^2 \qquad (2.19)$$

The corresponding contact line hysteresis at $r = r_c$ is

$$h^{(2)}(r_c, \varphi) = H_2 \cos 2((\varphi - \varphi_1)) \qquad (2.20)$$

The gradient of $h^{(2)}(r,\varphi)$ in the cylindrical coordinates is:

$$\nabla h^{(2)}(r, \varphi) = \left[\frac{dh}{dr}, \frac{dh}{rd\varphi}\right] = \left[-2H_2 \cos(2(\varphi - \varphi_1))\frac{r_c^2}{r^3}, -2H_2 \sin(2(\varphi - \varphi_1))\frac{r_c^2}{r^3}\right]$$

$$(2.21)$$

The self-energy of an individual particle due to the excess surface area of the interface deformation around the particle is given by:

$$\delta E = \gamma \delta S = \gamma \int \frac{(\nabla h)^2}{2} dS \qquad (2.22)$$

where $(\nabla h)^2 = \nabla h \cdot \nabla h = 4H_2^2 \frac{r_c^4}{r^6}$. Substituting this relation into Equation 2.22 yields $\delta E \approx \pi \gamma H_2^2$ for the quadrupole approximation.

The interaction energy between two particles (A and B) with the quadrupolar interface deformation is $\Delta E_{AB} = \gamma(\delta S_{AB} - \delta S_A - \delta S_B)$ where δS_{AB} is the interface area around the two particles, and δS_A and δS_B are the interface areas around each isolated particle.[35] Based on the superposition approximation of the interface height in long-range separations ($h_{AB} = h_A + h_B$) and the first Green identity theorem, the pair interaction energy becomes:

$$\delta E_{AB} = \frac{\gamma}{2} \int (\nabla(h_A + h_B)^2 - (\nabla h_A)^2 - (\nabla h_B)^2) dS$$

$$(2.23)$$

$$= \gamma \int \nabla h_A \cdot \nabla h_B dS = 2 \int_{C_B} h_B (\boldsymbol{n_B} \cdot \nabla h_A) dC$$

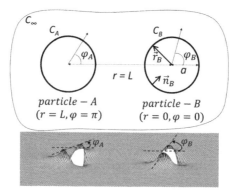

Figure 2.3   Schematic description of two particles with the quadrupole interface deformation.

where $\boldsymbol{n}_B = -[\cos\varphi,\ \sin\varphi]$ is the vector normal perpendicular to the boundary around the particle B (C_B) pointing toward the centre of the particle (Figure 2.3). The interface gradient caused by the particle A at the boundary around the particle B can be expressed using the Taylor expansion:

$$\nabla h_A|_{C_B} \approx \nabla h_A(r = L, \varphi = \pi) + r \cdot (\nabla \otimes \nabla) h_A(r = L, \varphi = \pi) \qquad (2.24)$$

where the position vector is $r = [r_c\cos\varphi,\ r_c\sin\varphi]$. The outer product of the two gradients is:

$$(\nabla \otimes \nabla) = \begin{pmatrix} \dfrac{d}{dr^2} & \dfrac{1}{dr}\left(\dfrac{d}{rd\varphi}\right) \\ \dfrac{1}{r}\dfrac{d^2}{d\varphi dr} & \dfrac{1}{r^2}\dfrac{d^2}{d\varphi^2} \end{pmatrix} \qquad (2.25)$$

The second term on the right-hand side in Equation 2.24 becomes:

$$r \cdot (\nabla \otimes \nabla) h_A(r = L, \varphi = \pi) = \begin{bmatrix} \dfrac{r_c^3 H_2}{L^4}\left(6\cos(2\varphi_A)\cos\varphi - 4\sin(2\varphi_A)\sin\varphi\right) \\[2mm] \dfrac{r_c^3 H_2}{L^4}\left(-6\sin(2\varphi_A)\cos\varphi - 4\cos(2\varphi_A)\sin\varphi\right) \end{bmatrix}$$

$$\approx \dfrac{r_c^3 H_2}{L^4}\begin{bmatrix} 6\cos(\varphi + 2\varphi_A) \\ -6\sin(\varphi + 2\varphi_A) \end{bmatrix} \qquad (2.26)$$

Therefore, the asymptotic form of the capillary interaction in Equation 2.23 is obtained to be:

$$\delta E_{AB} = -12\pi\gamma H_2^2 \cos(2(\varphi_A + \varphi_B))\sin^4\theta_c \dfrac{r_c^4}{r^4} \qquad (2.27)$$

The corresponding force caused by the quadrupolar interface deformation in long-range separations is:

$$F_{\text{quad}} = -\frac{\delta E_{AB}}{dr} = -48\pi\gamma H_2^2 \cos(2(\varphi_A + \varphi_B))\sin^4\theta_c \frac{r_c^4}{r^5} \qquad (2.28)$$

where $a \approx r_c$ when the contact angle is close to $\theta_c \approx 90°$. The prefactor, $H_2 = 0.5a\Delta\phi$, is the amplitude of the quadrupolar meniscus and $\Delta\phi$ is the contact line hysteresis. For typical colloids with $a \approx r_c = 1$ μm, $H_2 = 10$ nm, and $\theta_c = 90°$ at an oil–water interface *($\gamma \approx 50$ mN m^{-1})* when two particles are aligned with $\varphi_A = \varphi_B = 0$ and their separation is $r = 2a$, for instance, the capillary interaction is approximately $-300k_BT$, which is significantly larger than thermal energy k_BT. Therefore, these particles will attract one another.

Danov and coworkers extended this work to the interactions between two particles with the capillary multipoles of arbitrary orders (m_A and m_B).[30] They used bipolar coordinates to obtain an explicit expression for the capillary interactions, $\Delta E(r) = E(r) - E(r = \infty)$. The excess surface energy for the two particles at infinite separations is:

$$E(r=\infty) = \frac{\pi\gamma}{2}(m_A H_A^2 + m_B H_B^2) \qquad (2.29)$$

In finite separations, the surface energy is expressed as:

$$E(r) = \pi\gamma\left[H_A^2 S(\tau_A) + H_B^2 S(\tau_B) - H_A H_B G \cos(m_A\varphi_A + m_B\varphi_B)\right] \qquad (2.30)$$

$$S(\tau_Y) = \sum_{n=1}^{\infty} \frac{n}{2}\coth[n(\tau_A + \tau_B)]A^2(n, m_Y, \tau_Y) \ (Y = A \text{ or } B) \qquad (2.31)$$

$$G = \sum_{n=1}^{\infty} \frac{nA(n, m_A, \tau_A)A(n, m_B, \tau_B)}{\sinh[n(\tau_A + \tau_B)]} \qquad (2.32)$$

$$A(n, m_Y, \tau_Y) = m \sum_{k=0}^{\min(m,n)} \frac{(-1)^{m-k}(m+n-k-1)!}{(m-k)!(n-k)!k!}\beta^{m+n-2k} \qquad (2.33)$$

where $\beta = e^{-\tau_Y}$. From geometrical relations in the bipolar coordinate, the two contact lines, C_A and C_B, correspond to $\tau_A = \cosh^{-1}\left(\dfrac{r^2 + r_{c,A}^2 - r_{c,B}^2}{2rr_{c,A}}\right)$ and $\tau_B = \cosh^{-1}\left(\dfrac{r^2 + r_{c,B}^2 - r_{c,A}^2}{2rr_{c,B}}\right)$. For the quadrupolar deformation ($m_A = m_B = 2$) with the equal size ($\tau_c = \tau_A = \tau_B$), Equations 2.29–2.33 are simplified as:

$$E(r=\infty) = \pi\gamma(H_A^2 + H_B^2) \qquad (2.34)$$

$$E(r) = \pi\gamma\left[(H_A^2 + H_B^2)S(\tau) - H_A H_B G \cos(2(\varphi_A + \varphi_B))\right] \tag{2.35}$$

$$S(\tau_C) = \sum_{n=1}^{\infty} \frac{n}{2\pi^2} \coth(2n\tau_C)A^2(n,\tau_C) \tag{2.36}$$

$$G = \sum_{n=1}^{\infty} \frac{nA^2(n,\tau_C)}{\pi^2 \sinh(2n\tau_C)} \tag{2.37}$$

$$A(n,\tau_C) = 4\pi \sinh(\tau_C)[n\sinh(\tau_C) - \cosh(\tau_C)]e^{-n\tau_C} \tag{2.38}$$

An asymptotic equation in the case of the capillary multipoles with an arbitrary order can also be simplified as:

$$\Delta E(r) \approx -12\pi\gamma H_A H_B \cos(m_A\varphi_A + m_B\varphi_B)\frac{r_{c,A}^{m_A} r_{c,B}^{m_B}}{r^{(m_A+m_B)}} \tag{2.39}$$

Notably, Equation 2.39 for the quadrupole ($m_A = m_B = 2$) with equal size ($r_c = r_{c,A} = r_{c,B}$) is equivalent to Equation 2.27.

For highly charged particles, Oettel and Dominquez suggested that electrocapillary interactions caused by the dipolar electric field of the particles could lead to attractive interactions.[31,34] This electrostatic capillary force ($F_{cap,el}$) is coupled to the electrostatic repulsive force, $F_{el} \sim \epsilon_F F_{cap,el}$ ($a \ll r$) in which $\epsilon_F = -F_v/(2\pi a\gamma_{ow} \sin\theta_c)$ is the ratio of vertical forces (F_v) acting on the colloid by external sources (*e.g.* gravitational, electrostatic, hydrostatic, *etc.*) with respect to the interfacial tension force ($a\gamma_{ow}$). Notably, the scaling of the electrocapillary force is identical to that of the electrostatic repulsion ($\sim 1/r^4$). This implies that the magnitude of the repulsive force between two particles, for example, is decreased as much as the strength of the electrocapillary force, whereas the scaling of the net force remains as $\sim 1/r^4$.

The asymmetric shape of colloidal particles at fluid–fluid interfaces also deforms their surrounding interface to satisfy two conditions, the wetting boundary condition and the minimum surface free energy.[37–45] The equilibrium topology of the deformed interface can be analytically and numerically determined. For an isolated cylindrical particle trapped at a fluid–fluid interface, for example, when the three-phase contact angle is less than 90°, the interface deflects upward on the planar end surfaces and downward around the sides, respectively.[37,45] Such interface deformations around the particles essentially induce the capillary interactions with neighbouring particles and therefore particle assembly and the formation of microstructures.

2.3 Interactions and Microstructures of Homogeneous Particles

The structures and properties of materials on macroscopic length scales are determined by the interactions and microstructures at the individual constituent levels. For colloidal dispersions and suspensions, understanding such microscopic scale behaviours is critical to control and modify the material properties in higher levels of hierarchical assembly, such as the phase behaviour and rheology. In this section, the pair interactions and the interaction heterogeneity between particles are discussed, based on direct measurements of pair interaction forces at an oil–water interface using optical laser tweezers and, subsequently, these are related to bulk property measurements, such as the equilibrium radial distribution function (RDF) of 2D equilibrium suspensions, the time evolution of aggregate microstructures, and the micromechanics of colloidal aggregates at the oil–water interface.

Interaction forces between two particles at a fluid–fluid interface can be directly measured by using optical laser tweezers.[11,12,27,46] Time-shared optical traps are used to hold the two particles confined at the interface. One particle is translated toward another stationary particle. Displacement (Δx) of the fixed particle from its equilibrium position is monitored as the centre-to-centre separation (r) is decreased. The interaction force can be calculated by using the relation, $F(r) = \kappa_t \Delta x(r)$, in which κ_t is the calibrated optical trap stiffness (Figure 2.4).

Addition of additives (*i.e.* electrolyte and surfactants) in a fluid phase leads to a decrease in the Debye screening length and an alteration in the wettability of the particle at the interface. This enables control of their interactions and microstructures at the interface, as discussed next.

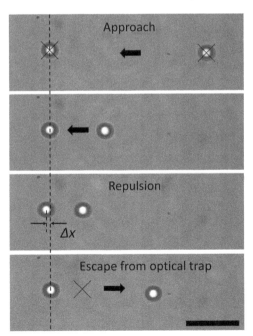

Figure 2.4 Interaction measurement between two spherical polystyrene particles with $2a = 3$ μm at the decane–water interface. The scale bar is 10 μm. Modified and reprinted from reference 27.

2.3.1 Pair Interactions at an Oil–Water Interface

The repulsive interaction force between two particles at an oil–water interface is directly measured using optical laser tweezers.[12,27] The magnitude of the repulsive interaction depends on the particle pairs that are measured. The pair interaction force over 32 pairs are measured, showing a broad distribution of the force profiles, as shown in Figure 2.5a. Using $F = \dfrac{3a_2 k_B T}{r^4}$, the fitted prefactor (a_2) varies between $1.2 < a_2 \times 10^{13} < 12.2$ m³ with a mean value of $\langle a_2 \rangle = 5.1 \pm 2.4 \times 10^{-13}$ m³. Notably, the force profiles for all curves exhibit the r^{-4} scaling behaviour, which is consistent with the proposed electrostatic interaction models (Equations 2.13 and 2.17). The corresponding histogram of a_2 values in Figure 2.5b clearly shows a skewed distribution of the interaction magnitude. This heterogeneous interaction can be characterized empirically by using a gamma distribution function with a shape parameter (k) and a scale parameter (θ):

$$f(a_2 ; k, \theta) = a_2^{k-1} \frac{e^{-a_2/\theta}}{\theta^k \Gamma(k)} \tag{2.40}$$

where $\Gamma(k)$ is the gamma function. The fitted parameter values in Figure 2.5b are $k = 4.42$ and $\theta = 3.87 \times 10^{-14}$ m³, which can be used for randomly generating the interaction potential for multiple pairs, as discussed in Section 2.3.2.

The pair interaction force also depends on the salt concentration in the aqueous sub-phase.[12] As shown in Figure 2.6a, the repulsive force decreases with the salt concentration, in which the measured force scales as r^{-4}. The dependence of the interaction force on the salt concentration suggests the effect of surface charge screening in the presence of counterions in the aqueous phase (Figure 2.6). The magnitude of the repulsive force is fitted using $F_{rep} = F_0 \left(\dfrac{a}{r}\right)^4$ and plotted as a function of Debye screening length (κ^{-1}) in Figure 2.6b. On average, over several pairs to consider the interaction

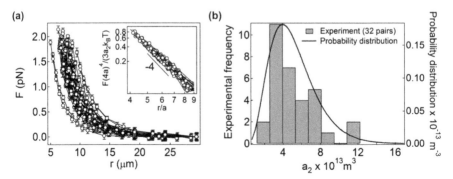

Figure 2.5 Heterogeneity of the repulsive pair interactions at the oil–water interface. Modified and reprinted from reference 27.

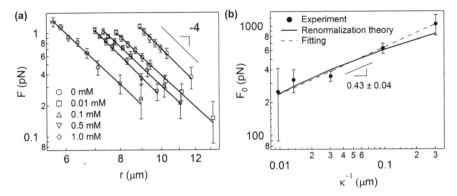

Figure 2.6 Effect of electrolyte on the pair interactions at the decane–water interface. (a) Repulsive force versus the particle separation with varying the salt concentration. (b) Magnitude of the repulsive pair interaction force as a function of Debye screening length.
Modified and reprinted with permission from reference 12. © 2008 American Chemical Society.

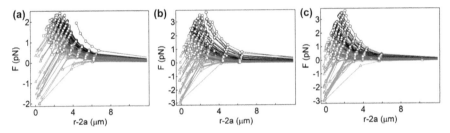

Figure 2.7 Heterogeneity of the pair interaction at the oil–water interface with addition of additives in each fluid phase: (a) 0.25 M NaCl in water, (b) 0.25 M NaCl and 0.1 mM SDS in water, and (c) 25 μM SPAN 85 in oil. Modified and reprinted from reference 11.

heterogeneity, the interaction force weakly depends on the screening length with a scaling exponent of 0.43 ± 0.04, which is consistent with that predicted from the charge renormalization model discussed in Section 2.2.2.1.[12,25] The experimental value of the interaction magnitude, however, is significantly lower than that predicted from the charge renormalization, as noted earlier.

Introducing different types of additives in each fluid phase also alters the interactions between particles at the oil–water interface.[11] In the presence of 0.25 M NaCl in the aqueous sub-phase, 0.25 M NaCl/0.1 mM SDS in the sub-phase, or 25 μM SPAN 85 in the decane super-phase, the measurements of the interaction force of approximately 100 particle pairs show that the interactions are heterogeneous (Figure 2.7), which is similarly observed in the interactions at the pure oil–water interface (Figure 2.5). In general, however, there is a significant decrease in the interaction magnitude, compared to the interactions at the pure interface, as shown in Figure 2.7. Some particle pairs show a reduced repulsive interaction (black and blue curves in Figure 2.7), whereas others exhibit a stronger attraction, resulting in the two particles attaching to each other (red curves in Figure 2.7).

The addition of NaCl in Figure 2.7a can effectively screen the surface charge in the aqueous sub-phase without changing the three-phase contact angle in which the Debye screening length is about $\kappa^{-1} \approx 0.6$ nm. Because the number of dissociated charges in the aqueous phase is constant, the reduction in the repulsive interaction is probably due to the charge screening effect in the aqueous phase. In the case of the addition of NaCl/SDS in water in Figure 2.7b, the interaction can be affected by increase in the three-phase contact angle ($\sim 20\%$) as well as decrease in the Debye screening length, resulting in significant reduction in the repulsive force. The addition of the non-ionic surfactant, SPAN 80, in the oil phase also increases the three-phase contact angle by $\sim$ 20%, but maintains the Debye screening length identical to the pure oil–water interface system ($\kappa^{-1} \approx 200$ nm for pure oil–water system at normal atmosphere with an ionic strength, $I \approx 2.2 \times 10^{-6}$ M). These last two results imply that the decrease in the number of dissociated charges in the aqueous phase leads to the reduction in the electrostatic repulsions. In short, the repulsive force between the two charged particles at the oil–water interface decreases by decreasing the Debye screening length and the number of surface charges in the aqueous sub-phase. The importance of the experimental results of these charged polystyrene systems is that the contribution of the charge effect on the electrostatic repulsive interaction is probably dominated by the charges dissociated in the aqueous sub-phase. Notably however, for hydrophobically modified silica particles, the interaction force increases as the portion immersed in the oil phase increases, suggesting the dominant contribution of the surface residual charge effect in the oil phase.[47,48]

The pair interactions of particles at the oil–water interface can change over time. Such time evolution in the pair interactions is observed when the aqueous phase contains a small amount of SDS.[12] As shown in Figure 2.8a,

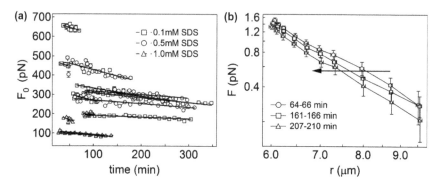

Figure 2.8   Time-dependent pair interactions at the oil–water interface in the presence of SDS in the aqueous sub-phase.
Modified and reprinted with permission from reference 12. © 2008 American Chemical Society.

the magnitude of the repulsive force for several pairs decreases with time in the presence of 0.1–1.0 mM SDS in the aqueous phase. The force profiles of a pair at the oil–water interface containing 0.1 mM SDS in water also clearly show that the repulsive interactions over three time intervals decrease (Figure 2.8b). The mechanism for this time evolution in the presence of surfactants is not completely understood. However, this time-dependent behaviour on the pair interactions can account for the macro-scopic aggregation behaviour that also depends on time, as discussed in Section 2.3.2.

2.3.2 Microstructure and Micromechanical Properties

Monte Carlo simulations of the 2D suspension can provide some insight into the effect of pair interactions on the microstructure, and especially the role of the pair interaction heterogeneity. The interaction potential between i and j particles is given by:[27]

$$\frac{U_{ij}}{k_B T} = \frac{(a_{2,i} + a_{2,j})}{r_{ij}^3} = \frac{a_{2,ij}}{r_{ij}^3} \tag{2.41}$$

Note that each particle carries its own half-pair potential ($a_{2,i}$ and $a_{2,j}$) and its sum ($a_{2,ij}$) indicates the pair interaction pre-factor, which is related to the measured force in Figure 2.5:

$$F = -\frac{dU}{dr} = \frac{3a_{2,ij} k_B T}{r_{ij}^4} \tag{2.42}$$

Based on pairwise additive potentials, $U_{tot} = \sum_j^N U_{ij}(i \neq j)$, the interaction heterogeneity can be introduced by using the two fitting parameters of the gamma distribution function, k and θ, in Equation 2.40. The radial distribution function (RDF) is generated to compare the microstructure between simulated and experimental 2D suspensions and is given by:

$$g(r) = \frac{\langle N_\lambda \rangle}{2\pi N \rho_N r_\lambda dr} \tag{2.43}$$

where N is the total number of particles, $\langle N_\lambda \rangle$ is the number of particles in the separation interval r_λ to $r_{\lambda+1} = r_\lambda + dr$, $\rho_N = \rho/\pi a_2$ is the particle number density per unit area, and ρ is the surface area fraction. As shown in Figure 2.9, the introduction of the interaction heterogeneity on the simulation based on the measured gamma distribution leads to excellent agreement with the experimental results. Pairwise potentials for the simulation are also used based on the average value ($a_2 = 5.1 \times 10^{-13}$ m^3) and lower limit ($a_2 = 0.5 \times 10^{-13}$ m^3) of the measured interaction. In this case, the interaction is homogeneous between particles. The resulting RDF in Figure 2.9 deviates significantly from the experimentally measured RDF. These results indicate that the heterogeneous interactions quantified by pair interaction measurements play an important role in forming the equilibrium microstructure.

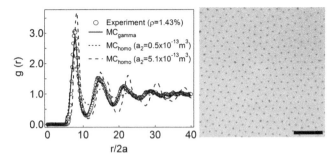

Figure 2.9   Effect of the heterogeneous pair interactions on the microstructure at the oil–water interface. The scale bar is 100 μm.
Modified and reprinted from reference 27.

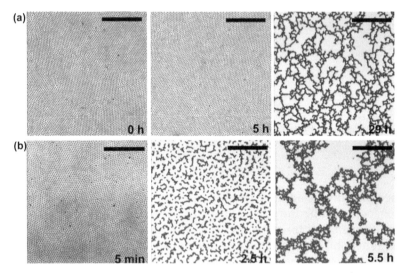

Figure 2.10   Time evolution of colloidal suspensions at the oil–water interface, containing (a) 0.1 M NaCl/0.1 mM SDS in the aqueous sub-phase and (b) 0.25 μM SPAN 80 in the oil super-phase. The scale bar is 100 μm.

The aggregation kinetics strongly depend on the composition of fluid phases upon the addition of salt and surfactant.[10] Two different conditions, the addition of SDS/NaCl in water and SPAN 80 in oil, have been studied. The particle aggregation kinetics in the suspension containing SDS/NaCl is considerably slower than that containing SPAN 80 (Figure 2.10). Note that similar time-dependent behaviour is observed in the pair interaction measurements in the presence of a small amount of SDS in the aqueous phase (Figure 2.9).[12] As noted previously, the mechanism for the time evolution is not completely understood, but the effects are immediately clear. Particle aggregation continues until a ramified, gel-like network forms. This is an important method for controlling the surface rheology.

More open microstructures develop when NaCl/SDS is used (Figure 2.10a). This suggests that rearrangements of the particles after forming aggregates is unlikely to occur, consistent with DLCA (diffusion-limited cluster aggregation) kinetics. In this case, the diffusion of particles or clusters plays an important role in the aggregate formation (*i.e.* high sticking probability). On the contrary, when SPAN 80 is used in the oil phase, the microstructure is denser due to the rearrangements of particles (Figure 2.10b). These kinetics are similar to the RLCA (reaction-limited cluster aggregation) model.

The micromechanics of colloidal aggregates have been characterized through measurements of their rupture mechanics and bending elasticity.[49] The deformation, rearrangements and rupture of aggregates are related to the elasticity and yield of 2D suspensions. Therefore, such studies provide an important connection between interactions on the particle scale to the mechanical properties of suspensions.

As previously shown in Figure 2.10, the microstructure of aggregates at the oil–water interface upon addition of different surfactants in each fluid phase exhibits a clear structural difference. To understand this, optical laser tweezers were used to impart compressive and tensile forces to colloidal aggregates and to measure the rigidity and rearrangements.[49] In the case of salt/SDS in Figure 2.11a, the approaching particle escapes from the translating trap without inducing significant structural deformation, suggesting a strong rigidity of the aggregate. In contrast, noticeable deformation and frequent rearrangements occur during the compression in the presence of SPAN 80 (Figure 2.11b). This result indicates that the colloidal bonds are significantly weaker in both the radial and tangential direction, compared to aggregates formed in the NaCl/SDS condition. The aggregate rigidity can be estimated by averaging the mechanical force during the

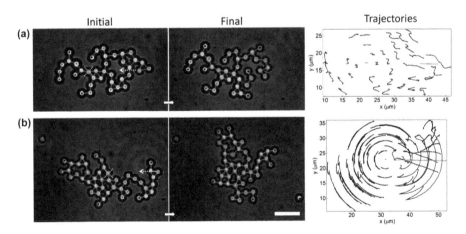

Figure 2.11   Compression of colloidal aggregates at the oil–water interface, containing (a) 0.1 M NaCl/0.1 mM SDS in the aqueous sub-phase and (b) 0.25 μM SPAN 80 in the oil super-phase. The scale bar is 10 μm.

deformation of aggregates and is found to be $\langle \kappa_a \rangle = 0.38 \pm 0.45$ mN m^{-1} for SPAN 80 and 5.7 ± 2.9 mN m^{-1} for NaCl/SDS. This suggests that the strong rigidity in NaCl/SDS suppresses the deformation and rearrangements, forming a more open percolated network, whereas the weaker rigidity in SPAN 80 causes frequent rearrangements, resulting in the denser structures.

2.3.3 Effect of Geometrical Anisotropy

The anisotropic shape of colloidal particles can impart unique properties on rheology and microstructure when they are dispersed at fluid–fluid interfaces. This is because such non-spherical particles inherently deform the interface surrounding them, which causes the attractive interaction to be dominant, compared to the electrostatic repulsive interactions. For ellipsoidal particles prepared using the mechanical stretching method,[50] the scaling of the interactions between two ellipsoids exhibits alignment dependence, *i.e.* the interaction force is found to be $F_{cap} \sim r^{-5}$ when the two particles approach in the tip-to-tip direction, whereas $F_{cap} \sim r^{-4}$ for the side-to-side approach.[39] Notably, the scaling exponent of the former case resembles the quadrupolar interaction between two spherical particles.[35] Indeed, it has been determined that the shape of the interface deformation around an ellipsoid is quadrupole in which the interface is depressed around the tip areas and is raised around the side regions.[39,40] It is also found that the assembly behaviours can be controlled by modifying the surface charge and wettability of the particles at the interfaces.[39,51] Exploiting this controllability on their lateral capillary interactions, Yunker *et al.*[52] reported that when a suspension drop containing ellipsoids is placed on a substrate, the particles form loosely packed structures at the air–water interface due to the existence of capillary interactions between the particles. This capillary force induced structure allows the particles to be distributed evenly at the air–water interface upon drying the solution, and thus prevents the non-uniform deposition around the edge of the suspension drop (*i.e.* coffee-ring effect). The particle shape-mediated deposition may provide a significant impact on practical applications, such as uniform coating, the suppression of cracking and the interface stabilization of foams and emulsions.[53]

Similar to the ellipsoidal particles, cylindrical particles confined at fluid–fluid interfaces also induce the static interface deformation around them.[37,38,41,42,45] The shape of the deformation is numerically determined, and found to be also quadrupole. These particles cause strong tip-to-tip capillary interactions, which scale as $F_{cap} \sim r^{-5}$ in long-range separations. Interestingly, the trajectories and orientation of the particles can be directed by controlling the interface curvature and gradients by using micro-fabricated posts.[54] The fact that the interface curvature acts as an external field for driving the particle assembly potentially provides a novel approach for engineering material structures and properties.[55]

2.4 Interactions and Conformations of Amphiphilic Particles

The surface activity of molecular amphiphiles and their assembly behaviours at fluid interfaces are strongly influenced by their molecular structure and affinity to each fluid phase (*i.e.* hydrophilic–lipophilic balance, HLB). Likewise, the configuration and interactions of particles at a fluid interface are likely to have significant impact on their efficacy as solid surfactants and also their ability to organize into novel colloidal assemblies.[56] Recently, a variety of techniques have been developed and utilized to prepare colloidal amphiphiles with chemical and geometric anisotropy.[57-63] Also known as Janus particles, these solid amphiphiles constitute a new class of colloids exhibiting a wide range of unique behaviours and novel applications.[64] For example, Janus particles, like molecular amphiphiles, have been shown to effectively stabilize multiphasic fluid mixtures such as emulsions and foams.[65,66] Recent studies have even demonstrated that Janus particles could potentially lead to the formation of thermodynamically stable emulsions due to the strong attachment of these particles to the interfaces.[67,68]

2.4.1 Equilibrium Configurations (Theoretical Approach)

To determine the equilibrium configurations of Janus particles at fluid–fluid interfaces, their attachment energy to the interface is calculated using Equation 2.10 as a function of the vertical displacement (d_v) and the orientation angle (θ_r) and, subsequently, is minimized to find the lowest energy state and the corresponding equilibrium displacement $(d_{v,eq})$ and orientation angle $(\theta_{r,eq})$.

The equilibrium configuration of amphiphilic (or Janus) particles is determined by a balance between two factors; one factor has to do with the tendency of Janus particles to maximize the contact area between one side and its preferred phase (the first two terms on the right-hand side, $S_{Ao} \cos\theta_A + S_{Po} \cos\theta_P$, in Equation 2.10) and the other has to do with the tendency of the system to maximize the displaced interfacial area due to the presence of the particle at the interface (the third term on the right-hand side, S_I, in Equation 2.10). These two factors compete with each other to minimize the attachment energy.

2.4.1.1 *Amphiphilic Spheres*

In general, the equilibrium configuration of a Janus sphere trapped at a fluid–fluid interface is such that each hemisphere is fully exposed to its preferred fluid phase and the wettability separation line (WSL or Janus boundary) between the two hemispheres coincides with the interface. This behaviour can be illustrated using a spherical Janus particle shown in Figure 2.12. The location of the WSL, as defined by α (the inclination angle)

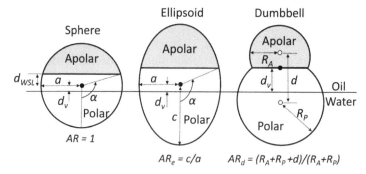

Figure 2.12 Geometries of Janus sphere, ellipsoid, at dumbbell at the oil–water interface.

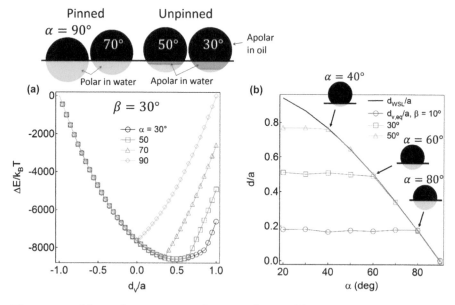

Figure 2.13 (a) Attachment energy of Janus spheres with $a = 10$ nm at an oil–water interface ($\gamma_{ow} = 50$ mN m^{-1}). (b) Critical vales of α at which the oil–water interface is pinned at the WSL.

or $|d_{WSL}|$ (the vertical distance between the WSL and the centre-of-mass of the particle), and the difference in wettability between the two surfaces, as defined by β (where $\beta = \theta_A - 90° = 90° - \theta_P$), control the equilibrium configuration (Figure 2.12).[13–15,69–71] For spherical Janus particles, the first factor ($S_{Ao}\cos\theta_A + S_{Po}\cos\theta_P$ in Equation 2.10) dominates the attachment energy, and thus the equilibrium configuration. For large values of $|d_{WSL}|$ and β, however, the interface could depin from the WSL because the second factor (S_I in Equation 2.10) becomes more significant than the first one.

As shown in Figure 2.13a, for Janus spheres with $\beta = 30°$ at an oil–water interface, the attachment energy becomes more asymmetric as the value of α decreases below $90°$ (apolar surface area increases). For $70° < \alpha < 90°$, the

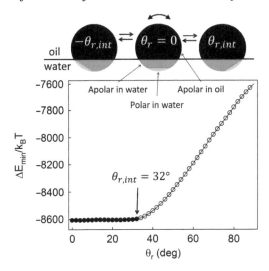

Figure 2.14 Free rotation of Janus sphere in which the WSL is unpinned at the oil–water interface; $a=10$ nm, $\alpha=\beta=30°$. The minimum attachment energy (ΔE_{min}) is the lowest energy among $\Delta E(d_v)$s at constant values of θ_r obtained from Equation 2.10.

oil–water interface stays pinned at the WSL of Janus particles at equilibrium. However, the interface no longer stays pinned at the WSL when α is between 30° and 50° (Figure 2.13a). The critical value of α (α_c), beyond which the interface stays pinned at the WSL (solid line in Figure 2.13b), increases as the value of β decreases because the energy penalty caused by non-preferred wetting becomes smaller for particles with small wettability difference. Interestingly, after the interface detaches from the WSL at $\alpha < \alpha_c$, the vertical displacement $(d_{v,eq})$ at equilibrium stays constant, regardless of α. In this regime, the equilibrium displacement is dominantly determined by the wettability of the apolar region (θ_A).

When the WSL of the Janus sphere detaches from the fluid interface, the particle can freely rotate over a range of angles, $|\theta_r| \leq \theta_{r,int}$. Due to the spherical geometry, the attachment energy does not change upon out-of-plane orientation (*i.e.* constant value of d_v and S_l) until the WSL makes contact with the interface, as shown in Figure 2.14.

2.4.1.2 Non-Spherical Amphiphilic Particles

In the case of particles with both geometric and chemical anisotropy, the effect of the two factors (*i.e.* the energy from the preferred wetting and the energy from the displaced area) on the particle configurations becomes more significant and strongly influences the equilibrium configurations. Ellipsoids and dumbbells (also known as dicolloids, dimer particles), which are two partially fused spheres, constitute excellent model systems to systematically study the effect of particle geometry and wettability on their configuration behaviours because these particles have been recently prepared and can potentially be used as effective solid surfactants.[65,66,72-75] As shown in Figure 2.12, the aspect ratio of an amphiphilic ellipsoid is defined as the ratio of major and minor axes, $AR_e = c/a$. The location of the WSL between the apolar and polar regions is parameterized by the inclination angle (α) in Figure 2.12. For the amphiphilic dumbbells consisting of two partially fused spheres with radii of R_A and R_P, the aspect ratio is $AR_d = (R_A + R_P + d)/(R_A + R_P)$ where d is the centre-to-centre distance between the two spheres (Figure 2.12). The WSL of the dumbbells corresponds to the boundary between the two spheres. Both particles are composed of apolar and polar regions, and the corresponding wettability can be characterized by the three-phase contact angles of the apolar (θ_A) and polar (θ_P) spheres at the fluid–fluid interface (Figure 2.2).

2.4.1.2.1 Amphiphilic Ellipsoids. The simplest case of amphiphilic ellipsoids is the one with geometric and chemical symmetry (*i.e.* the supplementary wetting (β) and $\alpha = 90°$).[16] In this case, the centre-of-mass of the particle always stays at the fluid–fluid interface, regardless the orientation angle. The attachment energy (ΔE) using Equation 2.10 is calculated as a function of AR_e and θ_r at a constant $\beta = 30°$ (Figure 2.15a), and as a function of β and θ_r at a constant $AR_e = 5$ (Figure 2.15c). The corresponding equilibrium orientation angles ($\theta_{r,eq}$) in both cases can be found as a function AR_e and β, respectively, by determining the lowest energy states (Figure 2.15b,d). Interestingly, for particular values of AR_e and β, secondary energy minima exist, which indicates that the particles can be kinetically trapped at metastable orientations (Figure 2.15b,d). Based on this approach, the orientation phase diagram for the amphiphilic ellipsoids with the supplementary wettability and $\alpha = 90°$ can be found and is shown as a function of AR_e and β in Figure 2.16a. In general, the upright orientation is observed for small AR_e and large β, whereas the tilted orientation becomes dominant for large AR_e and small β. Amphiphilic ellipsoids with more complicated geometries, such as non-supplementary wetting and $\alpha \neq 90°$, shows diverse orientation behaviours as shown in Figure 2.16b,c, indicating that the orientation behaviours depend strongly on the particle characteristics, such as geometric (α and AR_e) and chemical (β, θ_A, θ_P) parameters.[17]

2.4.1.2.2 Amphiphilic Dumbbells. The advantage of dumbbell geometry is the possibility of independently tuning the particle aspect ratio (AR_d) and the relative size (R_A/R_P) and wettability (β, θ_A, θ_P) of the two spheres *via* established synthetic protocols.[65,66,74-76] When used as colloid surfactants, these particle parameters can significantly affect the stability and properties of emulsions. As shown in Figure 2.17, a variety of orientation behaviours can be obtained by varying the particle characteristics.[16,17] Notably, when the difference in the wettability or the size of the two spheres is large, one sphere can completely detach from the interface. Due to the spherical geometry of each side, the attachment energy is identical over a

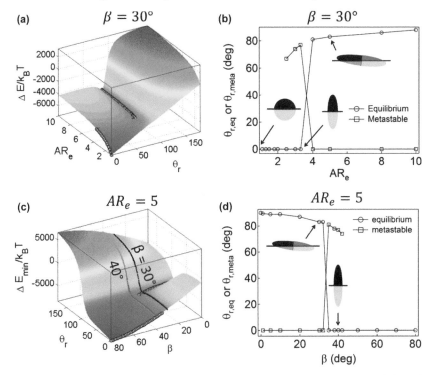

Figure 2.15 Attachment energy of amphiphilic ellipsoids with $\alpha = 90°$ and the effect of β and AR_e on the configuration behaviours.
Modified and reprinted with permission from reference 16. © 2012 American Chemical Society.

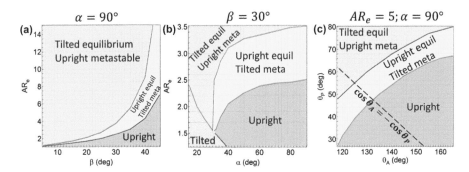

Figure 2.16 Orientation phase diagrams of amphiphilic ellipsoids depending the particle characteristics.
Modified and reprinted from reference 17.

range of orientation angles until the interface contacts the WSL (Figure 2.18) at $\theta = \theta_{r,int}$. This behaviour ('intermediate' regime in Figure 2.17) is similar to the case of spherical Janus particles with high asymmetry as shown in Figure 2.14.

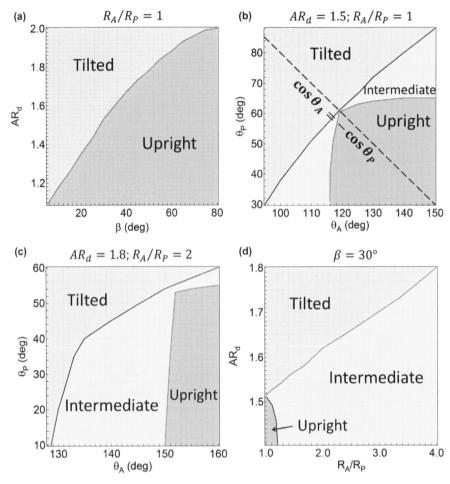

Figure 2.17 Orientation phase diagrams of amphiphilic dumbbells depicting the particle characteristics.
Modified and reprinted from reference 17.

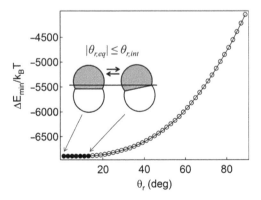

Figure 2.18 Intermediate state of amphiphilic dumbbells.
Modified and reprinted from reference 17.

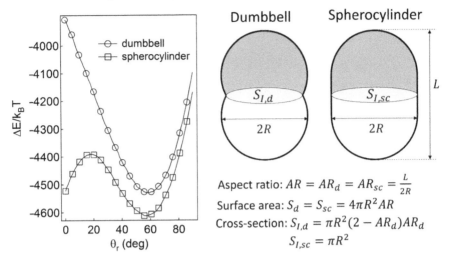

Figure 2.19   Comparison of the attachment energy between an amphiphilic dumbbell and an amphiphilic spherocylinder with $AR = 1.5$, $\beta = 15°$, $R = 10$ nm. Modified and reprinted with permission from reference 16. © 2012 American Chemical Society.

Unlike amphiphilic ellipsoids, amphiphilic dumbbells do not show any coexisting orientation regimes in the orientation phase diagrams (Figure 2.17).[16] The absence of metastable states is due to the presence of the thin 'waist' at the WSL of dumbbells. This rationale can be more clearly illustrated by calculating the attachment energies of an amphiphilic dumbbell and an amphiphilic spherocylinder with the same aspect ratio (same surface area). Figure 2.19 shows that the spherocylinder possesses the secondary energy minimum at the upright orientation, whereas only the primary energy minimum exists in the tilted orientation for the dumbbell. The smaller value of S_I when the dumbbell adopts the upright orientation increases the magnitude of attachment energy, suppressing the existence of a secondary energy minimum.

2.4.1.2.3   Effect of Particle Size.   The transition between the equilibrium and metastable configurations for large amphiphilic particles (> 10 nm) is unlikely to occur because the magnitude of the energy barrier between the two configurations is significantly larger than the thermal energy, $k_B T$ (Figure 2.20a). When the size of particles is sufficiently small, the particles with a metastable configuration can spontaneously adopt their corresponding equilibrium configuration.[16,17] The magnitude of the energy barrier between the equilibrium and metastable states decreases with the particle size at a constant aspect ratio and wettability, given by:

$$\Delta E_b(k) - \Delta E_{eq\,or\,meta}(k) = [\Delta E_b(k_0) - \Delta E_{eq\,or\,meta}(k_0)] \times (k/k_0)^2 \qquad (2.44)$$

where k is a linear dimension of the particle (*e.g.* a or c for ellipsoids, and R_A or R_P for dumbbells; see Figure 2.20b). Based on Monte Carlo simulations, it

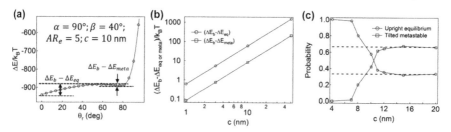

Figure 2.20   Size effect of the amphiphilic ellipsoid with $\alpha = 90°$, $\beta = 40°$ and $AR_e = 5$ on the configurations. (a) Attachment energy profile as a function of the orientation angle. (b) Magnitude of energy barriers as a function of the long axis length, c. (c) Probability change of the upright and tilted orientations with varying the size of the particle.
Modified and reprinted with permission from reference 16. © 2012 American Chemical Society.

is shown that as the particle size decreases, the frequency of the equilibrium orientation (*i.e.* upright orientation) increases (Figure 2.20c).

For larger particles ($c > 10$ nm), the frequency of each orientation can be estimated by using the value of $\theta_{r,b}$:

$$P(\text{upright}) = \theta_{r,b}/180° \text{ and } P(\text{tilted}) = (180° - \theta_{r,b})/180° \qquad (2.45)$$

as indicated by the dashed line in Figure 2.20c. In this approach, it is assumed that the initial orientation of particles is random, and the particles can continuously rotate to find an energy minimum.

2.4.2   Equilibrium Configurations (Experimental Approach)

Recent progress has enabled the fabrication of various types of amphiphilic particles with complex shapes.[57,61,63,65,66,72,75–81] For example, amphiphilic ellipsoids have been prepared using a mechanical stretching method in the presence of an external field.[72] Amphiphilic dumbbells comprising two partially fused spheres have been prepared using bulk synthesis as well as microfluidics using photopolymers.[74–76] These dumbbells are advantageous for independently controlling the size ratio of the two spheres and the wettability of each sphere. Amphiphilic cylinders have also been synthesized using a micro-moulding method.[61] These Janus cylinders possess a sharp and smooth WSL between the two sides, which is useful for systematic investigation of their interfacial behaviours.

The configuration of amphiphilic cylinders at a decane–water interface strongly depends on the aspect ratio (AR).[19,20] For particles with small aspect ratios of $AR = 0.9$ and 1.2 ($L_A/L_P = 1.3$; where the L_A and L_P represent the heights of apolar and polar regions in Figure 2.22d), they predominantly exhibit the upright orientation ($\sim 98\%$), and the WSL of these particles is pinned at the interface (Figure 2.21a,b). This experimental result is consistent with the attachment energy calculation in Figure 2.21e, indicating

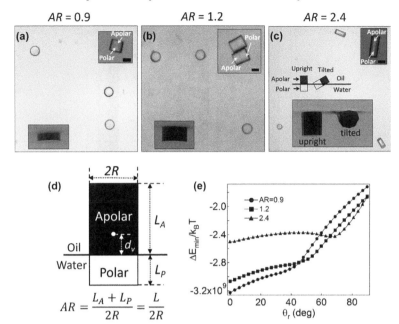

Figure 2.21 (a–c) Configurations of amphiphilic cylinders at the oil–water interface. The scale bar is 50 μm. (d) Schematic geometry of amphiphilic cylinder. (e) Minimum attachment energy profiles. Modified and reprinted from reference 19.

a single energy minimum at the upright orientation. Amphiphilic cylinders with a high aspect ratio of $AR = 2.4$ ($L_A/L_P = 1$) adopt both pinned-upright and tilted configurations (Figure 2.21c). The presence of a secondary energy minimum at a tilt angle in Figure 2.21e accounts for the coexistence of the two configurations and suggests that some particles can be kinetically trapped at the tilted configuration. The tilt angle is found to be $\theta_r = 64 \pm 6°$ for $AR = 2.4$, which also shows a good agreement with the calculation, $\theta_{r,\text{meta}} = 66°$.

The location of the energy barrier ($\theta_{r,b}$) between the primary and the secondary energy minima determines the frequency of each orientation (Equation 2.45).[19] For instance, the value of $\theta_{r,b}$ for $AR = 2.4$ in Figure 2.21e is found to be $\theta_{r,b} = 41$–$42°$. Using Equation 2.45, the frequencies of upright and tilted orientations result in $P_{\text{upright}} \approx 23\%$ and $P_{\text{tilted}} \approx 77\%$. These calculated values are consistent with the experimental observation, $P_{\text{upright}} \approx 19\%$ and $P_{\text{tilted}} \approx 81\%$ for $AR = 2.4$, as shown in Figure 2.22.

Orientation phase diagrams in Figure 2.23 summarize the effect of geometry (AR and L_A/L_P) and wettability (β and θ_A/θ_P) on the orientation behaviours of amphiphilic Janus cylinders at the oil–water interface. In each of the three phase diagrams in Figure 2.23, three distinct regions are observed. The upright-only region is generally observed in the low AR and strong wettability (large β, or large θ_A and small θ_P) ranges (pink in Figure 2.23).

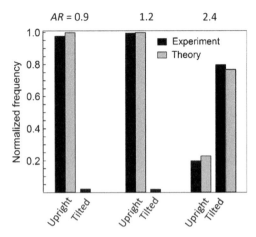

Figure 2.22 Histogram for orientation statistics of amphiphilic cylinders at the oil–water interface.
Modified and reprinted from reference 19.

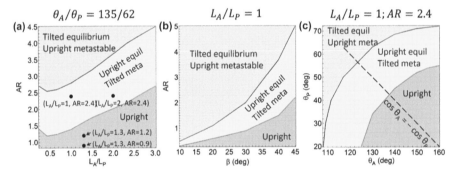

Figure 2.23 Orientation phase diagrams for amphiphilic cylinders at the oil–water interface.

Two coexisting regions appear: when *AR* is large and wettability difference is small, tilted equilibrium and upright metastable orientations coexist (light blue in Figure 2.23), whereas for moderate conditions of *AR* and wettability difference, coexistence of the upright equilibrium and tilted metastable orientations is observed (yellow in Figure 2.23). Notably, the tilted-only region is not shown in these phase diagrams; it may only exist under extreme conditions, such as sufficiently large *AR* and small β.[16]

An interesting feature of these amphiphilic cylinders is that their wettability becomes 'double hydrophilic' when they are placed at an air–water interface.[20] The contact angles of the two sides are $\theta_P \approx 62°$ (polar) and $\theta_A \approx 135°$ (apolar) when the particles are at the oil–water interface; however at the air–water interface, these two angles become $\theta_{SP} \approx 32°$ (strongly polar) and $\theta_{WP} \approx 80°$ (weakly polar); thus these are asymmetrically hydrophilic cylinders. It is interesting to note that this switchable wetting property between

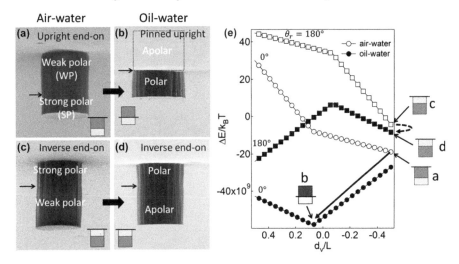

Figure 2.24 Wetting transition of double hydrophilic cylinders at the air–water interface to amphiphilic cylinders at the oil–water interface. Modified and reprinted with permission from reference 20. © 2013 American Chemical Society.

double hydrophilicity and amphiphilicity is analogous to double hydrophilic block copolymers (DHBCs), of which the surface activity (amphiphilicity) can be triggered by the presence of liquid–liquid or liquid–solid interfaces.[82]

To systematically understand the effect of wettability change upon the appearance of a new fluid–fluid interface, Janus cylinders with $AR = 1.2$ are initially dispersed at the air–water interface and a hydrocarbon oil is subsequently added on the water surface. At the air–water interface, double hydrophilic Janus cylinders adopt the upright end-on configuration ($d_v/L = -0.5$; $\theta_r = 0°$). When oil is added to the superphase, these particles become amphiphilic and adopt the pinned-upright configuration ($d_v/L \approx 0.08$; $\theta_r = 0°$) at the oil–water interface (Figure 2.24a,b). This pinned-upright configuration is such that the apolar and polar surfaces of the particle are fully exposed to their preferred fluid phases (*i.e.* apolar with oil and polar with water) and the WSL is pinned to the oil–water interface (Figure 2.24b). This out-of-plane transition is due to a decrease in the attachment energy, as indicated by the solid arrow in Figure 2.24e. A few particles that initially adopt the inverse end-on configuration ($d_v/L = -0.5$; $\theta_r = 180°$) at the air–water interface, however, does not translate upward into the oil phase (Figure 2.24c,d) because of the energy penalty caused by the unfavourable contact of the polar surface with the oil phase (Figure 2.24e).

2.4.3 Assemblies and Interactions

The assemblies and microstructures of amphiphilic particles at a decane–water interface are highly influenced by chemical and geometric properties

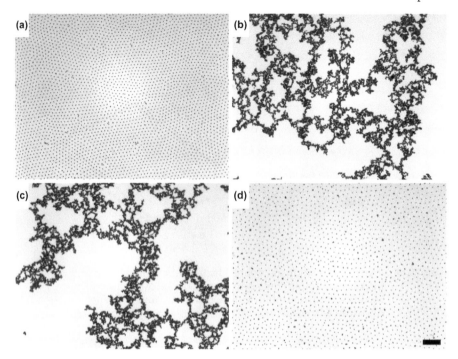

Figure 2.25   Microstructures of (a) unmodified PS, (b) Au-PS, (c) DDT (1-dodeca-
nethiol)-Au-PS and (d) MPA (3-mercaptopropionic acid)-Au-PS particles
at the oil–water interface. The scale bar is 100 µm.
Modified and reprinted from reference 71.

of the particles. As shown in Figure 2.25b,c, amphiphilic Janus spheres with
apolar and polar hemispheres (Au-PS and DDT-Au-PS) exhibit attractive
interactions, resulting in percolated aggregate networks.[71] In this case, an
undulating three-phase contact line on the particle surface due to the diffuse
Janus boundary leads to quadrupolar capillary attractions, which are confirmed by the pair interaction forces (see Section 2.2.2.2). In contrast, double
hydrophilic Janus spheres with two different polar hemispheres (MPA-Au-PS)
show repulsive interactions, forming a hexagonal microstructure shown in
Figure 2.25d. These structures are similar to those obtained with unmodified
PS as shown in Figure 2.25a. The similarity of the surface chemistry of the
two sides reduces the contact line undulation such that the electrostatic
repulsion becomes dominant.

The particle shape also has a significant impact on the lateral interactions
between Janus particles and can be clearly illustrated by the behaviour of
Janus cylinders at a fluid interface. The assemblies and interactions of
amphiphilic Janus cylinders are very different from those of chemically
homogeneous ellipsoids and cylinders, which exhibit highly deterministic
assembly behaviours. These non-spherical homogeneous particles dominantly exhibit quadrupolar capillary interactions ($F \sim r^{-5}$) with neighbouring
particles, caused by the quadrupolar shape of the interface meniscus

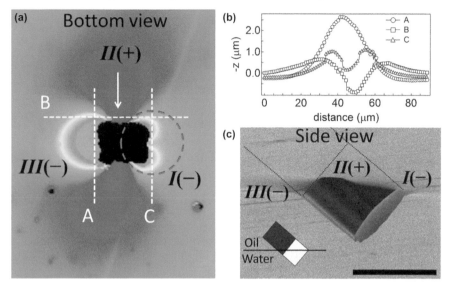

Figure 2.26   Quasi-quadrupolar interface deformation around a tilted Janus cylinder with $AR = 2.4$ ($L_A/L_P = 2$) at the oil–water interface. (a) Profilometry image. (b) Magnitude of the interface deformation corresponding to the dashed lines on panel (a). SEM image of a tilted particle on the PDMS slab. The scale bar is 30 µm.
Modified and reprinted from reference 19.

deformation around each particle.[37,39] This interaction force leads to deterministic assemblies between these anisotropic particles with either tip-to-tip or side-to-side attachment. In contrast, the interface deformation around tilted Janus cylinders at the oil–water interface is found to be more complex (Figure 2.26). The interface undulation around each particle consists of negative deformation around the apolar head (III), two positive deformations around two sides (II), and net negative deformation (I) including two small negative poles and one positive pole between them (see the one-dimensional deformation profile in Figure 2.26b).[19]

The presence of the asymmetric 'quasi-quadrupolar' interface deformation around individual amphiphilic cylinders at the oil–water interface significantly affects their lateral pair interactions.[19] As shown in Figure 2.27a, when two particles approach in a side-to-side alignment and come in contact at $t = t_{max}$, the value of exponent that best describes the following relationship $r \sim (t_{max} - t)^\lambda$ is about $\lambda = 0.167 \pm 0.003$. Based on a linear relationship between the interaction force ($F \sim r^{\xi-1}$) and the drift velocity $\left(v = \dfrac{dr}{dt} \right)$ in the low Reynolds number limit ($N_{Re} = \dfrac{2Rv\rho}{\mu} \ll 1$ where ρ and μ are the density and viscosity of the fluid, respectively), the power law exponent (ξ) of the interaction force can be related to the value of λ, yielding $\xi = 2 - 1/\lambda$. Thus, the corresponding value of ξ in the case of side-to-side approach is -4, indicating that the interaction can be described by the quadrupolar interface deformation.[35] When two particles approach in other

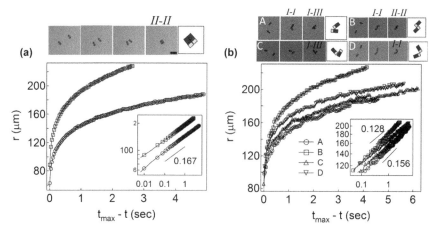

Figure 2.27 Lateral interaction between two tilted Janus cylinders with $AR = 2.4$ $(L_A/L_P = 2)$ when they approach (a) in side-to-side and (b) in arbitrary alignments. The scale bar is 50 μm.
Modified and reprinted from reference 19.

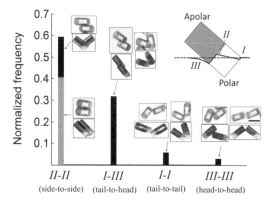

Figure 2.28 Statistics of lateral assemblies between two tilted amphiphilic cylinders with $AR = 2.4$ $(L_A/L_P = 2)$ at the oil–water interface. The scale bar is 50 μm.
Modified and reprinted from reference 19.

alignments, however, the obtained values of ξ vary from -5.8 to -4.4, as shown in Figure 2.27b. This broad distribution implies a partial contribution of hexapole characteristics to the interaction. Note that the power law exponents of the interaction force between two homogeneous spheres with quadrupole–hexapole and hexapole–hexapole deformations are reported to be $\xi = -5$ and -6, respectively.[30] In short, the dependence of interaction scaling behaviour on the pair alignments can be attributed to the presence of the asymmetric quasi-quadrupolar deformation.

The observed multipolar interface deformation around tilted Janus cylinders also accounts for diverse assembly behaviours.[19] A variety of possible combinations between multi-poles around each particle leads to the non-deterministic assembly structure, as shown in Figure 2.28. In general, the majority of particle pairs tend to assemble in side-to-side (II-II; 59%) and tail-to-head (I-III; 32%) alignments. Only a small number of pairs ($\sim$ 8%) with either the tail-to-tail (I-I) or head-to-head (III-III) assemblies are found, indicating that such geometries can be easily rearranged to form more energetically stable ones.

2.5 Concluding Remarks

The shape, wettability and size of particles significantly influence their interfacial phenomena, resulting in diverse microscopic pair interactions, assemblies and micromechanics. These parameters ultimately affect the bulk interfacial properties of particle-laden fluid interfaces. Understanding the colloidal behaviours at two-dimensional fluid–fluid interfaces from the micro- to the macroscopic levels potentially enables the bottom-up assembly of novel structures and the new pathways to designed functional materials.

Acknowledgements

D. Lee acknowledges funding from the American Chemical Society Petroleum Research Fund (ACS PRF) and NSF CAREER Award (DMR-1055594). B. J. Park acknowledges financial support from Basic Science Research Program through the National Research Foundation of Korea (NRF) funded by the Ministry of Science, ICT & Future Planning (NRF-2014R1A1A1005727).

References

1. B. P. Binks, *Curr. Opin. Colloid Interface Sci.*, 2002, **7**, 21–41.
2. B. P. Binks and T. S. Horozov, *Colloidal Particles at Liquid Interfaces*, Cambridge University Press, New York, 2006.
3. S. U. Pickering, *J. Chem. Soc. Trans.*, 1907, **91**, 2001–2021.
4. W. Ramsden, *Proc. R. Soc. Lond.*, 1903, **72**, 156–164.
5. M. Oettel and S. Dietrich, *Langmuir*, 2008, **24**, 1425.
6. H. Hoekstra, J. Vermant, J. Mewis and G. Fuller, *Langmuir*, 2003, **19**, 9134–9141.
7. E. J. Stancik, M. Kouhkan and G. G. Fuller, *Langmuir*, 2004, **20**, 90–94.
8. J. Fernández-Toledano, A. Moncho-Jordá, F. Martínez-López and R. Hidalgo-Alvarez, *Langmuir*, 2004, **20**, 6977–6980.
9. S. Reynaert, P. Moldenaers and J. Vermant, *Phys. Chem. Chem. Phys.*, 2007, **9**, 6463–6475.
10. S. Reynaert, P. Moldenaers and J. Vermant, *Langmuir*, 2006, **22**, 4936–4945.
11. B. J. Park and E. M. Furst, *Soft Matter*, 2011, **7**, 7676–7682.
12. B. J. Park, J. P. Pantina, E. M. Furst, M. Oettel, S. Reynaert and J. Vermant, *Langmuir*, 2008, **24**, 1686–1694.
13. B. P. Binks and P. D. I. Fletcher, *Langmuir*, 2001, **17**, 4708–4710.

14. S. Jiang and S. Granick, *J. Chem. Phys.*, 2007, **127**, 161102–161104.
15. T. Ondarçuhu, P. Fabre, E. Raphaël and M. Veyssié, *J. Phys. France*, 1990, **51**, 1527–1536.
16. B. J. Park and D. Lee, *ACS Nano*, 2012, **6**, 782–790.
17. B. J. Park and D. Lee, *Soft Matter*, 2012, **8**, 7690–7698.
18. Y. Hirose, S. Komura and Y. Nonomura, *J. Chem. Phys.*, 2007, **127**, 054707–054705.
19. B. J. Park, C.-H. Choi, S.-M. Kang, K. E. Tettey, C.-S. Lee and D. Lee, *Soft Matter*, 2013, **9**, 3383–3388.
20. B. J. Park, C.-H. Choi, S.-M. Kang, K. E. Tettey, C.-S. Lee and D. Lee, *Langmuir*, 2013, **29**, 1841–1849.
21. J. N. Israelachvili, *Intermolecular and Surface Forces*, 3rd edn, Academic Press, New York, 2011.
22. A. J. Hurd, *J. Phys. A: Math. Gen.*, 1985, **45**, L1055–L1060.
23. P. Pieranski, *Phys. Rev. Lett.*, 1980, **45**, 569–572.
24. J. Frank and H. Stillinger, *J. Chem. Phys.*, 1961, **35**, 1584–1589.
25. D. Frydel, S. Dietrich and M. Oettel, *Phys. Rev. Lett.*, 2007, **99**, 118302.
26. K. Masschaele, B. J. Park, E. M. Furst, J. Fransaer and J. Vermant, *Phys. Rev. Lett.*, 2010, **105**, 048303.
27. B. J. Park, J. Vermant and E. M. Furst, *Soft Matter*, 2010, **6**, 5327–5333.
28. R. Aveyard, B. P. Binks, J. H. Clint, P. D. I. Fletcher, T. S. Horozov, B. Neumann, V. N. Paunov, J. Annesley, S. W. Botchway, D. Nees, A. W. Parker, A. D. Ward and A. N. Burgess, *Phys. Rev. Lett.*, 2002, **88**, 246102–246104.
29. R. Aveyard, J. H. Clint, D. Nees and V. N. Paunov, *Langmuir*, 2000, **16**, 1969–1979.
30. K. D. Danov, P. A. Kralchevsky, B. N. Naydenov and G. Brenn, *J. Colloid Interface Sci.*, 2005, **287**, 121–134.
31. A. Dominguez, M. Oettel and S. Dietrich, *J. Phys.: Condens. Matter*, 2005, **17**, S3387–S3392.
32. R. Golestanian, M. Goulian and M. Kardar, *Phys. Rev. E*, 1996, **54**, 6725–6734.
33. P. A. Kralchevsky and K. Nagayama, *Langmuir*, 1994, **10**, 23–36.
34. M. Oettel, A. Dominguez and S. Dietrich, *Phys. Rev. E*, 2005, **71**, 051401–051416.
35. D. Stamou, C. Duschl and D. Johannsmann, *Phys. Rev. E*, 2000, **62**, 5263–5272.
36. N. D. Vassileva, D. van den Ende, F. Mugele and J. Mellema, *Langmuir*, 2005, **21**, 11190–11200.
37. E. P. Lewandowski, M. Cavallaro, L. Botto, J. C. Bernate, V. Garbin and K. J. Stebe, *Langmuir*, 2010, **26**, 15142–15154.
38. E. P. Lewandowski, P. C. Searson and K. J. Stebe, *J. Phys. Chem. B*, 2006, **110**, 4283–4290.
39. J. C. Loudet, A. M. Alsayed, J. Zhang and A. G. Yodh, *Phys. Rev. Lett.*, 2005, **94**, 018301–018304.
40. J. C. Loudet and B. Pouligny, *Europhys. Lett.*, 2009, **85**, 28003.

41. L. Botto, E. P. Lewandowski, M. Cavallaro and K. J. Stebe, *Soft Matter*, 2012, **8**, 9957–9971.
42. L. Botto, L. Yao, R. L. Leheny and K. J. Stebe, *Soft Matter*, 2012, **8**, 4971–4979.
43. H. Lehle, E. Noruzifar and M. Oettel, *Eur. Phys. J. E*, 2008, **26**, 151–160.
44. E. Van Nierop, M. Stijnman and S. Hilgenfeldt, *EPL (Europhysics Letters)*, 2005, **72**, 671.
45. E. P. Lewandowski, J. A. Bernate, A. Tseng, P. C. Searson and K. J. Stebe, *Soft Matter*, 2009, **5**, 886–890.
46. B. J. Park and E. M. Furst, *Langmuir*, 2008, **24**, 13383–13392.
47. A. D. Law, D. M. A. Buzza and T. S. Horozov, *Phys. Rev. Lett.*, 2011, **106**, 128302.
48. A. Law, T. Horozov and D. Buzza, *Soft Matter*, 2011, 7, 8923–8931.
49. B. J. Park and E. M. Furst, *Soft Matter*, 2011, 7, 7683–7688.
50. C. Ho, R. Ottewill, A. Keller and J. Odell, *Polym. Int.*, 1993, **30**, 207–211.
51. B. Madivala, J. Fransaer and J. Vermant, *Langmuir*, 2009, **25**, 2718–2728.
52. P. J. Yunker, T. Still, M. A. Lohr and A. Yodh, *Nature*, 2011, **476**, 308–311.
53. J. Vermant, *Nature*, 2011, **476**, 286–287.
54. M. Cavallaro, L. Botto, E. P. Lewandowski, M. Wang and K. J. Stebe, *Proc. Natl. Acad. Sci. U.S.A.*, 2011, **108**, 20923–20928.
55. E. M. Furst, *Proc. Natl Acad. Sci.*, 2011, **108**, 20853–20854.
56. A. Kumar, B. J. Park, F. Tu and D. Lee, *Soft Matter*, 2013, DOI: 10.1039/C1033SM50239B.
57. S. C. Glotzer and M. J. Solomon, *Nat. Mater.*, 2007, **6**, 557–562.
58. J. A. Champion, Y. K. Katare and S. Mitragotri, *J. Controll. Release*, 2007, **121**, 3–9.
59. A. Walther and A. H. E. Muller, *Soft Matter*, 2008, 4, 663–668.
60. S. Jiang, Q. Chen, M. Tripathy, E. Luijten, K. S. Schweizer and S. Granick, *Adv. Mater.*, 2010, **22**, 1060–1071.
61. C.-H. Choi, J. Lee, K. Yoon, A. Tripathi, H. A. Stone, D. A. Weitz and C.-S. Lee, *Angew. Chem. Int. Edit.*, 2010, **49**, 7748–7752.
62. T. Brugarolas, B. J. Park and D. Lee, *Adv. Funct. Mater.*, 2011, **21**, 3924–3931.
63. J. Hu, S. Zhou, Y. Sun, X. Fang and L. Wu, *Chem. Soc. Rev.*, 2012, **41**, 4356–4378.
64. P. G. de Gennes, *Rev. Mod. Phys.*, 1992, **64**, 645–648.
65. N. Glaser, D. J. Adams, A. Boker and G. Krausch, *Langmuir*, 2006, **22**, 5227–5229.
66. J.-W. Kim, D. Lee, H. C. Shum and D. A. Weitz, *Adv. Mater.*, 2008, **20**, 3239–3243.
67. S. Sacanna, W. K. Kegel and A. P. Philipse, *Phys. Rev. Lett.*, 2007, **98**, 158301.
68. R. Aveyard, *Soft Matter*, 2012, **8**, 5233–5240.
69. C. Casagrande, P. Fabre, E. Raphael and M. Veyssié, *Europhys. Lett.*, 1989, **9**, 251–255.
70. D. L. Cheung and S. A. F. Bon, *Soft Matter*, 2009, **5**, 3969–3976.

71. B. J. Park, T. Brugarolas and D. Lee, *Soft Matter*, 2011, **7**, 6413–6417.
72. O. Güell, F. Sagués and P. Tierno, *Adv. Mater.*, 2011, **23**, 3674–3679.
73. Y. Liu, W. Li, T. Perez, J. D. Gunton and G. Brett, *Langmuir*, 2012, **28**, 3–9.
74. E. B. Mock, H. De Bruyn, B. S. Hawkett, R. G. Gilbert and C. F. Zukoski, *Langmuir*, 2006, **22**, 4037–4043.
75. K.-H. Roh, D. C. Martin and J. Lahann, *Nat. Mater.*, 2005, **4**, 759–763.
76. J.-W. Kim, R. J. Larsen and D. A. Weitz, *J. Am. Chem. Soc.*, 2006, **128**, 14374–14377.
77. K. Fujimoto, K. Nakahama, M. Shidara and H. Kawaguchi, *Langmuir*, 1999, **15**, 4630–4635.
78. Y. Hong, N. M. K. Blackman, N. D. Kopp, A. Sen and D. Velegol, *Phys. Rev. Lett.*, 2007, **99**, 178103–178104.
79. D. Lee and D. A. Weitz, *Small*, 2009, **5**, 1932–1935.
80. T. M. Ruhland, A. H. Gröschel, A. Walther and A. H. E. Müller, *Langmuir*, 2011, **27**, 9807–9814.
81. Z. Zhang, P. Pfleiderer, A. B. Schofield, C. Clasen and J. Vermant, *J. Am. Chem. Soc.*, 2010, **133**, 392–395.
82. H. Cölfen, *Macromol. Rapid Comm*, 2001, **22**, 219–252.

CHAPTER 3

Polymer Colloidal Particles Prepared by Pickering Emulsion Polymerization or Self-Assembly Method

HANYING ZHAO* AND JIA TIAN

Department of Chemistry, Nankai University, Tianjin 300071, P. R. China
*Email: hyzhao@nankai.edu.cn

3.1 Introduction

The self-assembly of solid particles at liquid–liquid interfaces was first reported by Pickering[1] and Ramsden[2] about a century ago. Emulsions stabilized by solid particles are known as Pickering emulsions. When dispersed in a mixture of oil and water, the particles have a strong tendency to self-assemble at liquid–liquid interfaces, mainly because of the reduction in the total interfacial energy upon replacing part of the liquid–liquid interface by a liquid–particle interface.[3–5] The particles can be latex particles,[4–6] inorganic particles[7,8] or polymeric micelles.[9] As shown in Equation 3.1, in a given Pickering emulsion system, a decrease in the interfacial energy is related to the average size of the particles:[10]

$$E_0 - E_1 = \Delta E_1 = -\frac{\pi r^2}{\gamma_{o/w}} \left[r_{o/w} - (\gamma_{p/w} - \gamma_{p/o}) \right]^2 \tag{3.1}$$

RSC Soft Matter No. 3
Particle-Stabilized Emulsions and Colloids: Formation and Applications
Edited by To Ngai and Stefan A. F. Bon
Published by the Royal Society of Chemistry, www.rsc.org

where r is the radius of the solid particles, and $\gamma_{o/w}$, $\gamma_{p/w}$ and $\gamma_{p/o}$ represent interfacial tensions between oil and water, particle and water, as well as particle and oil. The reduction in the interfacial energy depends on r^2, and in comparison to large particles the assembly structure formed by smaller nanoparticles is less stable.[11] For microscopic particles, the decrease in the interfacial energy is much larger than the thermal energy ($k_B T$), and permanent confinement of the particles to the interface is achieved.[12] For nano-sized particles, the energy reduction is comparable to thermal energy, and the assembly of nanoparticles at liquid–liquid interfaces is dynamic, with particles adsorbing to and desorbing from the interfaces.[12]

The decrease in the interfacial energy is also related to the interfacial tensions between particle and water, and particle and oil as shown in Equation 3.1, which are dependent on the hydrophobicity of the solid particles. So the hydrophobicity of the solid particles plays an important role in determining the efficiency of the particular surfactants. If the particles are too hydrophobic, they tend to stay in the oil phase but at the interface; if the particles are too hydrophilic, they stay in the aqueous phase. Takahara and coworkers prepared silica particles with different hydrophobicity by reactions with an alkylsilane agent, and investigated the location of the particles in a mixture of water and oil.[13] They found that the bare hydrophilic silica particles dispersed only in water due to the surface hydroxyl groups. However, after fully reacting with an alkylsilane agent the hydrophobic particles dispersed well in toluene. Only silica particles after partial modification of the external surface can locate at the interface between oil and water.

Colloidal particles can be prepared by Pickering emulsion polymerization and particle-stabilized colloidal particles are obtained. Bon and coworkers prepared organic–inorganic hybrid hollow spheres by a Pickering emulsion method.[14] This type of TiO_2-stabilized hollow spheres can be used in the fields of drug release and photocatalytic applications. In order to prepare hybrid hollow spheres, TiO_2 nanoparticles were dispersed in water, and emulsification of the oil mixture of monomer/hexadecane containing the initiator was conducted. The TiO_2-stabilized colloidosomes were polymerized overnight. The diameter of the hybrid hollow spheres can be tailored by variation of the total relative amount of the oil phase. Bon and coworkers also prepared PS colloidal particles armoured with Laponite RD clay particles. The PS colloidal particles were made by a Pickering mini-emulsion polymerization process.[15] Voorn and coworkers reported surfactant-free inverse emulsion polymerization by using organically modified clay platelets as stabilizers.[8] Colloidally stable inverse Pickering emulsions of aqueous monomers in cyclohexane stabilized by hydrophobic Cloisite were also reported. With both oil-soluble and water-soluble free-radical initiators, inverse latexes in the size range of 700–980 nm were obtained.

Polymer brushes refer to an assembly of polymer chains that are tethered by one end to a surface or interface. The most widely studied polymer brushes are block copolymer brushes and mixed homopolymer brushes.[16–19]

One of the most important properties of the polymer brushes is their response to environmental conditions. For example, block copolymer brushes and mixed homopolymer brushes on a solid surface can make different nanopatterns through the treatments of selective solvents, and smart surfaces with different properties can be created.[20,21] Zhao and coworkers synthesized poly(styrene-*block*-methyl methacrylate) (PS-*b*-PMMA) block copolymer brushes on silicate substrates. They found that if the block copolymer brushes were immersed in cyclohexane, a solvent for PS and a precipitant for PMMA, PMMA blocks collapse from the solvent and PS blocks form a shield around the PMMA aggregates.[22] Polymer brushes can be prepared on many different solid substrates, and nanopatterns of the polymer brushes can be created in selective solvents. Zhao and Shipp prepared poly(styrene-*block*-butyl acrylate) (PS-*b*-PtBA) block copolymer brushes on the surface of clay layers, and they found the block copolymer brushes form worm-like surface aggregates after treatment in acetone.[23,24] Because of the response of the polymer brushes in different solvents, the polymer brush-modified solid particles present different surface properties in different solvents. Favourable interfacial energy reduction is achieved when the brush-modified particles are used in Pickering emulsions. In this chapter, applications of polymer brush-modified inorganic particles in Pickering emulsions are reviewed.

3.2 Applications of Polymer Brush-Modified Clay Layers in Pickering Emulsions

3.2.1 Clay Layers with Block Copolymer Brushes

Layered silicates consist of two silica tetrahedral sheets fused to an edge-shared octahedral sheet.[25] A lattice is formed from periodic stacking of the layers, and the stack held together by van der Waals forces. The silicate layers possess a net charge deficiency that is compensated by cations in the interlayer galleries. The interlayer spacing between the silicate layers is hydrophilic and can be modified by organic compounds or polymers through cation exchange reactions. Some groups reported preparation of clay-armoured polymer colloidal particles by specific interactions or Pickering emulsion polymerization.[8,15,26] For example, Antonietti and co-workers prepared 'armoured colloidal particles' based on electrostatic interaction between positively charged colloidal particles and negatively charged clay sheets.[26] The coating of the colloidal particles with clay sheets increases the surface roughness of the particles. However, the overall negative charges and the amphoteric rim of the clay particles are usually not enough to produce stable dispersions. Introducing polymer brushes on the surfaces of clay layers is an efficient way to improve the assembly of clay particles at a liquid–liquid interface and to obtain stable colloidal particles.

Yang and coworkers synthesized amphiphilic poly(methyl methacrylate-*block*-2-(dimethylamino)ethyl methacrylate) (PMMA-*b*-PDMAEMA) block copolymer brushes on the surface of clay layers by *in situ* atom transfer radical polymerization (ATRP).[27] The block copolymer brushes make different nanopatterns on the clay layers after treatment in different solvents. For example, after treatment in THF, a good solvent for both PMMA and PDMAEMA blocks, block copolymer brushes form lamella structures on the surface of clay layers, as shown in Figure 3.1a and b. The average thickness of the lamella strip is about 3 nm. On the surface of clay layers the block copolymer brushes have phase separation to avoid unfavourable interaction between PMMA blocks and PDMAEMA blocks. So the lamella structure on the surface of clay was formed. After treatment in water at a pH value of 5.6, surface micelles, as indicated by an arrow in Figure 3.1c, are formed. The diameter of surface micelles is in the range 3–5 nm. Water with a pH of 5.6 is a solvent for PDMAEMA blocks and a precipitant for PMMA blocks. In aqueous solution in order to avoid contact with water, PMMA blocks

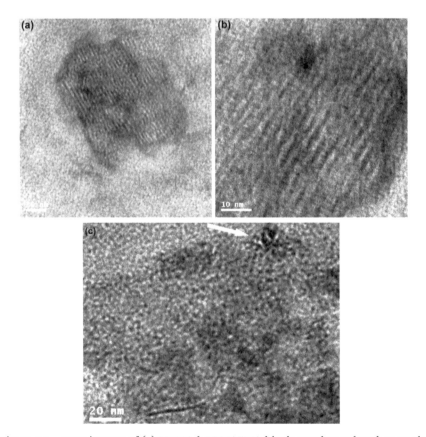

Figure 3.1   TEM images of (a) PMMA-*b*-PDMAEMA block copolymer brushes on the surface of clay layers after treatment in THF, (b) a magnification of a part of image a, (c) block copolymer brushes on the surface of clay layers after treatment in water at a pH of 5.6. All the samples were stained by OsO_4. Reproduced with permission from *Langmuir* 2007, **23**, 2867.

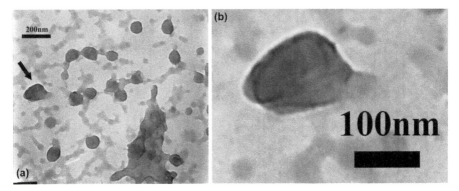

Figure 3.2 (a) A TEM image of PMMA colloid particles stabilized by clay/PMMA-*b*-PDMAEMA. (b) A magnified TEM image of a specific colloid particle indicated by an arrow in image a.
Reproduced with permission from *Langmuir* 2007, **23**, 2867.

collapse, forming the cores of surface micelles, and PDMAEMA blocks form layers around PMMA cores.

PMMA colloid particles stabilized by clay layers with block copolymer brushes were prepared by suspension polymerization. Clay/block copolymer nanocomposite was dissolved in acetone and added into water at a pH value of 5.6. After acetone was removed at a reduced pressure, MMA monomer, DVB cross-linker and BPO initiator were added into the solution. Figure 3.2a shows a TEM image of PMMA colloid particles and Figure 3.2b is a magnification of the particles indicated by an arrow on Figure 3.2a. The TEM images demonstrate that clay layers locate around the surface of colloid particles. MMA monomer is a solvent for PMMA and PDMAEMA, and water at pH 5.6 is a solvent for PDMAEMA. Clay layers with block copolymer brushes tend to locate at liquid-liquid interface so that the interfacial tension between two fluids can be reduced. After polymerization, PMMA colloid particles with clay layers on the surface were prepared.

3.2.2 Clay Layers with Homopolymer Brushes

Zhang and coworkers prepared a composite of clay and free-radical initiator with quaternary alkylammonium by ion exchange. PDMAEMA brushes were prepared by free-radical polymerization.[28] The conversion of DMAEMA monomer was kept at a low level by controlling the polymerization time. Because the polymerization time of DMAEMA is much short than the half-life of the free-radical initiators, the clay particles with PDMAEMA brushes and the leaving initiator molecules can be used in Pickering emulsion polymerization of MMA. PDMAEMA brushes on the surfaces of clay are hydrophilic, and the free-radical initiators on the clay are hydrophobic, so clay layers with polymer brushes and free-radical initiators are efficient stabilizers in Pickering emulsion polymerization of MMA. TEM images of

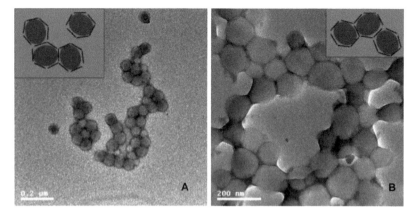

Figure 3.3 TEM images of PMMA colloidal particles stabilized by clay layers with PDMAEMA brushes.
Reproduced with permission from *J. Polym. Sci., Part A: Polym. Chem.* 2008, **46**, 2632.

PMMA colloidal particles stabilized by clay layers with PDMAEMA brushes are shown in Figure 3.3. Dark clay platelets on the surface of the particles are observed. It is worth noting that in comparison to the particles stabilized by organically modified clay layers, the colloidal particles stabilized by clay layers with PDMAEMA brushes are smaller, with narrower size distributions.

PDMAEMA is a pH-sensitive polymer, and the amine groups on the polymer chains are protonated under acidic conditions, so clay layers with PDMAEMA brushes have positive zeta potentials at low pH values (curve a in Figure 3.4) and the zeta potential decreases with the increase in pH value in the aqueous solution. The surfaces of the original clay particles are negatively charged, but the grafting of PDMAEMA changes the surface of the clay layers from negatively to positively charged. PMMA colloidal particles stabilized by clay layers with PDMAEMA brushes also have positive zeta potentials at low pH values (curve b in Figure 3.4). The isoelectric point of the colloidal particles is obtained at a pH value of about 7.3.

3.2.3 Clay Layers with Mixed Homopolymer Brushes

Two or more different polymers randomly grafted to the solid substrate form a mixed polymer brush layer.[29] Mixed homopolymer brushes also have environmental response, and can make different nanopatterns after treatment in selective solvents. Clay layers with mixed polymer brushes are expected to have interfacial activity and can be used as stabilizers in Pickering emulsions.

Yang and coworkers synthesized PS and PDMAEMA mixed polymer brushes on the surface of clay layers by *in situ* free-radical polymerization.[30] In order to prepare PS colloidal particles, clay layers with mixed polymer brushes were dispersed in THF and added to water at a pH value of 5.6. After removal of THF at reduced pressure, styrene monomer, DVB and free-radical initiator were added into the solution. Figure 3.5 represents SEM images of

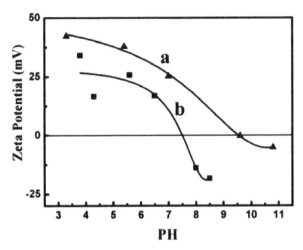

Figure 3.4 Plots of zeta potential versus pH values for clay layers with PDMAEMA brushes (a) and PMMA colloidal particles stabilized by clay layers with PDMAEMA brushes (b).
Reproduced with permission from *J. Polym. Sci., Part A: Polym. Chem.* 2008, **46**, 2632.

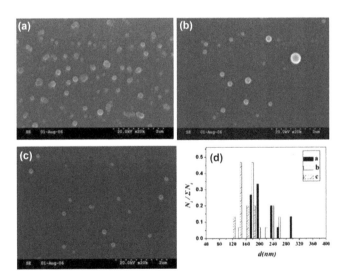

Figure 3.5 SEM images of PS colloidal particles stabilized by varying amounts of clay layers with mixed polymer brushes, (a) 0.5 wt%, (b) 1 wt%, (c) 1.5 wt%, and size distribution of particles (d).
Reproduced with permission from *J. Polym. Sci., Part A: Polym. Chem.* 2007, **45**, 5759.

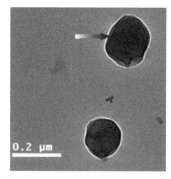

Figure 3.6   TEM image of PS colloidal particles stabilized by clay layers with mixed polymer brushes. The weight percentage of clay layers is 1 wt%. Reproduced with permission from *J. Polym. Sci., Part A: Polym. Chem.* 2007, **45**, 5759.

PS colloidal particles and the particle distribution. With an increase in the amount of clay layers added into the solution, the size of colloidal particles decreases and the size distribution becomes narrower. Figure 3.6 shows a TEM image of PS colloidal particles, in which clay layers on the surface of colloidal particles can be clearly seen. The size of the irregular clay layers is about 130 nm×150 nm. After polymerization the X-ray diffraction peak disappeared totally, indicating the interpenetration of PS chains into the gallery of clay layers and the intercalated structure was destroyed. For the clay-protected PS colloidal particles, positive zeta potentials were observed at low pH values due to the cationic nature of the PDMAEMA brushes.

3.2.4   Clay Layers with Hydrophilic Faces and Hydrophobic PS Brushes on the Edges

A clay particle is composed of layered plates, and the surface of a plate is negatively charged. In the galleries of clay layers there are inorganic positive counter ions. By exchanging these inorganic cations with organic long-chain aliphatic quaternary ammonium, many different types of polymer–clay nanocomposites were prepared.[31,32] Besides ion exchanging, another modification method is based on reactions of chloro- or alkoxysilanes with the silanol groups located on the edges of the clay sheets.[33,34] Wu and coworkers anchored ATRP initiator to the edges of clay particles, and prepared PS brushes on the edges of clay particles by ATRP (Figure 3.7).[35] The clay particles with hydrophilic faces and hydrophobic PS brushes on the edges can be used as efficient stabilizers in Pickering suspension polymerization of styrene. Figures 3.8a and b show TEM images of PS colloidal particles. In a magnified TEM image, clay particles on the surface of the colloidal particles can be observed (Figure 3.8b). The average diameter of PS colloidal particles is about 260 nm. Clay particles with hydrophilic faces and

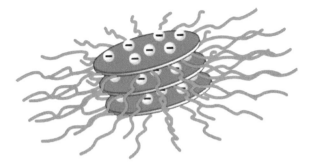

Figure 3.7 A cartoon representing the layered silicate particles with hydrophilic surface and hydrophobic PS brushes on the edges.
Reproduced with permission from *J. Polym. Sci., Part A: Polym. Chem.* 2009, **47**, 1535.

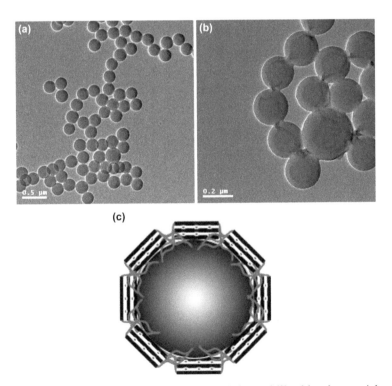

Figure 3.8 (a) A TEM image of PS colloidal particles stabilized by clay particles with hydrophilic faces and hydrophobic PS brushes on the edges, (b) a magnified TEM image of PS colloidal particles, (c) a schematic illustration of the PS colloidal particles stabilized by clay particles with amphiphilic structure.
Reproduced with permission from *J. Polym. Sci., Part A: Polym. Chem.* 2009, **47**, 1535.

hydrophobic PS brushes on the edges locate at the surface of PS colloidal particles with PS brushes penetrating into colloidal particles and negatively charged surfaces facing the medium, and the interfacial tension between colloidal particles and water is reduced. The structure is illustrated in Figure 3.8c. Because of the stabilization of the colloidal particles by negatively charged amphiphilic clay particles, the synthesized PS colloidal particles are negatively charged and further modification of the colloidal particles is also possible. Poly(2-vinylpyridine) (P2VP) was chosen to interact with PS colloidal particles in aqueous solution at pH 3.0. At this pH value, some of the P2VP units were protonated and the polymer chains were positively charged. The positively charged P2VP chains were adsorbed onto the surfaces of the colloidal particles, and the P2VP brushes could be used as a template in the preparation of metal nanoparticles.

3.3 Stabilization of Emulsions by Gold Nanoparticles (AuNPs) with Polymer Brushes

3.3.1 Amphiphilic AuNPs Formed at a Liquid–Liquid Interface

AuNPs are the most stable metal nanoparticles and have many fascinating properties. Colloidal particles with AuNPs on the surface find applications in catalysis and biotechnology. Many different methods have been employed in the synthesis of AuNPs on the surface of colloidal particles. An efficient and direct approach is based on emulsions stabilized by AuNPs. However, in a given Pickering emulsion system, a decrease in the interfacial energy is related to the size of the solid particles, and in comparison to large particles, the assembly structures formed by nano-sized particles are less stable. So the self-assembly of AuNPs with sizes of around 5 nm at a liquid–liquid interface is difficult. The hydrophilic surfaces of AuNPs could be partly modified by hydrophobic polymer chains, and the amphiphilic AuNPs are able to undergo self-assembly onto the liquid–liquid interface. Based on *in situ* modification and self-assembly of AuNPs at a liquid–liquid interface, emulsions stabilized by AuNPs and a variety of advanced structures can be prepared.

Tian and coworkers demonstrated that a stable emulsion could be obtained by mixing an aqueous dispersion of citrate-stabilized AuNPs and toluene solution of thiol-terminated PS (PS-SH).[36] A controlled experimental result shows that the average size of toluene droplets in the aqueous dispersion of AuNPs is much bigger than a mixture with dissolved PS-SH in the toluene phase. The decrease in toluene droplet size is attributed to the ligand exchange between PS-SH and citrate-stabilized AuNPs at the liquid–liquid interface and the grafting of PS chains to the AuNPs (Figure 3.9). The interfacial tension between toluene and water is reduced due to the location of the amphiphilic AuNPs at the liquid–liquid interface. Colloidal particles with PS cores and AuNPs coronae were prepared by adding the emulsion to a 5-fold excess of methanol. Upon addition of emulsions into the methanol,

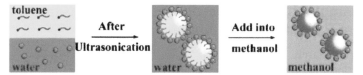

Figure 3.9 A scheme for the preparation of colloidal particles with PS cores and AuNPs coronae.
Reproduced with permission from *Langmuir* 2010, **26**, 8762.

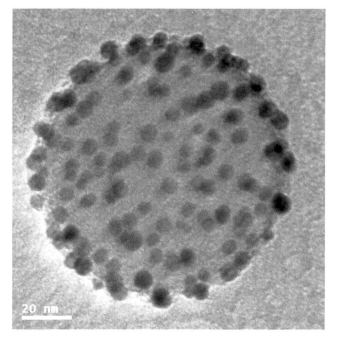

Figure 3.10 TEM image of a colloidal particle with PS cores and AuNPs corona. The weight ratio of PS-SH to AuNPs is about 3:1.
Reproduced with permission from *Langmuir* 2010, **26**, 8762.

PS chains collapse, forming the cores of the colloidal particles, and the hydrophilic AuNPs stay on the surface to stabilize the colloidal particles. A TEM image of a colloidal particle with PS cores and AuNPs corona is shown in Figure 3.10. The weight ratio of PS-SH to AuNPs plays an important role in controlling the morphology of colloidal particles. TEM and light scattering results indicate that the average size of the colloidal particles increases with the weight ratio. The surface of a nanoparticle in the corona is divided into two parts; one part is embedded in the PS phase because of the grafting of PS chains, and the other part is a citrate-protected surface and is hydrophilic. Because the AuNPs are negatively charged, the ζ-potential values of the colloidal particles are negative.

3.3.2 Amphiphilic Nanoparticles Complexes Formed at a Liquid–Liquid Interface

In other research, Tian and coworkers prepared PS brushes on Fe_3O_4 nanoparticles (Fe_3O_4NPs) by reversible addition-fragmentation chain transfer (RAFT) polymerization.[37] After a reduction reaction PS brushes with terminal thiol groups were synthesized on Fe_3O_4NPs (HS-PS-Fe_3O_4NPs). Citrate-stabilized AuNPs (5 nm) were dispersed in aqueous solution, and Fe_3O_4NPs with thiol-terminated PS brushes were dispersed in toluene. Upon mixing of the organic solution with the aqueous solution, AuNPs interacted with thiol terminal groups on PS brushes at the oil–water interface, and amphiphilic nanoparticle complexes (AuNPs-PS-Fe_3O_4NPs) were formed. A stable o/w emulsion was obtained due to the formation of the amphiphilic nanoparticle complexes at the liquid–liquid interface. Upon addition of the emulsion to excess methanol, the amphiphilic nanoparticle complexes self-assembled into colloidal particles with hydrophobic Fe_3O_4NPs cores and hydrophilic AuNPs coronae (Figure 3.11). When the colloidal particles were redispersed into THF, a good solvent for PS, core–shell structures can be observed on TEM, indicating the formation of AuNPs-PS-Fe_3O_4NPs nanoparticle complexes (Figure 3.12a). In THF, PS brushes on the surface of Fe_3O_4NPs stretch into the solution, and attach to the AuNPs on the ends *via* Au-S interaction. The colloidal particles with Fe_3O_4NPs cores and AuNPs coronae were well dispersed in methanol due to the stabilization of AuNPs. However, all of the colloidal particles were concentrated in a magnetic field. When the magnet was removed, the colloidal particles were redispersed in methanol, indicating the magnetic property of the colloidal particles.

3.3.3 Hollow Capsules Prepared by Interfacial Polymerization or Chemical Reaction

By using AuNPs-stabilized emulsions as templates, hybrid hollow capsules can be fabricated.[38,39] Hydrophilic citrate-stabilized AuNPs were dispersed in water, and a hydrophobic polymer with a disulfide group at the midpoint

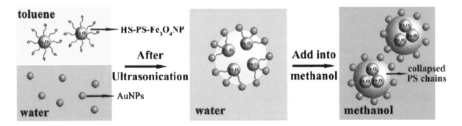

Figure 3.11 A scheme for the formation of amphiphilic nanoparticle complexes at a liquid–liquid interface and the preparation of colloidal particles with Fe_3O_4NPs cores and AuNPs coronae.
Reproduced with permission from *J. Phys. Chem. C* 2011, **115**, 3304.

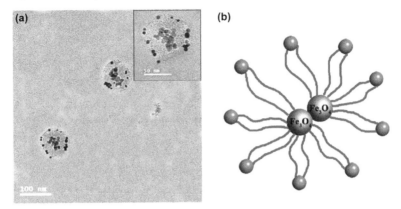

Figure 3.12   (a) TEM image of self-assembly structures formed by dispersion of colloidal particles into THF, (b) a schematic illustration of amphiphilic nanoparticle complexes formed *via* Au-S interaction.
Reproduced with permission from *J. Phys. Chem. C* 2011, **115**, 3304.

and methacrylate groups on the repeating units was dissolved in toluene.[38] Upon mixing of the two solutions, a stable emulsion was obtained. At the liquid–liquid interface, amphiphilic reactive AuNPs were prepared after ligand exchange between disulphide-containing polymer and citrate on AuNPs. The crosslinking polymerization of the reactive AuNPs was conducted at the liquid–liquid interface. Water-soluble free-radical initiator was used to initiate the polymerization. The decomposition of the initiator was in the aqueous phase, and only free radicals diffusing to the interface were able to initiate the polymerization of methacrylate on AuNPs. One-component hollow capsules with AuNPs on the surfaces were prepared by interfacial crosslinking polymerization of methacrylate (Figure 3.13). The amphiphilic AuNPs at the liquid–liquid interface were used as crosslinkers and stabilizers in the preparation of crosslinked hollow capsules. TEM images of the hollow capsules are shown in Figure 3.14. The average size of the hollow capsules can be controlled by controlling the volume ratio of oil to water.

Interfacial copolymerization of acrylamide (AM) and amphiphilic AuNPs at the liquid–liquid interface was conducted and multicomponent hollow capsules were prepared (Figure 3.13). Figure 3.15 shows TEM images of the multicomponent hollow capsules at two different magnifications. The density of the AuNPs in white domains is much lower than the deep domains, indicating the nano-sized phase separation on the surface of the multi-component hollow capsules. The white domains represent PAM-rich phases on the hollow capsules, and the deep domains represent AuNPs-rich phases.

In other research, Tian and coworkers prepared hybrid hollow capsules by using UV photodimerization.[39] A hydrophobic random copolymer with a disuldife group at the midpoint and pendant anthracene groups on the

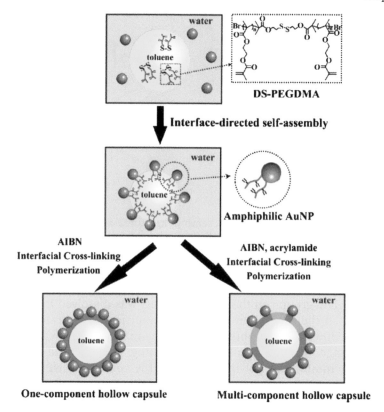

Figure 3.13 A scheme for the preparation of one-component and multi-component hollow capsules.
Reproduced with permission from *Langmuir* 2012, **28**, 9365.

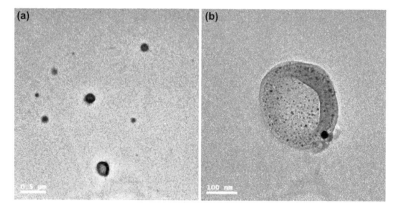

Figure 3.14 TEM images of hollow capsules prepared by interfacial free-radical polymerization, (a) at low magnification and (b) at high magnification.
Reproduced with permission from *Langmuir* 2012, **28**, 9365.

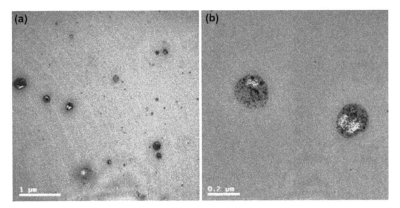

Figure 3.15 (a, b) TEM images of multicomponent hollow capsules with AuNPs and poly(acrylamide) on the surface.
Reproduced with permission from *Langmuir* 2012, **28**, 9365.

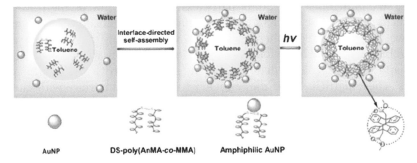

Figure 3.16 A scheme for the preparation of nanocapsules with AuNPs on the surface based on ligand exchange and interfacial photodimerization of anthrance.
Reproduced with permission from *Polym. Chem.* 2013, **4**, 1913.

repeating units was synthesized. The hydrophobic random copolymer was dissolved in toluene and hydrophilic citrate-stabilized AuNPs were dispersed in water. Upon mixing of the two solutions, amphiphilic AuNPs were produced at the liquid–liquid interface. The disulphide-containing random copolymer chains were anchored to the surface of AuNPs *via* ligand exchange. Hollow capsules with AuNPs on the surface were synthesized after interfacial photodimerization of anthracene under UV irradiation (Figure 3.16). The hybrid nanocapsules show effective catalytic activity in the reduction of 4-nitrophenol.

3.4 Stabilization of Emulsions by Janus Particles

Janus particles are compartmentalized colloidal particles that possess two sides of different chemistry or polarity. If one side is hydrophobic and the other side is hydrophilic, the particles are called amphiphilic Janus particles.

One of the most important applications of Janus particles is their use as solid surfactants for the stabilization of emulsions and foams.[40,41] Amphiphilic Janus particles could locate at the interface of water–oil with hydrophobic hemispheres immersed in the oil phase and hydrophilic hemispheres in the aqueous phase. Just like small molecular surfactants or amphiphilic block copolymers, amphiphilic Janus particles can enable self-assembly at the liquid–liquid interface and can be used as surfactants in water-based emulsions. It was shown that Janus particles have significantly higher interfacial activity than their homogeneous counterparts, and effectively lower the interfacial tension between oil and water.[42] Binks and coworkers conducted a theoretical study of the adsorptions of homogeneous particles and Janus particles at the oil–water interface.[43] Their study shows that homogeneous particles are strongly surface active but are not amphiphilic, whereas Janus particles are both surface active and amphiphilic. The surface activity of the particles increases with the amphiphilicity. Janus particles retain their strong adsorption even for average contact angles of 0 and 180°, indicating that Janus particles with either low or high average contact angles will be efficient emulsion stabilizers. Janus dumb-bells made of two partially fused spheres of opposing wettability have shown considerable surfactant behaviours,[44] which indicates that shape factor could play an important role in determining the surface activity of the Janus particles. These particles have the tendency to cover the interface with closely packed arrangements. Based on dynamic surface tension measurements and computer simulations, the assembly behaviours of Janus particles with different shapes were studied.[45] It was found that the adsorption kinetics and packing behaviours of Janus particles at the liquid–liquid interface strongly depend on their geometry. For Janus spheres, Brownian diffusion to the interface is responsible for the rapid decrease in the surface tension in the early stages. As the surface coverage of the particles grows with time, the pre-existing particles at the interface tend to prevent the additional adsorption of particles and the reduction in surface tension slows down. For Janus discs, the particles attach to the interface with either upright or inverted orientation in the initial stage, and subsequently the re-orientation or adsorption process of the inverted particles accounts for the slower evolution in the surface tension.

Synthesis and self-assembly of amphiphilic Janus disks at the oil–water interface were reported. Liu and coworkers developed a facile and versatile method for the synthesis of amphiphilic Janus Laponite disks (Figure 3.17).[46] Positively charged PS colloidal particles were prepared by ATRP emulsion polymerization. Negatively charged Laponite disks were added into PS emulsions, and the nanosized disks were adsorbed onto the surface of PS particles *via* electrostatic interaction. One side of a Laponite disk touches the surface of a colloidal particle and the other side faces the medium. After addition of positively charged polymeric micelles or quaternized poly(2-(dimethylamino)ethyl methacrylate) (q-PDMAEMA) chains into the aqueous dispersions of the colloidal particles, the micelles or

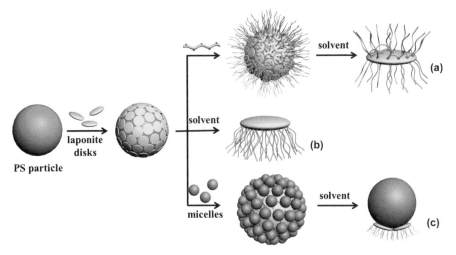

Figure 3.17 Schematic outline for the synthesis of Janus Laponite disks, (a) a Laponite disk with hydrophobic PS brushes on one side and hydrophilic quaternized PDMAEMA on the other side, (b) a Laponite disk with PS brushes one side, (c) a polymeric micelle with Laponite disks. Reproduced with permission from *Macromolecules* 2013, **46**, 5974.

polymer chains were immobilized onto the Laponite disks, and the amphiphilic Janus disks were produced on particle templates. In a mixture of oil and water, large oil droplets are formed because of the high interfacial tension (Figure 3.18a). On addition of a small amount of amphiphilic Janus Laponite disks into the emulsion, the sizes of the oil droplets are reduced and the size distribution of the oil droplets becomes much narrower (Figure 3.18b). Figure 3.18c shows a TEM image of an o/w emulsion containing Janus Laponite disks. In the TEM image, spherical self-assembled structures of Janus disks can be observed, which indicates that in the emulsion the Janus disks locate at the liquid–liquid interface and the spherical self-assembled structures are created after evaporation of liquids.

Colloidal particles stabilized by Janus particles were also prepared. Müller and coworkers prepared amphiphilic Janus particles by selectively cross-linking the spherical polybutadiene (PB) microdomains within the lamella-sphere morphology of a microphase-separated template of a PS-PB-PMMA triblock terpolymer and subsequent hydrolysis of PMMA to poly(methacrylic acid).[47] They studied emulsion polymerizations of three monomers by using Janus particles as stabilizers. The emulsion polymerizations can be conducted in a facile fashion and do not need any additives or mini-emulsion polymerization techniques. The resulting latex dispersions show very well-controlled particle sizes with low polydispersities.

Faria and coworkers explored metal-supporting Janus particles as interfacial catalysts.[48] Two types of Janus particles, one with catalyst on both faces and the other with catalyst only on the hydrophobic side, were used to stabilize o/w emulsions with the oil phase containing benzaldehyde and the

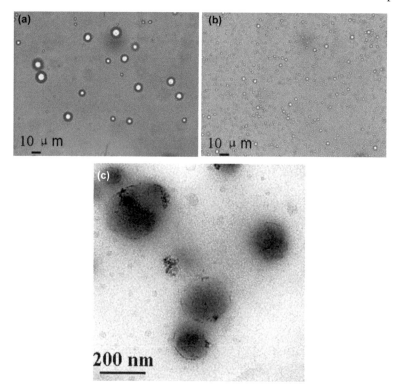

Figure 3.18   Optical microscope images of (a) toluene droplets dispersed in water
and (b) emulsions with Janus Laponite disks. (c) TEM image of dried
o/w emulsion showing aggregation of Janus Laponite disks at the
liquid–liquid interface.
Reproduced with permission from *Macromolecules* 2013, **46**, 5974.

aqueous phase containing glutaraldehyde. Their experimental results revealed that Janus particles sit at the oil–water interface, and the particles
with catalyst just on the hydrophobic face had better phase selectivity than
the particles with catalyst on both sides.

3.5   Concluding Remarks

Clay layers with polymer brushes grafted to the surface can be used as efficient stabilizers in Pickering emulsions, and colloidal particles with clay
layers on the surface can be prepared by emulsion polymerization. Amphiphilic AuNPs can be prepared at the liquid–liquid interface *via* ligand
exchange, and the AuNPs are able to self-assemble at the liquid–liquid
interface. Based on self-assembly of the AuNPs, advanced structures, including spheres and hollow structures, can be fabricated. Amphiphilic Janus
particles with one side hydrophilic and the other side hydrophobic are able
to locate at the interface of oil and water with their hydrophilic side in the

water phase and their hydrophobic side in the oil phase. The Janus structures are also able to be used as stabilizers for emulsions and colloidal particles. Interfacial tension between water and oil is reduced after addition of Janus structures. Functional emulsions and particles can be prepared by using Janus particles as stabilizers.

References

1. S. U. Pickering, *J. Chem. Soc.*, 1907, **91**, 940.
2. W. Ramsden, *Proc. R. Soc. London*, 1903, **72**, 156.
3. B. P. Binks and S. O. Lumsdon, *Langmuir*, 2000, **16**, 2539.
4. O. D. Velev, K. Furusawa and K. Nagayama, *Langmuir*, 1996, **12**, 2374.
5. O. D. Velev, K. Furusawa and K. Nagayama, *Langmuir*, 1996, **12**, 2385.
6. O. D. Velev and A. M. Lenhoff, *Curr. Opin, Colloid Interface Sci.*, 2000, **5**, 56.
7. S. Melle, M. Lask and G. Fuller, *Langmuir*, 2005, **21**, 2158.
8. D. J. Voorn, W. Ming and A. M. Van Herk, *Macromolecules*, 2006, **39**, 2137.
9. S. Fujii, Y. Cai, J. V. M. Weaver and S. P. Armes, *J. Am. Chem. Soc.*, 2005, **127**, 7304.
10. P. Pieranski, *Phys. Rev. Lett.*, 1980, **45**, 569.
11. A. Böker, J. He, T. Emrick and T. P. Russell, *Soft Matter*, 2007, **10**, 1231.
12. Z. Niu, J. He, T. P. Russell and Q. Wang, *Angew. Chem. Int. Ed.*, 2010, **49**, 10052.
13. Y. K. Takahara, S. Ikeda, S. Ishion, K. Tachi, K. Ikeue, T. Sakata, T. Hasegawa, H. Mori, M. Matsumura and B. Ohtani, *J. Am. Chem. Soc.*, 2005, **127**, 6271.
14. T. Chen, P. J. Colver and S. A. F. Bon, *Adv. Mater.*, 2007, **19**, 2286.
15. S. Cauvin, P. J. Colver and S. A. F. Bon, *Macromolecules*, 2005, **38**, 7887.
16. P. Uhlmann, H. Merlitz, J. Sommer and M. Stamm, *Macromol. Rapid Commun.*, 2009, **30**, 732.
17. R. C. Advincula, W. J. Britain, K. C. Caster and J. Rühe (eds), *Polymer Brushes*, Wiley-VCH, Weinheim, 2004.
18. W. J. Brittain and S. Minko, *J. Polym. Sci., Part A: Polym. Chem.*, 2007, **45**, 3505.
19. B. Zhao and W. J. Brittain, *Prog. Polym. Sci.*, 2000, **25**, 677.
20. B. Zhao, W. J. Brittain, W. Zhou and S. Z. D. Cheng, *J. Am. Chem. Soc.*, 2000, **122**, 2407.
21. B. Zhao, W. J. Brittain, W. Zhou and S. Z. D. Cheng, *Macromolecules*, 2000, **33**, 8821.
22. B. Zhao and W. J. Brittain, *Macromolecules*, 2000, **33**, 8813.
23. H. Zhao and D. A. Shipp, *Chem. Mater.*, 2003, **15**, 2693.
24. H. Zhao, B. P. Farrell and D. A. Shipp, *Polymer*, 2004, **45**, 4473.
25. T. J. Pinnavaia, *Science*, 1983, 220.
26. B. Putlitz, K. Landfester, H. Fischer and M. Antonietti, *Adv. Mater.*, 2001, **13**, 500.

27. Y. Yang, L. Liu, J. Zhang, C. Li and H. Zhao, *Langmuir*, 2007, **23**, 2867.
28. J. Zhang, K. Chen and H. Zhao, *J. Polym. Sci., Part A: Polym. Chem.*, 2008, **46**, 2632.
29. I. Luzinov, S. Minko and V. V. Tsukruk, *Prog. Polym. Sci.*, 2004, **29**, 635.
30. Y. Yang, J. Zhang, L. Liu, C. Li and H. Zhao, *J. Polym. Sci., Part A: Polym. Chem.*, 2007, **45**, 5759.
31. Y. Yang, D. Wu, C. Li, L. Liu, X. Cheng and H. Zhao, *Polymer*, 2006, **47**, 7374.
32. P. Viville, R. Lazzaroni, E. Pollet, M. Alexandre and P. Dubois, *J. Am. Chem. Soc.*, 2004, **126**, 9007.
33. P. A. Wheeler, J. Wang, J. Baker and L. J. Mathias, *Chem. Mater.*, 2005, **17**, 3012.
34. P. A. Wheeler, J. Wang and L. J. Mathias, *Chem. Mater.*, 2006, **18**, 3937.
35. Y. Wu, J. Zhang and H. Zhao, *J. Polym. Sci., Part A: Polym. Chem.*, 2009, **47**, 1535.
36. J. Tian, J. Jin, F. Zheng and H. Zhao, *Langmuir*, 2010, **26**, 8762.
37. J. Tian, F. Zheng and H. Zhao, *J. Phys. Chem. C*, 2011, **115**, 3304.
38. J. Tian, L. Yuan, M. Zhang, F. Zheng, Q. Xiong and H. Zhao, *Langmuir*, 2012, **28**, 9365.
39. J. Tian, G. Liu, C. Guan and H. Zhao, *Polym. Chem.*, 2013, **4**, 1913.
40. S. Jiang, Q. Chen, M. Tripathy, E. Luijten, K. S. Schweizer and S. Granick, *Adv. Mater.*, 2010, **22**, 1060.
41. B. P. Binks, *Curr. Opin. Colloid Interface Sci.*, 2001, **7**, 21.
42. N. Glaser, D. J. Adams, A. Boker and G. Krausch, *Langmuir*, 2006, **22**, 5227.
43. B. P. Binks and P. D. I. Fletcher, *Langmuir*, 2001, **17**, 4708.
44. J. W. Kim, D. Lee, H. C. Shum and D. A. Weitz, *Adv. Mater.*, 2008, **20**, 3239.
45. T. M. Ruhland, A. H. Gröschel, N. Ballard, T. S. Skelhon, A. Walther, A. H. E. Müller and S. A. F. Bon, *Langmuir*, 2013, **29**, 1388.
46. J. Liu, G. Liu, M. Zhang, P. Sun and H. Zhao, *Macromolecules*, 2013, **46**, 5974.
47. A. Walther, M. Hoffmann and A. H. E. Müller, *Angew. Chem. Int. Ed.*, 2007, **120**, 723.
48. J. Faria, M. P. Ruiz and D. E. Resasco, *Adv. Synth. Catal.*, 2010, **352**, 2359.

CHAPTER 4

Pickering Suspension, Mini-Emulsion and Emulsion Polymerization

STEFAN A. F. BON

Department of Chemistry, University of Warwick, Coventry CV4 7AL, UK
Email: s.bon@warwick.ac.uk

4.1 Introduction: How to Define Suspension, Mini-Emulsion and Emulsion Polymerization

The terms emulsion polymerization and mini-emulsion polymerization can lead to confusion in the literature.[1] An emulsion is defined as a dispersion of droplets of a liquid phase (for example oil) into a continuous liquid phase (for example water). Note that the inverse scenario, whereby water droplets are dispersed into an oil phase, is commonly referred to as an inverse emulsion. When the oil droplets consist of a monomer that can be polymerized, we do *not* speak of an emulsion polymerization. This heterogeneous polymerization process, whereby monomer-containing droplets are converted into polymer beads, is referred to as a suspension polymerization.[1] In essence one can envisage each droplet as an individual microscopic bulk/solution polymerization reactor. In contrast, the loci of the (radical) polymerization in an emulsion polymerization process are polymer particles that are newly formed in the reaction mixture upon initiation in the water phase. Particle formation is generally confined to the initial stages of the emulsion polymerization process, and occurs through either micellar[2,3]

RSC Soft Matter No. 3
Particle-Stabilized Emulsions and Colloids: Formation and Applications
Edited by To Ngai and Stefan A. F. Bon
© The Royal Society of Chemistry 2015
Published by the Royal Society of Chemistry, www.rsc.org

or homogeneous nucleation.[4,5] The monomer droplets simply can be seen as temporary storage containers for excess amounts of monomer and serve to replenish the drop in monomer concentration in the growing particles. Transport of monomer from the droplets through the water phase into the particles, or direct refill of the monomer concentration in the particles upon reversible collision with monomer droplets, therefore plays an important role in an emulsion polymerization reaction. The final dispersion of polymer particles in the liquid reaction medium (read water) is also known as a polymer latex. One may ask oneself why the monomer droplets do not polymerize into polymer beads in an emulsion polymerization process? The answer lies in the low probability of a growing radical oligomer in the water phase entering a monomer droplet, as the total surface area of the monomer droplets is low in comparison to the combined surface area of the polymer latex particles. One could, however, increase the probability of entry of radicals into the monomer droplets by decreasing the size of the droplets (to sub-micron diameters), thereby shifting the locus of polymerization in favour of droplet polymerization.[6] This is the concept of a mini-emulsion polymerization,[7] which in an ideal scenario can be seen as a miniaturized version of a suspension polymerization. Key to the mini-emulsion polymerization process is that Ostwald ripening of the mini-emulsion droplets, *i.e.* smaller droplets disappear in favour of growth of larger ones by monomer diffusion through the water phase, is efficiently suppressed over the time period of the mini-emulsion polymerization process by making use of small amounts (typically a few wt%, based on monomer) of so-called hydrophobes, for example *n*-hexadecane or stearyl methacrylate. In the ideal mini-emulsion polymerization process all of the monomer droplets are converted into polymer particles. The synthesis of polymer particles by mini-emulsion polymerization alleviates the need for transport through the water phase and allows the individual mini-emulsion droplet 'reactors' to be pre-loaded with materials, opening the pathway to hybrid and nanocomposite latexes and capsules.[8] In reality, one should consider all kinetic events of emulsion polymerization in a mini-emulsion polymerization process, keeping in mind the one key difference. Whereas in emulsion polymerization droplet nucleation can be neglected, in mini-emulsion polymerization it is the dominant pathway in the formation of polymer particles. One should, however, always be aware that particles can be formed through homogeneous nucleation, and potentially through micellar nucleation. In the case of a mini-emulsion polymerization experiment where the droplets were pre-loaded with, for example, quantum dots, the formation of particles through these two latter 'undesired' but often unavoidable pathways leads to a fraction of polymer particles that do not contain quantum dots, thereby imparting the ideological concept of the mini-emulsion polymerization process.

In both emulsion and mini-emulsion polymerizations, molecular surfactants are commonly used to maintain colloidal stability of the latex particles throughout and after the polymerization process. Examples include small molecules such as sodium dodecyl sulfate or sodium bis(2-ethylhexyl) sulfosuccinate, or macromolecular compounds such as poly(vinyl alcohol) or hydroxyethyl cellulose.

This chapter focuses on Pickering mini-emulsion and emulsion polymerization processes, which were developed over the last decade. The Pickering heterogeneous polymerization processes differ from their conventional analogues in that molecular surfactants have been replaced by nanoparticles, the latter effectively serving as Pickering stabilizers.[9-11] We will, however, start with some historical comments on the origin of Pickering stabilization and the use of insoluble emulsifiers in suspension polymerization, after which we will summarize some recent developments in the use of Pickering suspension polymerization as a method to synthesize hybrid and composite microspheres and capsules.

4.2 Early Studies on Pickering Stabilization and its Use in Suspension Polymerization

The phenomenon that solid particles can adhere to soft and thus deformable interfaces (*i.e.* liquid–liquid or liquid–gas interfaces), and thus have the ability to stabilize dispersed oil droplets in water has been coined Pickering stabilization, after Spencer Umfreville Pickering.[10] When solid particles are adhered to the interface of emulsion droplets they provide a barrier which protects the droplets from coalescence. Hence the use of the word stabilization. In his 1907 paper Pickering describes a detailed investigation into the use of insoluble emulsifiers such as CaO, CuO, $CuSO_4 \cdot 5H_2O$ and $FeSO_4 \cdot 7H_2O$, for the stabilization of paraffin oil emulsions in water. In the first paragraph he does acknowledge the work on emulsification by Ramsden published in 1903,[9] who essentially drew the same conclusions. Hilderband and coworkers concisely described the basics of Pickering stabilization in their 1923 paper on the theory of emulsification (Figure 4.1).[11]

They said that obviously the powder must collect at the interface in order to be effective and that this would only occur if the solid particles were wetted by both liquids. They stated that in general the particles had a preference for one liquid over the other, which meant that the particles

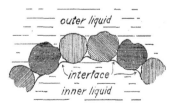

Figure 4.1   Schematic drawing of the interface of a water-in-benzene emulsion droplet stabilized by solid lampblack particles as observed by microscopic examination.
Reprinted with permission from reference 11. © 1923 American Chemical Society.

would be drawn more into the former. They described that the accumulation of particles onto the liquid interface could cause the interface to bend in the direction of the more poorly wetting liquid, facilitating it to become the dispersed phase. They concluded that the type of emulsion, *i.e.* oil-in-water or water-in-oil, could be predicted on the basis of wettability and thus on the basis of the contact angle of the interface with the solid.

One could argue, however, that the origin of a description for Pickering stabilization lies further back, in the area of (froth) flotation. William Haynes filed a patent in 1860 in which he described mixing coal tar and resin with crushed ore to which metallic particles adhered, while the gangue could be removed with water.[12] In 1877 the Bessel brothers filed a patent describing the process for the flotation of graphite from ores by adding 1.0–10% of non-polar oils. This slurry was added to water and brought to boiling point. The patent describes in detail that graphite flakes attached themselves to bubbles, rose to the surface, and could be skimmed off.[13]

The use of solid particles as stabilizers in heterogeneous emulsion-based polymerization techniques was first described by Röhm and Trommsdorff in 1934/5 using various inorganic stabilizers such as barium sulfate, talc, kaolin clay and aluminum oxide in the suspension polymerization of a variety of monomers.[14] Hohenstein and coworkers described the suspension polymerization of styrene and dichlorostyrene in presence of bentonite and talc,[15] and also reported the use of tricalcium phosphate.[16] Other early examples include works by Mayne[17] who reported the use of bentonite, and Haward and Elly who used titanium dioxide[18] as inorganic solids stabilizers in suspension polymerizations.

The onset of polymerization in a suspension polymerization results in an increase in the viscosity of the emulsion droplets. Whereas this is good for resistance to distortion and droplet break-up due to viscous drag, it makes the droplets sticky and increases their tendency to aggregate and coalesce. The latter phenomenon is minimized by the presence of solid particles at the surface, as they forms a protective barrier. Winslow and Matreyek carried out suspension polymerizations of a mixture of isomers of ethylbenzene and divinylbenzene in the presence of inorganic solid stabilizers such as bentonite and tricalcium phosphate.[19] They stated that their use was associated with the production of opaque polymer beads, partly due to surface roughness and that the inorganic particles were undesired contaminants difficult to remove. The coarsening of the emulsion droplets through coalescence, or better, the prevention of that, clearly is a function of the amount and the shape and particle size of the Pickering stabilizer. This effect of limited coalescence, a process in which after a certain time a stable set of solids-armoured liquid droplets is obtained, was first described by Hardy in 1928.[20] In 1954, Wiley carried out a series of experiments for the emulsification of styrene in water using Dowex 50 ion exchange resins and bentonite clay platelets as Pickering stabilizers in combination with small amounts of calcium chloride and gelatin as adhesion promotors.[21] He derived an equation to estimate the diameter of the styrene droplets, assuming full coverage with a square packing arrangement, and complete adhesion of all the colloidal particles onto the interface supporting his experimental findings. In 1956 von Wenning published a detailed account of solids stabilizers for suspension polymerization, focusing hereby on the use and physical behaviour of barium sulfate as Pickering stabilizer for emulsions of styrene and water in the presence of various amounts of octa-decane sulfonic acid as conventional molecular surfactant.[22] As part of this work he showed that small additions of octadecane sulfonic acid (from 0 to 0.3%) improved the emulsification of styrene in water, with optimum performance at 0.3% for a styrene-in-water emulsion. Further increase in the ratio of molecular surfactant to barium sulfate particles up to *ca.* 30% led to the formation of the inverse water-in-styrene emulsion. A further increase in the relative amount of octadecane sulfonic acid caused the emulsion to invert back to a dispersion of styrene droplets in water.

In 1987, Deslandes[23] studied the surface morphology of polymer beads made by suspension copolymerization of styrene and butadiene using hydroxyapatite as a Pickering stabilizer in conjunction with alkanol XC (isopropylated naphthalenesulfonic acid, sodium salt). He found by cross-sectional TEM analysis that the polymer beads had a surface layer of two distinct parts (Figure 4.2).

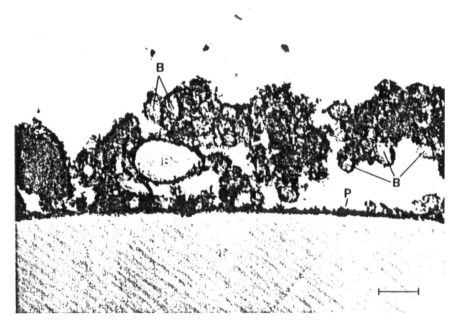

Figure 4.2 Transmission electron micrograph of a cross-section of a poly(styrene-*co*-butadiene) bead showing the two distinct layers that make the suspension stabilizer layer. Legend: I: bead interior, P: thin uniform layer of primary particles of hydroxyapatite, T: thicker outer layer made of hydroxyapatite aggregates, B: small polymer bead within the hydroxyapatite layer (see text for explanation). Scale bar is 1 μm.
Reprinted with permission from reference 23. © 1987 John Wiley & Sons, Inc.

The inner layer was a densely packed monolayer of hydroxyapatite particles. This is in line with the concept of Pickering stabilization. However, this layer of adhered hydroxyapatite particles was surrounded by a thicker, more loose, layer of aggregates of hydroxyapatite containing smaller polymer particles, interestingly also coated with a monolayer of hydroxyapatite.

In the above studies the focus remained on the fabrication of polymer beads or pearls using inorganic particles as effective stabilizers, thereby disregarding the opportunity to fabricate composite polymer particles deliberately armoured with a layer of solid, potentially functional, particles.[24]

4.3 Pickering Suspension Polymerization as a Tool to Fabricate Armoured Microstructures

Interest in using suspension polymerization as a tool to fabricate micron-sized armoured particles and capsules emerged from the renewed interest in Pickering stabilization of emulsion droplets initiated by the 1996 triptych of Velev *et al.*,[25–27] in which they described the assembly of particles into supracolloidal structures using emulsion droplets as template. These permeable structures, created by fully covering emulsion droplets with colloidal building blocks, were coined 'colloidosomes' by Dinsmore and coworkers.[28] The following set of examples is by no means complete, but it summarizes noticeable developments.

Yongjun He demonstrated the preparation of polyaniline microspheres by oxidative suspension polymerization of droplets of a mixture of aniline and toluene dispersed in water and stabilized with ZnO,[29] $Cu_2(OH)_2CO_3$[30] and CeO_2 nanoparticles.[31] Ge *et al.*[32] prepared cage-like permeable hollow capsules from droplets of methyl methacrylate or vinyl acetate in water and templated with sulfonated polystyrene particles (Figure 4.3). After a 72-hour swelling period and subsequent polymerization by gamma irradiation the intricate morphologies were obtained. The authors suggest that the original polystyrene particles 'pop out' of the structure, creating the cage-like morphology.

Bon and coworkers combined the phenomenon of Pickering stabilization and the concept of an emulsion droplet armoured with particles being a supracolloidal 'colloidosome' structure, with heterogeneous polymerization techniques, such as suspension and mini-emulsion polymerization.[33,34] They used colloidosomes made from poly(methylmethacrylate-*co*-divinylbenzene) microgel latexes as micron-sized capsules (Figure 4.4), and demonstrated that the flexibility of the capsule wall could be tailored by varying the monomer composition in the suspension polymerization.[34]

Bon *et al.* used titanium dioxide nanoparticles in the suspension polymerization of mixtures of styrene, divinylbenzene and *n*-hexadecane to fabricate hybrid microspheres and capsules that were armoured with a layer of TiO_2 (the use of the words Pickering emulsion polymerization in the title and throughout the manuscript is incorrect).[35] Du and coworkers also used

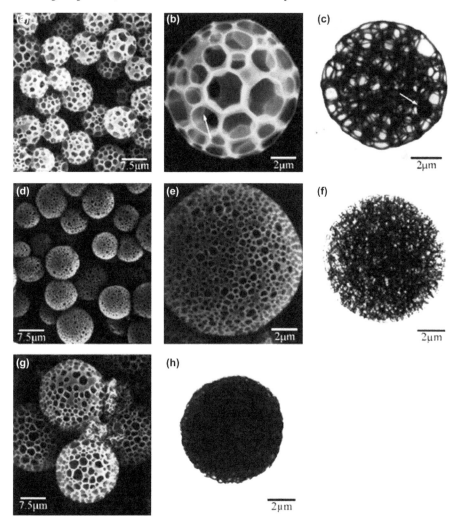

Figure 4.3 Morphological characterizations of the polymer microspheres. SEM (a and b) and TEM (c) micrographs of PMMA microspheres; SEM (d and e) and TEM (f) micrographs of PVA microspheres prepared under the same conditions; (g) SEM micrograph of PMMA microspheres using 2.94 μm PS latex particles as emulsion stabilizers; (h) TEM micrograph of PMMA microspheres from emulsion without being reserved for a certain time. Reprinted with permission from reference 32. © 2005 American Chemical Society.

titania as Pickering stabilizer in the preparation of armoured polymer microspheres.[36] Nie *et al.*[37] showed that control of the size of Pickering emulsion droplets could be achieved by using droplet-based microfluidics in which the solid particles were added to the dispersed phase to prevent device clogging. Bon and Chen demonstrated that the concept of using armoured emulsion droplets as polymerization vessels could be used as a fabrication

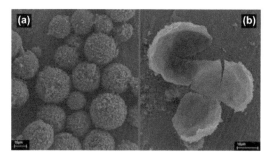

Figure 4.4 Field emission scanning electron microscopy (FE-SEM) images showing (a) a collection of supracolloidal interpenetrating network reinforced capsules made using PMMA microgel colloidosomes as radical polymerization vessels. Weight ratios of *n*-hexadecane–monomer 4 : 1, styrene–divinylbenzene 1 : 1. (b) A burst scaffolded supracolloidal capsule.
Reprinted with permission from reference 34. © 2006 Royal Society of Chemistry.

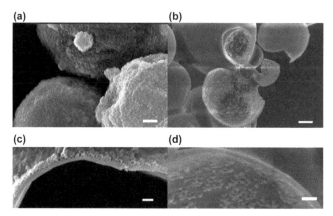

Figure 4.5 Field emission scanning electron microscopy (FE-SEM) images of complex nanopatterned silica microstructures. (a) Siliceous microstructures with clay nanocapsules on the outer surface after calcination at 600 °C for 6 h (scale bar = 1 µm). (b) Siliceous microstructures with clay nanocapsules on the inner surface after calcination at 600 °C for 6 h (scale bar = 2 µm). (c) Microtomed section of a single microcapsule with clay nanocapsules on the outside (scale bar = 400 nm). (d) Microtomed section of a single microcapsule with clay nanocapsules on the inside (scale bar = 400 nm).
Reprinted with permission from reference 38. © 2007 American Chemical Society.

tool for complex silica-based microcapsules with nanopatterned hollow features, in that the walls of the siliceous capsules were decorated on either the outside, or inside, with nanocapsules composed of Laponite clay (Figure 4.5).[38]

Hasell and coworkers showed that iron oxide nanoparticles could be used as Pickering stabilizers in the suspension polymerization of methyl methacrylate to yield Fe_3O_4 armoured magnetic microspheres.[39]

Inverse Pickering suspension polymerization (one could argue that this is an inverse Pickering mini-emulsion polymerization without added Ostwald ripening retarder, *e.g.* salt, to the water phase), that is the free radical polymerization of solids-stabilized monomer containing water droplets, was reported by van Herk *et al.*[40] who dispersed and polymerized aqueous solutions of acrylamide and 2-hydroxyethyl methacrylate in cyclohexane stabilized by hydrophobic Cloisite 20A clay platelets. Cryo-TEM analyses of the poly(acrylamide) latexes clearly showed that the clay was present at the surface of the latex particles. Wu and coworkers synthesized and characterized thermoresponsive poly(*N*-isoacrylamide) microspheres armoured with silica particles by inverse Pickering suspension polymerization.[41]

Tong and coworkers reported the fabrication of dual nanocomposite multihollow polymer microspheres using the principle of multiple Pickering emulsion droplets.[42] A water-in-oil-in-water (w/o/w) Pickering suspension polymerization was carried out in which small water droplets were dispersed in styrene using silica particles, after which this inverse emulsion was dispersed in water using magnetic iron oxide particles followed by subsequent polymerization of the monomer. The inverse system, in which small toluene droplets were dispersed in an aqueous salt containing (bis)acrylamide phase using Laponite clay discs, with this being subsequently dispersed into toluene using hydrophobic silica particles, was also included.

The idea of miniaturization of the Pickering suspension polymerization toward the use of smaller droplets armoured with a layer of solid nanoparticles brings us to the area of Pickering mini-emulsion polymerization.

4.4 Pickering Mini-Emulsion Polymerization

The first study that raised the concept of Pickering stabilization in mini-emulsion polymerization was reported by Landfester *et al.* in 2001.[43] They demonstrated the fabrication of polymer latexes armoured with silica nanoparticles by mini-emulsion polymerizations of styrene in the presence of auxiliary comonomer 4-vinylpyridine, hexadecane as hydrophobe to suppress Ostwald ripening, and deionized silica nanoparticles (Ludox TMA). They took inspiration from the fabrication of silica-armoured latexes by dispersion polymerization using 4-vinylpyridine, as reported by Armes and coworkers.[44,45] Optimal results were obtained from experiments carried out at pH 10 with 16.7 wt% of 4-vinylpyridine (Figure 4.6). The influence of the particle morphology upon addition of small amounts of anionic, non-ionic and cationic surfactants was also reported. When cetyl trimethyl ammonium chloride was used at amounts that fully covered the silica nanoparticles, a raspberry morphology rather than armoured supracolloidal structures were obtained, with the nanoparticles embedded and dispersed throughout the latex particle. Interestingly, and worth noting, is that the silica sol was added *after* the emulsification step through sonication, which raises the question of whether the silica particles will adhere to the emulsion droplets prior to

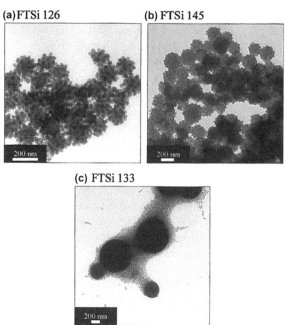

Figure 4.6 TEM pictures of latexes using silica particles as stabilizer for monomer droplets. As monomers, a mixture of styrene and 4-vinylpyridine is used. (a) Represents a latex with a monomer to silica weight ratio of 1 : 0.72. (b) Represents a latex with a monomer-to-silica ratio of 1 : 1.08. (c) Shows the imperfect stabilization effect if 4-vinylpyridine is not employed. Reprinted with permission from reference 43. © 2001 American Chemical Society.

polymerization, or whether they adhere to polymer particles formed during the polymerization process.

Bon and coworkers were the first to generate mini-emulsion droplets through sonication in the presence of Laponite clay discs as Pickering stabilizers, which was followed by carrying out the actual Pickering mini-emulsion polymerization.[33] Laponite clay particles are discotic platelets with an approximate lateral diameter of 25 nm, and a 1 nm thickness.[46,47] They have the ability to self-exfoliate in water into individual colloidal particles. In the experiments sodium chloride was added (0.1 M) to induce slight colloidal instability leading to clay particle flocculation, which greatly enhances the ability of the clay discs to serve as Pickering stabilizers.[48] Bon and coworkers showed that polystyrene colloids armoured with Laponite particles can be prepared by Pickering mini-emulsion polymerization and that an estimate for the size of the nanocomposite latexes could be obtained from packing considerations whereby the Laponite clay discs dictate the total interface generated. The first mechanistic study on Pickering mini-emulsion polymerization using Laponite clay discs as stabilizers was reported by Colver and Bon in 2007.[49] Particle morphologies were studied, showing that indeed polymer latexes armoured with clay discs were fabricated, and a model was developed to estimate the average particle size of the latex particles in Pickering mini-emulsion polymerizations, making use of partitioning (during the emulsification step) of the clay discs between the water phase and the surface of the monomer droplets, and the overall polymerization rates were studied and discussed.

For the Pickering mini-emulsion polymerizations to be successful it is conceptually easy to grasp that the particles used as Pickering stabilizer need to be able to armour the mini-emulsion droplet. In other words they have to be able to adhere to and cover (parts of) the surface of the emulsion droplet, or when flexible, be able to wrap themselves around the droplet. Thus we have to take into account the curvature of the surface of the droplet. For example, it seems impossible to wrap a rigid (square) clay sheet of 350 nm in diameter around a mini-emulsion droplet of 200 nm in diameter whereas theoretically 1.03 clay sheets are required to (fully) cover the surface of the droplet. This obviously leads to geometric constraints on the Pickering stabilizers. Because mini-emulsion droplets typically have diameters in the range of 50 nm to 1.0 μm, the solid particles to be used as stabilizer need to be nano-sized, and commonly have geometric measures of less than 200 nm. When flexible particles are used as Pickering stabilizer, for example graphene oxide sheets, larger dimensions can be allowed for but the bending rigidity of these colloidal objects needs to be taken into consideration. A wide range of nanoparticles have been employed in Pickering mini-emulsion polymerizations (not necessarily by free radical polymerization), including clay discs,[33,49–52] silica nanoparticles,[53–56] iron oxide,[57] cerium oxide,[58] zinc oxide,[59,60] cellulose nano-whiskers,[61] and graphene (oxide) sheets.[62–65] When the dimensions of Pickering stabilizers become comparable or exceed those of the mini-emulsion droplets only a few or one of these particles will adhere to the interface, or the inverse behaviour can be observed, in which multiple monomer droplets (and after polymerization polymer particles/blobs) are adhered to a single Pickering stabilizer. This is nicely illustrated in the work of Bonnefond *et al.*,[66] who performed Pickering mini-emulsion copolymerizations of *n*-butyl acrylate and methyl methacrylate in the presence of montmorillonite clay sheets, *i.e.* a reactive organoclay ((2-methacryloylethyl)hexadecyldimethylammonium modified montmorillonite, CMA16). They observed clay platelets 'sandwiched' between two polymer particles.

4.4.1 Particle/Droplet Size Distributions in Pickering Mini-Emulsions

The total amount of Pickering stabilizers used, taking into account their efficacy to adhere to the mini-emulsion droplets and their effective surface-to-volume ratio, has an influence on the droplet size distribution, and later the particle size distribution of the armoured polymer latexes. As mentioned earlier, a model to correlate the average particle size from Pickering mini-emulsion polymerizations with the amount of Pickering stabilizers used in relation to the total amount of monomer (and water) was developed by Colver and Bon.[49] They showed from Pickering emulsion polymerizations of styrene in presence of Laponite clay discs that there was a linear correlation between the added amount of clay and the calculated excess amount of Laponite that remained in the water phase and was not adhered to the surface of the particles. The latter was calculated from the data obtained for the average particle size as measured with dynamic light scattering. In other words the total amount of clay discs needed to cover the total surface area of the latex particles was calculated and this amount was subtracted from the analytical amount added in the Pickering mini-emulsion polymerizations. The linear correlation showed that the number of Laponite clay particles added dictated the total interface generated and thus determined the average particle size of the armoured latexes obtained. They stressed that the sonication allowed the clay discs to interact with the mini-emulsion droplet interface in a reversible manner (high energy allows for particle detachment of the interface) and that this reversibility allowed for the partitioning process between water and the water–droplet interface of the Laponite clay discs to reach equilibrium.

This reversibility upon sonication comes in handy when we would like to study the packing patterns of Pickering stabilizers onto the armoured droplets or latex particles, as it would allow for the particles to arrange themselves on the interface in a minimum energy configuration. This is indeed the case and was demonstrated by experimental determination of the packing patterns of (spherical) silica nanoparticles on polystyrene latexes made by Pickering emulsion polymerization and correlating this with results from metropolis Monte Carlo (MC) simulations. Bon and coworkers showed that excellent agreement could be obtained (Figure 4.7).[53] The ability to predict the arrangement of the particles on the surface of the latex particles in addition showed that an increase in dispersity of the particle size distribution of the nanoparticles would distort the clarity of the 12-point dislocations, commonly observed when monodisperse particles are packed onto a sphere,[67–69] thereby fading out the grain boundary scars.

In a paper by Zgheib *et al.*,[70] the partitioning of CeO_2 nanoparticles between the water phase and the surface of methyl methacrylate mini-emulsion droplets stabilized using octadecylacrylate as hydrophobe was studied by means of measuring the average droplet size, upon addition of a range of small amounts of methacrylic acid to tune the wettability of the CeO_2 nanoparticles. Indeed using small amounts of methacrylic acid enhanced the ability of the particles to adhere to the interface, allowing for a greater surface area to be stabilized, which resulted in a larger number of smaller mini-emulsion droplets.

They demonstrated by means of (cryo)-TEM analysis that indeed the CeO_2 particles adhered to the interface of the methyl methacrylate droplets (Figure 4.8). Interestingly they showed that some of the droplets showed signs of buckling (deflating shape deformation), which potentially indicates

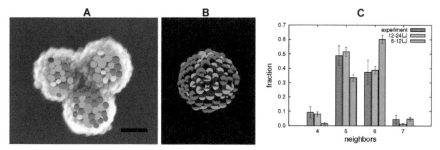

Figure 4.7 Packing patterns of nanosized silica particles on the surface of polystyrene latex particles: (A) Experimental micrograph taken by SEM (scale bar 100 nm). (B) Simulation snapshot of one equilibrium configuration with the 12–24 LJ potential (205 particles). The colours represent the number of neighbours (see text) of each silica nanoparticle: yellow (four), green (five), blue (six), red (seven). (Colours in the micrograph have been added manually.) To compare the experimental result and the simulated data calculated with both the 6–12 LJ and 12–24 LJ potentials, the fraction of neighbours are compared (C). The histogram shows excellent agreement between the experiment and the simulation with the 12–24 LJ potential. The data has been averaged over the two batches: cc-1-114 and cc-1-199. Five simulations have been run for each batch.
Reprinted with permission from reference 53. © 2009 American Chemical Society.

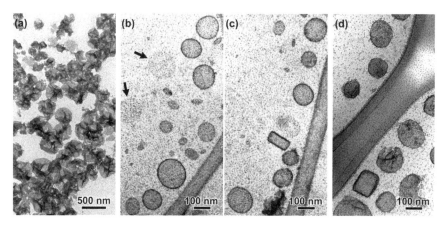

Figure 4.8 TEM images of MMA droplets prepared in the presence of 35 wt% CeO_2 (ME8 sample, Table 1): (a) dry specimen; (b, c) ice-embedded droplets from the quench-frozen suspension immediately after the preparation of the mini-emulsion (the arrows in (b) point to droplets with a very low contrast); (d) ice-embedded droplets from the same suspension after 1 week of aging.
Reprinted with permission from reference 70. © 2012 American Chemical Society.

that some (initial) Ostwald ripening could have occurred. Certain armoured droplets even appeared to be of 'rectangular' shape. The nanoceria armoured mini-emulsion droplets of methyl methacrylate were stable for at least 24 hours at room temperature, with no substantial increase in average droplet size diameter.

4.4.1.1   Ostwald Ripening or Coarsening in Mini-Emulsion Droplets

This brings us to the question of whether Pickering stabilizers have the ability to retard, or in the ideal scenario, arrest Ostwald ripening. Ashby and Binks showed that for emulsion droplets (mean diameter > 10 μm) of toluene in water stabilized by Laponite clay discs in the presence of sodium chloride, limited non-linear Ostwald ripening was observed, after which a plateau value for the mean diameter was observed, indicating arrest of coarsening.[48] This behaviour is explained in that in essence the particles are irreversibly adhered to the interface (although this is highly questionable upon severe buckling), which would lead to a change in interfacial tensions for shrinking smaller (decreasing) and expanding larger droplets (increasing), balancing out the Laplace pressures and thereby reaching a quasi-equilibrium. In addition, deformation of the buckling smaller droplet can lead to a surface rheological component on top of the Laplace pressure, slowing down further buckling and thus transport from molecules from one droplet to the other by means of diffusion. The latter process is known from studies on coarsening of bubbles in aqueous foams,[71] such as in beer.[72] To the best of our knowledge papers on Pickering mini-emulsion polymerizations include the use of hydrophobes to retard the Ostwald ripening process. Experiments in our laboratory omitting the use of a hydrophobe have thus far failed to be successful in producing stable armoured latexes.

4.4.1.2   Homogeneous/Secondary Nucleation and Limited Coagulation in Mini-Emulsion Polymerization

It has been stressed, for example in the early work of Bon and coworkers,[33] that it is important to use oil-soluble initiators in order to avoid/limit co-agulation in the Pickering mini-emulsion process. A plausible reason for this is that the use of water-soluble initiators can lead to particle formation by homogeneous nucleation. These secondary latex particles will not necessarily be covered in Pickering stabilizers, and upon growth they become unstable, which leads to coagulation. Indeed Asua and coworkers showed that the use of water-soluble initiators had to be avoided as it led to complete coagulation in the surfactant free and high-solids (50 wt%) mini-emulsion copolymerization of vinyl acetate with Veova-10 using silica nanoparticles which were modified with 2-[2-ethoxy(poly(ethyleneoxy)$_{9-12}$)propyl]-trimethoxy silane.[56] The use of relatively low amounts of Pickering stabilizers furthermore resulted in limited coagulation, *i.e.* cluster formation, during the mini-emulsion polymerization process. The phenomenon of limited coagulation at low concentrations of Pickering stabilizers was previously reported in the Pickering emulsion polymerization of vinyl acetate using silica nanoparticles as stabilizers.[73] This obviously has an impact on the particle size distribution, but could potentially be advantageous as multimodality in the particle size distribution of the armoured dispersion can have a beneficial effect in lowering the overall viscosity of the dispersion. One could ask why this happens exclusively upon polymerization of the mini-emulsion in the experiments performed by Asua *et al.*[74] and not prior to polymerization. Their data showed that the mini-emulsions did not coarsen upon storage before polymerization. The answer most probably can be found in the high water solubility of vinyl acetate, which is 4.27 wt% at 60 °C.[75] This (i) increases the probability of homogeneous nucleation, most probably through radical exit after transfer, and (ii) creates 'monomer-starved' conditions at the end of the polymerization process, which means that newly formed polymer deposits itself near the outer surface of the particles, potentially inducing limited destabilization as the mobility of the Pickering stabilizers to relocate can be restricted.

4.4.2 Rates of Polymerization in Pickering Mini-Emulsion Polymerizations

The overall rate of (radical) polymerization in heterogeneous polymerization techniques, such as mini-emulsion and emulsion polymerization, are often higher than in the analogous solution and bulk free radical polymerizations. This can be ascribed to the phenomenon of compartmentalization, where two radicals each present in a different particle do not have the ability to undergo bimolecular termination. In the case of Pickering mini-emulsion polymerizations, the armour of Pickering stabilizers present at the surface of the interface can and will have a profound effect on the polymerization kinetics. Phase transfer events, such as radical entry and exit, for example could be retarded and less efficient due to the presence of the impermeable solid nanoparticles at the surface of the latex particle. Moreover, as discussed before, for Pickering mini-emulsion polymerization to succeed the best approach is to use oil-soluble initiators. The majority of these will generate radicals that are capable of propagation in pairs. Initiator efficiency in small droplets (say < 80 nm in diameter) therefore could be compromised as a direct result of the confined and restricted total volume of the droplet. Detailed kinetic studies on Pickering emulsion polymerization are scarce. One interesting observation made by Colver and Bon[49] is that profound compartmentalization effects are found in the Pickering mini-emulsion polymerization of styrene in the presence of Laponite clay discs as stabilizers. The observed relative overall rates of polymerization (in comparison to the equivalent bulk polymerization, by taking the ratio of the two, tentatively named φ) increase when the droplet sizes and thus particle sizes of the latexes stabilized by Laponite clay discs are decreased. Relative enhancements of the rates of polymerization observed as a plateau value were 7.17, 5.74, 4.43, 3.45, 1.76, 1.54 and 1.34 for particle diameters (DLS) of 234.8, 287.7, 391.5, 495.7, 643.2, 607.9 and 658.3 nm, respectively. More striking, however, was that this compartmentalization effect on the overall polymerization rate showed large retardation effects, which extended to higher monomer conversion when latex particles were smaller, up to 40% monomer conversion. This is possibly due to the presence of the armoured layer of nanoparticles.

Pickering mini-emulsion polymerization requires pre-emulsification of the monomer using high-shear devices, or in a laboratory environment, sonication is popular. A downside of most Pickering stabilizers is that they are hard particles, which are abrasive. In addition, high-shear environments can lead to coagulation of the Pickering stabilizers and this complicates scale-up. This was a driver in developing the Pickering emulsion polymerization process, which alleviates these issues.

4.5   Pickering Emulsion Polymerization

For a thorough understanding of (free radical) emulsion polymerization the reader is referred to excellent books by Blackley,[76] Fitch,[77] Gilbert[78] and van Herk.[79] A standard recipe for a free radical emulsion polymerization contains the following ingredients: water as continuous phase, potentially containing a buffer to regulate pH; hydrophobic monomer(s) (*e.g.* styrene, butadiene, ethylene, isoprene, (meth)acrylates, vinyl acetate or veova analogues); (small) amounts of hydrophilic monomer(s) to aid colloidal stability and provide surface functionality of the latex particles (*e.g.* (meth)acrylic acid, acrylamide, sodium styrene sulfonate); low molar mass emulsifiers (*e.g.* sodium dodecyl sulfate, aerosol surfactants, anionic or non-ionic disponil surfactants, non-ionic Brij surfactants) and/or polymer surfactants (*e.g.* poly(vinyl alcohol), poly(vinyl pyrrolidone), Natrosol hydroxyethyl cellulose); and a water soluble free radical initiator (*e.g.* ammonium persulfate, or hydrogen peroxide in combination with ascorbic acid as redox system). Note that emulsion polymerizations can operate in the absence of surfactants, and are then referred to as soap-free or emulsifier-free emulsion polymerizations. Let us stress again, as already addressed in the introduction (see Section 4.1) that emulsion polymerization is *not* the polymerization of emulsion droplets.

In essence the goals of Pickering emulsion polymerizations are to replace molecular surfactants with solid, often nanosized, particles as stabilizer as well as to fabricate nanocomposite armoured polymer latexes. Again, as in the case of Pickering mini-emulsion polymerization (see Section 4.4), a range of Pickering stabilizers have been explored, such as amphiphilic polymer Janus particles,[80] silica nanoparticles[73,81–84] and nanosized clay discs.[85,86]

4.5.1 Adding Nanoparticles to Emulsion Polymerizations: The Dawn of Pickering Emulsion Polymerization

The concept of adding nanoparticles when synthesizing polymer colloids of submicron-sized dimensions dates back several decades. The development of Pickering emulsion polymerization as a technique evolved from these experiments. By no means to be seen as a complete overview, some key works are described below.

Solc nee Hajna filed a patent in 1980 that described the incorporation of nanosized iron oxide (ferrofluid) but also titania, calcium carbonate, zinc oxide, clay and silica into polymer latexes through emulsion polymerization.[87] Iron oxide nanoparticles or commercial ferrofluids were dispersed in water with the aid of potassium oleate and the dihexyl ester of sodium sulfosuccinic acid, after which conventional emulsion copolymerization reactions with styrene and butyl acrylate were carried out. Nanocomposite latexes that contained the iron oxide nanoparticles of up to 25 wt% solids were reported. Yanase and coworkers[88] carried out emulsion polymerizations of styrene in the presence of a commercial ferrofluid (Ferricolloid W-75; Fc). These magnetite particles are stabilized with a monolayer of sodium oleate and a large excess of sodium dodecylbenzene sulfonate (SDBS), with added ethylene glycol and glycerol as anti-freeze agents. They showed that using the ferrofluid directly as a waterborne dispersion led to polystyrene latexes with the iron oxide nanoparticles located at the interface. Purification of the ferrofluid prior to polymerization, ready removal of excess (SDBS) surfactant, on the contrary led to nanocomposite magnetic polymer colloids where the nanoparticles were dispersed throughout the polymer matrix (see Figure 4.9).

Long and coworkers described in 1991 the emulsifier-free emulsion polymerization of methyl methacrylate in the presence of silica and alumina nanosols, respectively.[89] Potassium persulfate was used as anionic free radical initiator and polymerizations were carried out at pH 5 at 60 °C. They successfully fabricated nanocomposite latex particles with encapsulation of the Al_2O_3 nanoparticles throughout the polymer matrix. They reasoned that the high encapsulation efficiency was due to the opposite charges of the polymer chains (sulfate end groups) and the (at pH 5) cationic alumina sol. Encapsulation of the silica nanoparticles was not efficient. From the TEM data presented it is hard to see whether in this case bare poly(methyl methacrylate) particles, or armoured and thus Pickering stabilized latex particles were obtained.

Armes and coworkers reported in 1999 the dispersion polymerization of 4-vinylpyridine (5.0 mL) in water (pH 10, 100 mL) in the presence of a waterborne sol of 20 nm silica particles (Nyacol 2040, 8.0 g) at 60 °C using 50 mg of ammonium persulfate as radical initiator.[44] They reported that nanocomposite particles could be made with the silica nanoparticles dispersed throughout the poly(4-vinylpyridine) matrix. Reactions in the absence of the silica nanoparticles resulted in coagulation, thereby clearly demonstrating that the silica particles played a role in colloid stabilization.

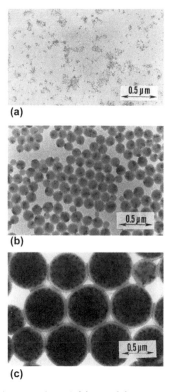

Figure 4.9 TEM photomicrographs of (a) Fc, (b) magnetic poly(St) latex particles
obtained by the use of Fc (**f** = 14%), and (c) magnetic poly(St) latex
particles prepared by the use of the ultrafiltered Fc (**f** = 20%).
Reprinted with permission from reference 88. © 1993 Wiley.

Modification of the monomer composition to methylmethacrylate : 4-vinyl-
pyridine (80 : 20 wt ratio) and styrene : 4-vinylpyridine (90 : 10 wt ratio),
thereby effectively turning the polymerization into an emulsion polymerization, yielded nanocomposite stable latexes of similar morphology. Wang
and coworkers showed that emulsion copolymerization of styrene and
4-vinylpyridine yields stable latexes under similar conditions and in the
absence of a silica sol.[90]

Xue and Wiese filed a patent in 2001 in which they described the synthesis
of polymer nanocomposite latexes with Nyacol 2040 silica nanoparticles
dispersed throughout the polymer matrix, at 40 wt% overall solids with no
excess of silica nanoparticles.[91] The recipe was based on conventional
emulsion polymerization using a combination of quaternary nitrogen
compounds, *e.g.* *N*-cetyl-*N,N,N*-trimethylammonium bromide (CTAB) as
co-surfactant, and small amounts of reactive and functional siloxane
comonomers, *e.g.* methacryloyloxypropyltrimethoxysilane, to modify the
surface of the silica nanoparticles in order to promote full encapsulation
and dispersion of the silica nanoparticles throughout the polymer

latex particles. One of the target applications was scratch-resistant water-borne coatings.

Wu and coworkers described in 2005 the synthesis of poly(methyl methacrylate) latexes armoured with silica nanoparticles.[83] They used a cationic auxiliary monomer, 2-(methacryloyl)ethyltrimethylammonium chloride (MTC), and postulated that electrostatic attraction between the negatively charged silica particles and the positively charged MTC was responsible for anchoring the silica particles onto the surface of the latex particles. They carried out a key experiment in which the silica particles were omitted. This led to coagulation, which demonstrated that the silica nano-particles played a key role in achieving colloidal stability. They proposed that the silica particles, modified with MTC, acted as Pickering stabilizers in the emulsion polymerization. In a way the wettability of the silica sol was tuned with MTC to promote adhesion to the surface of the growing latex particles.

Yu and Zhang reported in 2007 the emulsion copolymerization of butyl acrylate : methyl methacrylate : 2-hydroxyethyl methacrylate : acrylic acid at (41.50 : 32.25 : 23.13 : 3.12 wt ratio) in the presence of a 12 nm diameter silica sol (Klebosol 30R12, up to 7 wt% in total).[92] However, the fate and distribution of the silica nanoparticles, whether encapsulated throughout, present on the surface of the latex particles, or remaining in the water phase, was unclear as the TEM images provided were unconvincing.

Walther and coworkers reported at the end of 2007 the emulsion polymerizations of styrene and *n*-butyl acrylate using amphiphilic anisotropic particles as stabilizer.[80] These Janus-type nanoparticles were prepared by selectively crosslinking spherical polybutadiene microdomains within the lamella–sphere morphology of a microphase-separated template of a polystyrene-*block*-polybutadiene-*block*-poly(methyl methacrylate) triblock terpolymer (PS-PB-PMMA) and subsequent hydrolysis of the PMMA block into the hydrophilic poly(methacrylic acid). Results showed a clear decrease in the average diameter of the latex particle produced upon increasing the Janus particle concentration. The dispersity of particle size distributions of all latexes was low.

Armes and coworkers reported in 2008 the Pickering emulsion polymerization of styrene and styrene : butyl acrylate (1 : 1 weight ratio) in the presence of a glycerol-modified silica sol (Bindzil CC40, average diameter of 19 nm) to prepare successfully polystyrene latexes armoured with a layer of modified silica nanoparticles.[84] A typical recipe comprised 5.4 g of silica sol (eq. 2.0 g of dry silica), 5 g of monomer, 37.6 g of water, and 50 mg of the cationic free radical initiator 2,2-Azobis(isobutyramidine) dihydrochloride (AIBA) dissolved in 4.0 g of water. The Pickering emulsion polymerizations were carried out at 60 °C. They stated that use of the cationic initiator was essential as experiments using anionic ammonium persulfate as initiator induced colloidal unstability. They also said that the use of glycerol-modified silica was essential, as experiments carried out with unmodified silica nanoparticles (Bindzil 2040, average diameter of 20 nm) failed to produce well-defined armoured polystyrene latex particles.

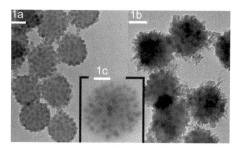

Figure 4.10   TEM images (scale bar = 100 nm) of (a) poly(methyl methacrylate) latex armoured with silica nanoparticles obtained by Pickering emulsion polymerization. Multilayered nanocomposite polymer colloids with (b) a 'hairy' outer layer of poly(acrylonitrile) and (c) a soft shell of poly(*n*-butyl acrylate).
Reprinted with permission from reference 81. © 2008 American Chemical Society.

In the same year Bon and coworkers showed that Pickering emulsion polymerization of methyl methacrylate (and ethyl methacrylate) using un-modified silica nanoparticles (Ludox TM-40, approximate average diameter 25 nm), and without the need for auxiliary monomers, surfactants or cat-ionic initiators, was possible under batch conditions, up to an overall solids content of 45 wt% (see Figure 4.10).[81]

Optimum results were obtained when the pH of the silica sol, once added to the water phase, was adjusted to 5.5. The use of styrene or *n*-butyl methacrylate led to the production of bare latex particles with the silica nanoparticles remaining in the water phase. Moreover, they showed that when the armoured latex particles were used as seed and that when *n*-butyl acrylate was used as monomer, the silica nanoparticles migrated to the newly created outer surface, whereas when acrylonitrile was polymerized intricate hybrid particle morphologies were obtained (Figure 4.10).

4.5.2   Pickering Emulsion Polymerization: A Mechanistic Approach

By 2008 it was generally accepted that Pickering emulsion polymerization could be carried out and that hybrid latex particles armoured with nano-particles, which served as Pickering stabilizers in the emulsion polymer-ization process, can be synthesized. However, detailed mechanistic insight into how exactly the Pickering emulsion process works is still limited. Bon *et al.*[73,81,85] proposed the following mechanistic insights. For convenience in this discussion we will subdivide the Pickering emulsion process into two stages: (i) particle formation, and (ii) particle growth.

4.5.2.1   *Particle Formation in Pickering Emulsion Polymerization*

It is well known that in soap-free emulsion polymerizations the particle formation process occurs through homogeneous nucleation, often with limited coagulation, commonly referred to as the Hansen–Ugelstad–Fitch–Tsai (HUFT) model of nucleation.[77] Primary latex particles are created from the water phase by successive initiation (for example by a sulfate radical), growth (through monomer consumption and thus propagation in the water phase) and collapse (*i.e.* phase separation from the water phase) of a single (growing) polymer chain, then identified as a primary particle. Swelling (with monomer) and further growth of these primary loci of polymerization leads to colloidal instability. Essentially a growing primary particle that is stabilized with a single sulfate group is unsustainable as a colloidal entity. This causes the primary particles to coalesce with one another (*i.e.* undergo self-assembly into clusters and rearrange into one new larger particle, as they are swollen with monomer and hence soft) until the decreasing surface-to-volume ratio of these newly formed larger particles warrants colloidal stability (in this case through electrostatic repulsion as a result of the surface charges caused by the sulfate end groups of the polymer chains). This assembly process, or limited coagulation event, therefore eventually generates a stable number of 'mature' growing latex particles. This constant number of growing latex particles marks the end of the particle formation phase and is achieved when new aqueous-phase radical species can have no other fate but to terminate or to enter existing particles. Typically, for soap-free emulsion polymerizations, this particle formation or nucleation process is fast relative to the overall timescale of polymerization (*i.e.* to reach complete monomer conversion). This difference in timescales between the particle nucleation period and the consecutive particle growth period (one could argue that the particles are also growing during particle formation, but let us leave that aside) basically means that particle size distributions of low dispersity will be obtained (secondary nucleation or coagulation during polymerization are absent).

The question now is how this nucleation process will be altered when we add nanoparticles to the water phase, in other words when we carry out a Pickering emulsion polymerization. It is logical to assume that growing waterborne oligomers can interact with a nanoparticle when wetting of the nanoparticle with the polymer chain is favourable. During physisorption and thus adhesion of a growing polymer chain onto a nanoparticle (Pickering stabilizer) a hybrid primary particle is formed. We believe this event is crucial for the success of a Pickering emulsion polymerization. When wetting of the growing waterborne oligomers is unfavourable and in the extreme case does not take place at all, an ordinary emulsifier-free emulsion polymerization takes place, leading to a blend of 'naked' ordinary latex particles with the nanoparticles remaining in the water phase (note that the presence of such non-interacting nanoparticles can potentially still have an effect on the particle size distribution of the polymer latex due to depletion interactions). Pickering emulsion polymerizations of pure styrene and *n*-butyl acrylate,[81] and of vinyl pivalate (see also Figure 4.12a)[73] using silica nanoparticles as Pickering stabilizer. were indeed unsuccessful in that binary blends of dispersed bare polymer particles and silica nanoparticles

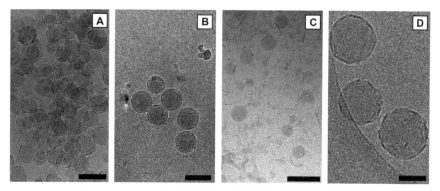

Figure 4.11 Cryo-TEM images of polymer latex particles: (A) poly(MMA-*co-n*-(BA)/ Laponite, (B) poly(Sty-*co*-BA)/Laponite, (C) poly(Sty-*co*-2-EHA)/Laponite, (D) poly(Sty-*co*-2-EHA)/Laponite with methacrylic acid. Scale bars of 100, 100, 200 and 50 nm, respectively.
Reprinted with permission from reference 85. © 2011 American Chemical Society.

were obtained. The silica nanoparticles are inherently too hydrophilic for the waterborne oligomers to adhere to irreversibly. This wettability issue can be overcome when small amounts of hydrophilic comonomers are used. Examples include the use of an auxiliary poly(ethylene glycol) mono-methylether methacrylate (PEGMA) macromonomer in the Pickering emulsion polymerization of styrene in the presence of silica particles as proposed by Sheibat-Othman and Bourgeat-Lami,[82] as well as its use in the Laponite-clay stabilized Pickering emulsion copolymerization of styrene and *n*-butyl acrylate.[86] Bon and coworkers showed that Pickering emulsion copolymerization of the hydrophobic monomers styrene and 2-ethylhexyl methacrylate using Laponite clay as stabilizers was possible when small amounts of methacrylic acid were used as comonomer (see Figure 4.11).[85]

High charge density of Pickering stabilizers, for example of the Ludox TM-40 silica sol in water indicated by a large negative ζ-potential (approximately −50 mV) also leads to bare latex particles, even in the case of the Pickering emulsion polymerization of the more hydrophilic methyl methacrylate, presumably due to combined effects of charge repulsion and effective hydration of the silica particle surface.[81]

Now let us go back to the scenario where a growing waterborne oligomer successfully adheres to a nanoparticle (Pickering stabilizer). The newly formed primary hybrid particles grow further and take part in a similar coagulative nucleation process, as in the case of emulsifier-free radical polymerization. They are in essence an additional player in this particle formation process, which now is an interplay between Pickering stabilizers, hybrid primary particles, bare primary latex particles and (growing) waterborne polymer chains. It means that the amount of Pickering stabilizers used in Pickering emulsion polymerization can have an effect on the total number of particles, and thus the final particle size distribution. Absence or

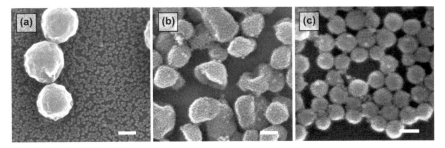

Figure 4.12 FEG-SEM images of latexes prepared in batch Pickering emulsion polymerization in the presence of Ludox TM-40 colloidal silica. (a) Poly(vinyl pivalate) latex with a silica/monomer ratio of 0.44. (b) Poly(vinyl acetate) latex with a silica/monomer ratio of 0.44. (c) Poly(vinyl acetate) latex with a silica/monomer ratio of 1.04. Scale bars are 200 nm. Reprinted with permission from reference 73. © 2010 American Chemical Society.

low amounts of Pickering stabilizers at any stage of the polymerization process could lead to limited (see Figure 4.12b)[73] or full coagulation.[81,85]

Obviously the amount of latex particles is driven in a way by the frequency of radical generation (*i.e.* the frequency of production of waterborne oligomers that reach a critical chain length at which they become surface active, and/or adhere to a Pickering stabilizer, or reach the critical chain length at which they collapse on themselves to form a primary particle). This means that the effect an increase in the number density of Pickering stabilizers in Pickering emulsion polymerizations has on the final particle size distribution of the armoured latex particles fades out. Indeed this is observed by Bon and coworkers when the amount of Laponite XLS clay is gradually increased in Pickering emulsion copolymerizations of styrene and *n*-butyl acrylate.[85]

4.5.2.2 Particle Growth in Pickering Emulsion Polymerization

The second stage in the Pickering emulsion polymerization is particle growth. The increase in volume of polymer (and monomer) increases the interfacial area of the growing particles. For a fixed number of nanoparticles present at the surface of the growing latex this means that the surface coverage drops, exposing areas of bare interface that are generally hydrophobic. A high fraction of bare interface is bad news, in that it induces colloidal instability and triggers coagulation events. It is here that the Pickering stabilizers present in the water phase play an important role. We argue that when a latex particle grows it will heterocoagulate with a nanoparticle. Upon collision the nanoparticle can adhere to the interface, acting as a Pickering stabilizer and, additionally, will provide extra charge to secure (temporarily) a sufficient electrostatic barrier between polymer latex particles, warranting colloidal stability of the system. Indeed TEM analysis carried out at different times throughout the Pickering emulsion polymerization of methyl methacrylate in the presence of a silica sol (Ludox TM-40) shows a gradual increase in the number of nanoparticles on the surface of the growing latex particles.[81]

Bon and coworkers were able to quantitatively monitor the amount of silica nanoparticles in the water phase throughout the Pickering emulsion polymerization process by disc centrifugation.[73] They developed a model to calculate the amount of silica nanoparticles remaining in the water phase from the average particle size of the armoured latexes made by Pickering emulsion polymerization as a function of monomer conversion,[73] taking into account the packing patterns of the silica nanoparticles onto the surface.[53] Excellent agreement was obtained between experimental data on the amount of Pickering stabilizers and the model. As noted before, the concentration of Pickering stabilizers in the water phase is important, as too low amounts can induce (partial) coagulation when the latex particles are still growing. When the extent of this coagulation event is limited, armoured particles form non-spherical clusters (see Figure 4.12b).

4.6   Outlook

This review hopefully has triggered your imagination over what can be achieved when the established areas of suspension, mini-emulsion and emulsion polymerization as ways of making polymer dispersions are combined with the phenomenon of Pickering stabilization. Pickering suspension polymerization was the subject of scientific interest in the 1930s to 1950s, but the sole targeted focus as a stabilization method to manufacture polymer beads and no sparked interest in what beneficial properties armoured micron and millis-sized polymeric objects could have, put Pickering stabilization for heterogeneous polymerization techniques in a dormant state until the late 1990s where through a drive for nanostructures and leaps in innovation in the microscopic imaging of colloids sparked a renewed interest.

Whereas great efforts are being made in the synthesis of armoured structures by Pickering suspension, mini-emulsion and emulsion polymerization (note: also Pickering dispersion polymerization, but this was not discussed in this review), it is important that a mechanistic, kinetic and physical understanding of these Pickering heterogeneous polymerization techniques is developed, and of equal importance, used in an appropriate manner for scientific explanations of experimental results. No doubt the years ahead will give us greater insight into Pickering stabilized polymerization processes. Technological applications of armoured nanocomposite polymer latexes look promising, but are still in their infancy. Whereas polymer latexes with nanoparticles throughout their matrix are used commercially as scratch-resistant coatings, armoured particles still need exploration. Part of the reason why armoured latexes have not found many applications is that until recently their synthesis was not straightforward. The developments in Pickering emulsion polymerization alleviate this. One example that shows promise was reported by Wang *et al.*,[52] who demonstrated that addition of small amounts of soft armoured latexes (2.7 wt%) to waterborne adhesives induces a marked increase in tack adhesion energy and hence performance (70% increase).

References

1. S. Slomkowski, J. V. Alemán, R. G. Gilbert, M. Hess, K. Horie, R. G. Jones, P. Kubisa, I. Meisel, W. Mormann, S. Penczek and R. F. T. Stepto, *Pure Appl. Chem.*, 2011, **83**, 2229–2259.
2. W. D. Harkins, *J. Am. Chem. Soc.*, 1947, **69**, 1428–1444.
3. W. D. Harkins, *J. Polym. Sci.*, 1950, **5**, 217–251.
4. R. M. Fitch, *Off. Dig., J. Paint Tech. Eng.*, 1965, **37**, 32.
5. C. P. Roe, *Ind. Eng. Chem.*, 1968, **60**, 20–23.
6. J. Ugelstad, M. S. El-Aasser and J. W. Vanderhoff, *J. Polym. Sci. Polym. Lett. Ed.*, 1973, **11**, 503–513.
7. F. J. Schork, Y. Luo, W. Smulders, J. P. Russum, A. Butté and K. Fontenot, *Adv. Polym. Sci.*, 2005, **175**, 129–255.
8. K. Landfester, *Angew. Chem. Int. Ed. Engl.*, 2009, **48**, 4488–4507.
9. W. Ramsden, *Phil. Trans. R. Soc. London*, 1903, **72**, 156–164.
10. S. U. Pickering, *J. Chem. Soc. Trans.*, 1907, **91**, 2001–2021.
11. P. Finkle, H. D. Draper and J. H. Hildebrand, *J. Am. Chem. Soc.*, 1923, **45**, 2780–2788.
12. W. Haynes, 1860, BP 488. February 23.
13. G. Bessel, 1877, Berlin Patent 42. July 2.
14. O. Röhm and E. Trommsdorff, 1939, US 2,171,765.
15. W. P. Hohenstein and H. Mark, *J. Polym. Sci.*, 1946, **1**, 127–145.
16. W. P. Hohenstein, 1950, US 2,524,627.
17. J. E. O. Mayne, 1947, GB 594,653.
18. R. N. Haward and J. Elly, 1954, GB 710,498.
19. F. H. Winslow and W. Matreyek, *Ind. Eng. Chem.*, 1951, **43**, 1108–1112.
20. W. P. Hardy, in *Colloid Symposium Monograph VI*, 1928, p. 8.
21. R. M. Wiley, *J. Colloid Sci.*, 1954, **9**, 427–436.
22. V. H. Wenning, *Die Makromol. Chemie*, 1956, **20**, 196–213.
23. Y. Deslandes, *J. Appl. Polym. Sci.*, 1987, **34**, 2249–2257.
24. R. F. A. Teixeira and S. A. F. Bon, *Adv. Polym. Sci.*, 2010, **233**, 19–52.
25. O. D. Velev, K. Furusawa and K. Nagayama, *Langmuir*, 1996, **12**, 2374–2384.
26. O. D. Velev, K. Furusawa and K. Nagayama, *Langmuir*, 1996, **12**, 2385–2391.
27. O. D. Velev and K. Nagayama, *Langmuir*, 1997, **13**, 1856–1859.
28. A. D. Dinsmore, M. F. Hsu, M. G. Nikolaides, M. Marquez, A. R. Bausch and D. A. Weitz, *Science*, 2002, **298**, 1006–1009.
29. Y. He, *Powder Technol.*, 2004, **147**, 59–63.
30. Y. He, *Mater. Lett.*, 2005, **59**, 2133–2136.
31. Y. He, *Mater. Chem. Phys.*, 2005, **92**, 134–137.
32. X. He, X. Ge, H. Liu, M. Wang and Z. Zhang, *Chem. Mater.*, 2005, **17**, 5891–5892.

33. S. Cauvin, P. J. Colver and S. A. F. Bon, *Macromolecules*, 2005, **38**, 7887–7889.
34. S. A. F. Bon, S. Cauvin and P. J. Colver, *Soft Matter*, 2007, **3**, 194–199.
35. T. Chen, P. J. Colver and S. A. F. Bon, *Adv. Mater.*, 2007, **19**, 2286.
36. Y. Zhao, H. Wang, X. Song and Q. Du, *Macromol. Chem. Phys.*, 2010, **211**, 2517–2529.
37. Z. Nie, J. Il Park, W. Li, S. A. F. Bon and E. Kumacheva, *J. Am. Chem. Soc.*, 2008, **130**, 16508–16509.
38. S. A. F. Bon and T. Chen, *Langmuir*, 2007, **23**, 9527–9530.
39. T. Hasell, J. Yang, W. Wang, J. Li, P. D. Brown, M. Poliakoff, E. Lester and S. M. Howdle, *J. Mater. Chem.*, 2007, **17**, 4382.
40. D. J. Voorn, W. Ming and A. M. van Herk, *Macromolecules*, 2006, **39**, 2137–2143.
41. L. Duan, M. Chen, S. Zhou and L. Wu, *Langmuir*, 2009, **25**, 3467–3472.
42. Q. Gao, C. Wang, H. Liu, Y. Chen and Z. Tong, *Polym. Chem.*, 2010, **1**, 75–77.
43. F. Tiarks, K. Landfester and M. Antonietti, *Langmuir*, 2001, **17**, 5775–5780.
44. C. Barthet, A. J. Hickey, D. B. Cairns and S. P. Armes, *Adv. Mater.*, 1999, **11**, 408–410.
45. M. J. Percy, C. Barthet, J. C. Lobb, M. A. Khan, S. F. Lascelles, M. Vamvakaki and S. P. Armes, *Langmuir*, 2000, **16**, 6913–6920.
46. E. Balnois, S. Durand-Vidal and P. Levitz, *Langmuir*, 2003, **19**, 6633–6637.
47. P. Mongondry, J. F. Tassin and T. Nicolai, *J. Colloid Interface Sci.*, 2005, **283**, 397–405.
48. N. P. Ashby and B. P. Binks, *Phys. Chem. Chem. Phys.*, 2000, **2**, 5640–5646.
49. S. A. F. Bon and P. J. Colver, *Langmuir*, 2007, **23**, 8316–8322.
50. P. J. Colver, T. Chen and S. A. F. Bon, *Macromol. Symp.*, 2006, **245–246**, 34–41.
51. Y. Yang, Z. Liu, D. Wu, M. Wu, Y. Tian, Z. Niu and Y. Huang, *J. Colloid Interface Sci.*, 2013, **410**, 27–32.
52. T. Wang, P. J. Colver, S. A. F. Bon and J. L. Keddie, *Soft Matter*, 2009, **5**, 3842–3849.
53. S. Fortuna, C. A. L. Colard, A. Troisi and S. A. F. Bon, *Langmuir*, 2009, **25**, 12399–12403.
54. F. Tiarks, K. Landfester and M. Antonietti, *Langmuir*, 2001, **17**, 5775–5780.
55. M. F. Haase, D. O. Grigoriev, H. Möhwald and D. G. Shchukin, *Adv. Mater.*, 2012, **24**, 2429–2435.
56. K. González-Matheus, G. P. Leal, C. Tollan and J. M. Asua, *Polymer*, 2013, **54**, 6314–6320.
57. Q. Xiao, X. Tan, L. Ji and J. Xue, *Synth. Met.*, 2007, **157**, 784–791.
58. N. Zgheib, J.-L. Putaux, A. Thill, F. D'Agosto, M. Lansalot and E. Bourgeat-Lami, *Langmuir*, 2012, **28**, 6163–6174.
59. J. H. Chen, C.-Y. Cheng, W.-Y. Chiu, C.-F. Lee and N.-Y. Liang, *Eur. Polym. J.*, 2008, **44**, 3271–3279.

60. J. Jeng, T.-Y. Chen, C.-F. Lee, N.-Y. Liang and W.-Y. Chiu, *Polymer*, 2008, **49**, 3265–3271.
61. A. Ben Mabrouk, M. Rei Vilar, A. Magnin, M. N. Belgacem and S. Boufi, *J. Colloid Interface Sci.*, 2011, **363**, 129–136.
62. X. Song, Y. Yang, J. Liu and H. Zhao, *Langmuir*, 2011, **27**, 1186–1191.
63. M. M. Gudarzi and F. Sharif, *Soft Matter*, 2011, **7**, 3432.
64. S. H. C. Man, N. Y. M. Yusof, M. R. Whittaker, S. C. Thickett and P. B. Zetterlund, *J. Polym. Sci. Part A: Polym. Chem.*, 2013, **51**, 5153–5162.
65. G. Yin, Z. Zheng, H. Wang, Q. Du and H. Zhang, *J. Colloid Interface Sci.*, 2013, **394**, 192–198.
66. A. Bonnefond, M. Micusik, M. Paulis, J. R. Leiza, R. F. A. Teixeira and S. A. F. Bon, *Colloid Polym. Sci.*, 2013, **291**, 167–180.
67. J. J. Thomson, *Philos. Mag. Ser. 6*, 1904, 7, 237–265.
68. P. Lipowsky, M. J. Bowick, J. H. Meinke, D. R. Nelson and A. R. Bausch, *Nat. Mater.*, 2005, **4**, 407–411.
69. A. R. Bausch, M. J. Bowick, A. Cacciuto, A. D. Dinsmore, M. F. Hsu, D. R. Nelson, M. G. Nikolaides, A. Travesset and D. A. Weitz, *Science*, 2003, **299**, 1716–1718.
70. N. Zgheib, J.-L. Putaux, A. Thill, F. D'Agosto, M. Lansalot and E. Bourgeat-Lami, *Langmuir*, 2012, **28**, 6163–6174.
71. A. D. Ronteltap, B. R. Damsté, M. De Gee and A. Prins, *Colloids and Surfaces*, 1990, **47**, 269–283.
72. A. D. Ronteltap and A. Prins, *Colloids and Surfaces*, 1990, **47**, 285–298.
73. C. A. L. Colard, R. F. A. Teixeira and S. A. F. Bon, *Langmuir*, 2010, **26**, 7915–7921.
74. K. González-Matheus, G. P. Leal, C. Tollan and J. M. Asua, *Polymer*, 2013, **54**, 6314–6320.
75. X.-S. Chai, F. J. Schork, A. DeCinque and K. Wilson, *Ind. Eng. Chem. Res.*, 2005, **44**, 5256–5258.
76. D. C. Blackley, *Emulsion Polymerization: Theory and Practice*, Springer, New York, 1975.
77. R. M. Fitch, *Polymer Colloids: A Comprehensive Introduction*, Academic Press, San Diego and London, 1997.
78. R. G. Gilbert, *Emulsion Polymerization: A Mechanistic Approach*, Academic Press, London, 1995.
79. A. M. van Herk (ed.), *Chemistry and Technology of Emulsion Polymerisation*, 2nd edn, Wiley, Chichester, 2013.
80. A. Walther, M. Hoffmann and A. H. E. Mueller, *Angew. Chem. Int. Ed.*, 2008, **47**, 711–714.
81. P. J. Colver, C. A. L. Colard and S. A. F. Bon, *J. Am. Chem. Soc.*, 2008, **130**, 16850.
82. N. Sheibat-Othman and E. Bourgeat-Lami, *Langmuir*, 2009, **25**, 10121–10133.
83. M. Chen, S. Zhou, B. You and L. Wu, *Macromolecules*, 2005, **38**, 6411–6417.
84. A. Schmid, J. Tonnar and S. P. Armes, *Adv. Mater.*, 2008, **20**, 3331–3336.

85. R. F. A. Teixeira, H. S. McKenzie, A. A. Boyd and S. A. F. Bon, *Macromolecules*, 2011, **44**, 7415–7422.
86. E. Bourgeat-Lami, T. R. Guimarães, A. M. C. Pereira, G. M. Alves, J. C. Moreira, J.-L. Putaux and A. M. Dos Santos, *Macromol. Rapid Commun.*, 2010, **31**, 1874–1880.
87. J. Solc nee Hajna, 1983, US 4,421,660A.
88. N. Yanase, H. Noguchi, H. Asakura and T. Suzuta, *J. Appl. Polym. Sci.*, 1993, **50**, 765–776.
89. F. Long, W. Wang, Y. Xu and T. Cao, *Tianjin Daxue Xuebao (Journal Tianjin Univ.)*, 1991, **4**, 10–15.
90. Y. Wang, L. Feng and C. Pan, *J. Appl. Polym. Sci.*, 1999, **74**, 1502–1507.
91. Z. Xue and H. Wiese, 2006, US 7094830 B2.
92. F.-A. Zhang and C.-L. Yu, *Eur. Polym. J.*, 2007, **43**, 1105–1111.

CHAPTER 5

Emulsions Stabilized by Soft Microgel Particles

ZIFU LI[a] AND TO NGAI*[b]

[a] Department of Chemical and Materials Engineering, University of Alberta, Edmonton, Alberta T6G 2G6, Canada; [b] Department of Chemistry, The Chinese University of Hong Kong, Shatin, N. T., Hong Kong
*Email: tongai@cuhk.edu.hk

5.1   Introduction

Emulsions are mixtures of immiscible liquids where one liquid is dispersed throughout another liquid in the form of droplets.[1] They occur as end products in a wide range of areas including the food, cosmetic, pharmaceutical and petroleum industries. Because high specific surface areas resulting from the dispersion process are not energetically favoured, emulsions are thermodynamically unstable and will separate into two phases over a period of time. Emulsions are often kinetically stabilized through the adsorption of low molecular weight surfactants or amphiphilic polymers at the liquid–liquid interfaces. The adsorbed molecules and polymers favour stabilization by inhibiting the coalescence and Ostwald ripening of droplets.[2]

In addition to surfactants, it has been known that solid particles of colloidal size can self-assemble at the interface and provide excellent long-term kinetic stability to both simple and multiple emulsions.[3] Such particle-stabilized emulsions, today referred to as 'Pickering emulsions', were described a century ago, first by Ramsden[4] and a few years later by Pickering.[5] It was found that the particle wettability is the key factor that dictates the emulsion type and stability.[6] Hydrophilic particles, like metal oxides, with a

RSC Soft Matter No. 3
Particle-Stabilized Emulsions and Colloids: Formation and Applications
Edited by To Ngai and Stefan A. F. Bon
© The Royal Society of Chemistry 2015
Published by the Royal Society of Chemistry, www.rsc.org

contact angle less than 90° (measured through the water phase), tend to stabilize oil-in-water (o/w) emulsions, while hydrophobic particles, like carbon, with larger contact angles can stabilize water-in-oil (w/o) emulsions.[6] Different types of colloidal particles, including silica, clays and latex particles, have been used as particulate emulsifiers to form either o/w or w/o emulsions.[3,6] Nowadays, this class of Pickering emulsion is a fundamental component of many modern products and formulations. However, the exploitation of this Pickering emulsion for the manufacture of new functional materials has only recently become the subject of intense investigation.[3]

Recently, a new class of materials, responsive microgels, has been successfully used to stabilize emulsion droplets.[7,8] Microgels are colloidal particles that consist of chemically crosslinked three-dimensional polymer networks. Hence, they combine properties of typical colloids with the responsiveness of gels.[9] Most of the previously studied microgel systems are based on poly(N-isopropylacrylamide) (PNIPAM), which shows a well-known volume phase transition temperature (VPTT) at around 32 °C.[10–13] In recent years, PNIPAM-based microgels at the oil–water interface have also received intense attention.[14–31] One of the driving forces for these explorations is that these microgels can be used to stabilize stimuli-responsive emulsions. The stability of microgel-stabilized emulsions can be triggered by changing the environmental conditions such as temperature or pH, thus offering a novel way to control emulsion stability.[15,17] For example, emulsions stabilized with poly(N-isopropylacrylamide-co-methacrylic acid) (PNIPAM-co-MAA) microgels have been shown to be stable at low temperature and high pH, but become unstable at high temperature and low pH.[7,8] This peculiar tunable stabilizing property is especially desirable in industrial applications such as biocatalysis, fuel production and oil transportation, where the emulsions can be prepared and broken on demand.[14]

Despite microgels as stabilizers having been well demonstrated, the stabilization and destabilization mechanism involved using such microgels as stabilizers for emulsions are far from completely understood.[15,17] For conventional, rigid, solid Pickering stabilizers, it is well documented that the key parameter for controlling emulsion type and stability is the contact angle that the particle makes with the oil–water interface.[3,6] The contact angle is related to the particle's wettability through a force balance on the contact line, which yields Young's equation. The underlying assumption for applying Young's equation is that the rigid particle quickly approaches the equilibrium position at the oil–water interface. However, recent studies showed that this underlying assumption might be unrealistic.[32] Due to particle surface roughness and heterogeneities, the relaxation of rigid solid particles at the interface to their equilibrium positions is an extremely slow process, appearing in logarithmic time, and thus the rigid solid particles at the oil–water interface are an issue out of equilibrium.

For PNIPAM microgels, their interfacial behaviour might be even more complex.[15,17] First, PNIPAM microgels are soft, porous and significantly deformed at the oil–water interface.[17,19,22] The deformation of microgels probably suggests that the traditional concept of contact angle may be no longer applicable for the analysis of microgels at the oil–water interface. Second, the complex nature of PNIPAM microgels further complicates their behaviour at the oil–water interface.[9] Structurally, PNIPAM microgels are shown as a core-shell structure with a highly crosslinked core surrounded by a shell of less crosslinked dangling polymer chains[33] due to the reactivity difference between crosslinker and NIPAM.[34] With these two distinct differences, it might be more difficult for microgels to reach their equilibrium positions at the oil–water interface as compared to rigid particles.

Here, we provide a review of the recent studies highlighting the unique properties of microgel-stabilized emulsions, and compare the specific properties of microgels as stabilizers for emulsions with classical solid rigid particles. We first describe the preparation and characterization of the PNIPAM-based microgel particles. After that, we outline the factors such as solution temperature, pH, deformability of particles, and interface microstructure affecting emulsion formation and stability. We then discuss the parameters that control the adsorption kinetics of microgels onto the oil–water interface. Lastly, we demonstrate how microgel-stabilized emulsions can serve as a template for the fabrication of novel materials including Janus particles, microcapsules and porous materials, which might open the door to new applications.

5.2 Microgel Particles: Synthesis and Characterization

Microgels are colloidal particles and have recently attracted much attention in soft matter research.[9,35–39] They are a chemically crosslinked network of polymer chains swollen in a solvent, like normal bulk gel.[40] The microgel particles exhibit a full reversible conformational transition in response to changing solvent quality, which may be affected by a number of external stimuli such as temperature, pH and ionic strength. During this transition, the microgel will adopt a more compact conformation in order to minimize polymer–solvent interactions, only returning to its original conformation when solvent conditions become more favourable.[12] Much research has illustrated the applicability of microgels in many different areas, such as in wound repair,[41] catalysts,[10] functional membranes and drug delivery.[9]

One of the most studied microgel systems is composed of the environmentally responsive polymer poly(N-isopropylacrylamide) (PNIPAM), which was first reported by Pelton and Chibante in 1986.[42] PNIPAM-based microgel particles are usually prepared by precipitation polymerization. Generally, in the presence of a surfactant such as sodium dodecyl sulfate (SDS), the hydrodynamic diameter of the PNIPAM microgels prepared by emulsion polymerization is less than 300 nm. However, there is a problem with completely removing the surfactants from the resulting microgel particles.[43] PNIPAM microgels are able to undergo a volume-phase transition (VPT) at around 32 °C. Figure 5.1 shows the temperature dependence of the mean

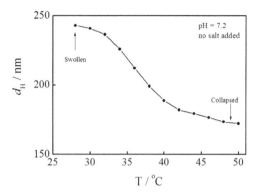

Figure 5.1 Temperature dependence of the mean hydrodynamic diameter d_H of PNIPAM-*co*-MAA microgel particles in an aqueous dispersion at pH 7.2. Reprinted with permission from reference 8. © 2006 American Chemical Society.

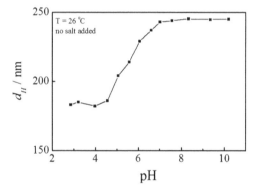

Figure 5.2 pH dependence of the mean hydrodynamic diameter d_H of PNIPAM-*co*-MAA microgel particles in an aqueous dispersion at room temperature. Reprinted with permission from reference 8. © 2006 American Chemical Society.

hydrodynamic diameter d_H, for the PNIPAM-*co*-MAA microgel particles in water at a constant pH of 7.2.[7,8] The microgel particles gradually shrink into an increasingly collapsed state when the temperature is raised to 50 °C. The decrease in d_H is expected, because the increase in temperature weakens the hydration and the PNIPAM chains gradually prefer each other over water, leading to the collapse of the microgel particles. On the other hand, the synthesized microgels are also sensitive to pH and ionic strength due to the copolymerization with ionic comonomer MAA. Figure 5.2 confirms the pH-responsive behaviour of the synthesized microgel particles.[7,8] At room temperature, the size of the microgel particles increases continuously from pH 4.0 to pH 8.0, and plateaus at higher pH. This is because the pH increase first induces dissociation of carboxylic acid groups (-COOH) on the network chains, leading to an increase in the charge density on the network.

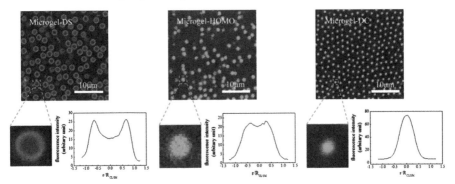

Figure 5.3 Confocal images of three different micron-sized, temperature and pH-responsive PNIPAM microgels with dense-shell (DS), dense-core (DC) or homogeneous (HOMO) structure.
Reprinted with permission from reference 45. © 2013 American Chemical Society.

The concomitant increase in mobile counterion content of the network increases the internal osmotic pressure, which induces the swelling of microgel particles. It is these adjustable properties that make microgel particles desirable for many applications.[9,35,38]

Recently, by combining semi-batch and temperature-programmed[44] surfactant-free precipitation polymerization, we have successfully developed a novel approach for the preparation of micron-sized ($\sim$ 2.5–5.0 μm), temperature and pH-responsive PNIPAM-based microgels with dense-shell (DS), dense-core (DC) or homogeneous (HOMO) structure.[45,46] We have investigated the interaction between the synthesized microgels and some fluorescent dye molecules using confocal laser scanning microscopy (CLSM). Our results have qualitatively revealed that the crosslinker and the functional carboxylic groups (-COOH) could be homogeneously distributed, predominately localized inside the core, or concentrated near the surface of the synthesized microgels (Figure 5.3). This large diameter of PNIPAM microgels thus offers the possibility to directly visualize the morphology of individual, isolated microgel particles by conventional optical microscopy. Note that this size is not traditionally obtainable by batch precipitation polymerization or by dispersion polymerization. We thereby expect that the synthesized microgels with improved control of crosslinker and functional group distributions would be helpful for making more advanced microgel structures for specific applications.

5.3 Factors Affecting Responsive Emulsions Stabilized by Microgel Particles

5.3.1 Influence of Solution Temperature

Using both thermal- and pH-sensitive PNIPAM-MAA microgel particles as stabilizers, Ngai *et al.* first prepared a novel type of emulsion responsive to

pH and temperature changes.[7,8] Octanol-in-water (O/W) emulsions stabilized by these PNIPAM-MAA microgel particles were stable at high pH and
low temperature; the emulsions, however, could be broken either by decreasing the pH (Figure 5.4A) or raising the temperature above the VPTT of
PNIPAM microgel particles (Figure 5.4B). These results clearly demonstrate
that microgel particles are very interesting stabilizers for emulsions as they
can offer unprecedented control of emulsion stability, well in excess of what
can be achieved by using small molecule surfactants or conventional solid
particles. To account for the observed pH- or temperature-triggered emulsion destabilization, they followed the general rules that govern the stability
of Pickering emulsions; namely the wettability of particle is crucial for the
emulsion type and stability.[3,6] They proposed that as the pH is lowered or
temperature is elevated, PNIPAM-MAA microgel particles become more
hydrophobic. As a result, the interfacial microgel particles probably move

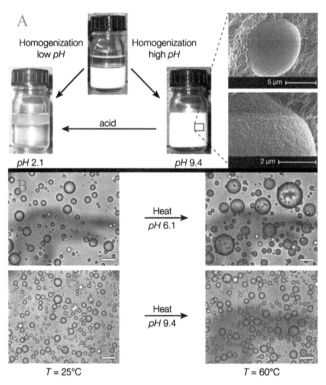

Figure 5.4 (A) Influence of the solution pH on the formation of octanol-in-water
emulsions after 48 hours at $T = 25$ °C. (B) Optical micrographs illustrating the temperature dependent stability of an octanol-in-water emulsion stabilized by microgel particles at pH 6.1 and 9.4 respectively (the
scale bar was 20 μm).
Reprinted with permission from reference 7. © 2005 The Royal Society of
Chemistry.

deeper into the oil phase, resulting in reducing the coverage of the oil–water interface and eventually breaking the emulsions (Figure 5.5).

Tsuji *et al.* have employed pure PNIPAM microgel and PNIPAM-carrying particles as stabilizers for making responsive Pickering emulsions using various oils, such as heptane, hexadecane, trichloroethylene and toluene.[28] The oil–water interfaces were found to be highly covered by microgel particles and the formulated emulsions were stable for more than 3 months when they were stored at room temperature. However, when the emulsions were heated to 40 °C, phase separation occurred (Figure 5.6). The authors argued that as temperature was raised, the microgels adsorbed at the interfaces shrank, reducing the coverage of the interfaces, and eventually emulsion droplets ripened or coalesced, and ultimately phase separation occurred.

Monteux *et al.* also prepared PNIPAM-based microgel particles as stabilizers for emulsions by copolymerization of NIPAM with dimethylamino ethyl methacrylate (DMAEMA) at various concentrations to control the size and size distribution of the resulting PNIPAM microgel particles.[24] The pendant drop technique was employed to measure the interfacial tension as a

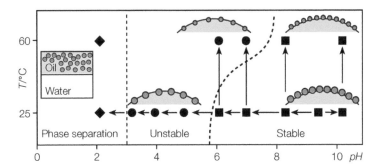

Figure 5.5   The stabilizing efficiency of PNIPAM-MAA microgel particles for octanol-in-water emulsions as a function of pH and temperature.
Reprinted with permission from reference 7. © 2005 The Royal Society of Chemistry.

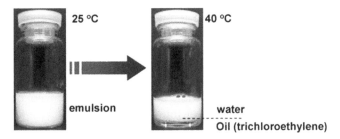

Figure 5.6   Photographs of the phase separation of trichloroethylene (TCE) emulsions stabilized by hairy particles.
Reprinted with permission from reference 28. © 2008 American Chemical Society.

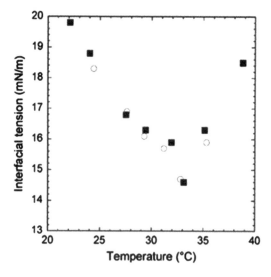

Figure 5.7 Variation of the interfacial tension (n-dodecane/water) as a function of temperature for PNIPAM microgel particles synthesized in the presence of DMAEMA.
Reprinted with permission from reference 24. © 2010 American Chemical Society.

function of temperature. Over the whole range of temperature (20–45 °C), the interfacial tension remains low, indicating the simultaneous adsorption of microgels to the interface (Figure 5.7). Interestingly, Figure 5.7 also shows that the interfacial tension always passes through a minimum (12 mN m^{-1}) at a temperature around the VPTT of microgel particles. They argued that below the VPTT, the decrease in interfacial tension with temperature is caused by the adsorption of dense layers of microgels because of the decrease in the excluded volume interactions. When the temperature was elevated above the VPTT, the increase in interfacial tension with temperature comes from the loosely packed PNIPAM microgel particles. Moreover, they investigated the effect of temperature on the stability of emulsions. Microgel particle stabilized emulsions, which formed at ambient temperature, were found to become unstable as the temperature was raised above the VPTT of microgels. They thus suggested that temperature-triggered emulsion destabilization is mainly caused by the adsorption of microgel aggregates above the VPTT and not from an important desorption of microgel particles or the reduced coverage at the oil–water interfaces. The physical origin of the temperature-triggered destabilization of emulsions stabilized by microgel particles is nonetheless a matter of intense debate.

5.3.2 Influence of Solution pH

Richtering's group has extensively employed PNIPAM-based microgels as stabilizers for making responsive emulsions, especially examining the influence of solution pH on emulsion stability.[14,17–19,21,25,27,29–31] They first prepared PNIPAM-MAA microgel particles under both acidic and basic conditions and investigated the influence of existing polymer residues on emulsion stability. It was found that raw microgel particles synthesized under basic conditions could stabilize octanol-in-water emulsions, whereas purified, they became inefficient stabilizers and were almost insensitive to pH.[29,30] They found that when the preparation of the microgel particles is conducted at high pH, only a small fraction of MAA comonomers can be incorporated into the microgel particles, rendering them less dependent on the degree of protonation. However, for microgel polymerization at low pH, the MAA content can be easily incorporated and tuned. This study has thus illustrated that microgel particles need to carry a certain amount of charges to be efficient stimuli-responsive emulsion stabilizers.[30]

Such temperature- and pH-sensitive PNIPAM-MAA microgel stabilized emulsion droplets were very stable at high pH and room temperature when the microgels are significantly swollen.[29,30] However, the emulsions can be broken by lowering the pH and elevating the temperature, and the sensitivity of the emulsion stability with respect to these parameters strongly depends on the polarity of oil. It has been demonstrated that complete separation into oil and water can be achieved, in which the microgels flocculate. This thus allows simple recycling of the microgel particles, an important feature for many industrial applications. The interfacial dilatational rheology was also employed to investigate the sensitivity of the emulsions with respect to these parameters, and a strong relationship between the interfacial elastic and loss moduli with pH as well as temperature was found.[29]

Brugger and Richtering have further reported two very interesting results, mainly focusing on PNIPAM-MAA microgel particles at heptane–water interfaces.[27] First, the interfacial tension is lowered by the microgel particles at both pH 3 and 9, but the emulsions formed at low pH are less stable, indicating that the emulsion destabilization at pH 3 is not caused by desorption of microgel particles from the water–oil interface (Figure 5.8).

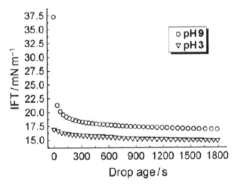

Figure 5.8 Decay of interfacial tension of a heptane–water interface as microgels adsorb at pH 3 and 9.
Reprinted with permission from reference 27. © 2009 Wiley.

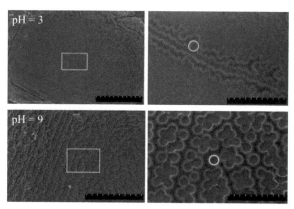

Figure 5.9 Cryo-SEM picture of the surface of a heptane drop covered with PNIPAM-*co*-MAA microgels at pH 3 and 9. Scale bars: 20 μm (left image) and 5 μm (right image).
Reprinted with permission from reference 27. © 2009 Wiley.

Second, by means of cryogenic scanning electron microscopy (cryo-SEM) to visualize the emulsion interface, they observed significantly different interfacial microstructures depending on pH: at pH 3, microgel particles formed dense two-dimensional crystalline arrays whereas at pH 9, when microgel particles are highly charged, they are packed in clustered crystalline patches with voids (Figure 5.9). Therefore, they suggested that the pH dependence of the emulsion stability is probably due to a variation in the interfacial viscoelastic properties rather than the wettability of the microgel particles.[27] The reason is that at low pH, the dense packing of microgel particles would lead to brittle interfaces that cannot withstand the mechanical forces during collisions of emulsion droplets, and eventually leading to breakage of the emulsions. However, at high pH, the presence of clusters would indicate a balance between attractive and repulsive forces, resulting in soft gel-like interfaces displacing higher elasticity. Hence, they concluded that in microgel-stabilized emulsions, the stabilization of droplets is not due to electrostatic repulsion between the interfacial microgels. Instead, the viscoelastic properties of the interface seem to play a dominant role in determining emulsion droplet stability.[25]

In order to clarify the influence of the spatial distribution of charges on emulsion stabilization, Schmidt *et al.* synthesized two different core-shell structured microgel particles: one with neutral PNIPAM core and MAA-bearing shell (MS) and the other with an MAA-bearing core and neutral PNIPAM-shell (MC) microgel particles as stabilizers.[21] Emulsions were formed with these two types of microgel particles under different conditions, but the presence of charges on the microgel particle was important for obtaining stable emulsions. However, the location of charges was found not to be relevant. Based on this result, they thus argued that stability of emulsions stabilized by pH-sensitive microgel particles is not purely due to electrostatic repulsion. Moreover, they also observed that microgel particles were strongly deformed at the oil–water interfaces and highlighted an important feature for using the soft and porous nature of the microgel particles in controlling emulsion stability: when the microgel particles were not strongly swollen and less deformable, no stable emulsions were formed; however, when the swelling and the deformability was enhanced, stable emulsions were obtained. This study thus links the stabilization of the emulsions to the swelling and structure of the microgel particles at the oil–water interface, which in turn highly correlates to particle morphology and properties.[21]

Very recently, by using two micrometre-sized, oppositely charged microgel particles (one anionic PNIPAM-MAA microgel particle and the other cationic poly(*N*-isopropylacrylamide-*co*-2-aminoethyl methacrylate) PNIPAM-AEM microgel particles) as stabilizers, Liu *et al.* have prepared stable emulsions with oppositely charged droplets.[18] Interestingly, droplets stabilized by oppositely charged microgel particles remain stable upon mixing, and no coalescence of droplets occurred. This result again seems to show that electrostatic interactions between droplets do not determine their stability and further demonstrate the peculiar properties of soft microgel particles with respect to producing emulsions.

To further explore the behaviour of microgel particles at the liquid interfaces, Geisel *et al.* studied the self-assembly of pH-sensitive microgels at a flat heptane–water interface by means of freeze-fracture shadow-casting cryo-SEM (FreSCa) in order to reveal the 3D information on the particle position relative to the interface.[17,19] It was found that microgel particles unexpectedly only slightly protruded into the oil phase; however, a significant deformation of the particles, comprising pronounced flattening and stretching of the polymer corona at the interface, was found (Figure 5.10). Surprisingly, they did not observe any influence of pH on the size of interfacial microgel particles or on the penetration depth into the oil phase, despite the fact that the microgel particle size in water and the emulsion stability depend on pH. The authors have accounted for the microgel behaviours at the interface by the balance among the salvation of the microgel in the two liquid phases, its interfacial activity, and internal elasticity of the microgel particles.

5.3.3 Influence of Particle Deformability

Destribats and Schmitt also published a series of papers on emulsions stabilized by PNIPAM microgels.[20,22,23] These emulsions were prepared through the technique of limited coalescence and had the advantage that the number of excess microgels in the continuous phase is greatly reduced. The most interesting feature observed in their study is related to the microstructure of microgels at the oil–water interface. As shown in Figure 5.11, the microgel particles are not rigid particles at the interfaces; instead they are strongly deformed and linked by filaments. This agrees with the observations reported by Richtering's group (Figure 5.10) that microgels stabilize emulsions

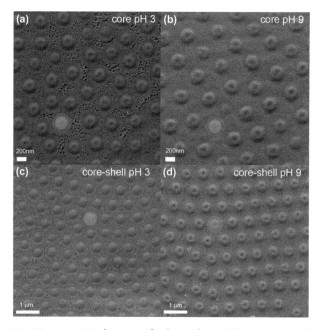

Figure 5.10   FreSCa cryo-SEM images of microgel particles at water/n-heptane interfaces imaged from the oil side. Colouring highlights the core-corona
morphology at the interface for all samples.
Reprinted with permission from reference 19. © 2012 American
Chemical Society.

better when they are charged because they are more swollen as compared to
when they are in the uncharged state.[17,19,21,27] They compared microgel
particles with different contents of crosslinker and demonstrated that increasing the degree of crosslinking reduced the stabilization efficiency. They
argued that the most deformable microgels tended to form 2D connected
networks, featured by significant overlapping of the peripheral parts.
Therefore, they proposed that the emulsion stability was mainly determined
by the softness or deformability of the microgel particles.

5.4   Adsorption Kinetics of Microgel Particles at the Fluid–Fluid Interface

Experimentally, both cryo-SEM[21,22,25,27] and confocal laser scanning microscopy (CLSM)[18,22] have clearly shown that PNIPAM microgels can adsorb to
and densely pack at the oil–water interface. However, the questions regarding how PNIPAM microgels adsorb to the oil–water interface, and what
parameters control such adsorption processes, have not been studied in
detail. Understanding the adsorption behaviours of microgels to the oil–
water interface has become increasingly important both in terms of fundamental science and applications of microgels as multi-stimuli responsive

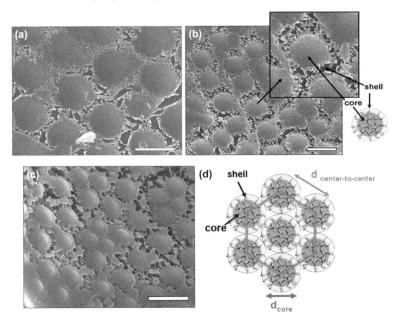

Figure 5.11 Cryo-SEM image of the interface of a heptane-in-water emulsion drop cover by: (a and b) 2.5 mol% BA crosslinked microgel particles after sublimation (front view), (c) 5 mol% BA crosslinked microgel particles after sublimation (sidelong view), scale bars are 1 μm; (d) scheme of the particle structure and arrangement at the interface.
Reprinted with permission from reference 22. © 2011 The Royal Society of Chemistry.

emulsion stabilizers. In the next section we will discuss the use of pendant drop tensiometry to trace the evolution of oil–water interfacial tensions under different conditions. We mainly investigated two PNIPAM microgels with 3.2 wt% (3.2%BA microgels) and 12.8 wt% (12.8%BA microgels) crosslink density as well as poly(styrene-*co*-NIPAM) core-shell particles with different styrene contents.[47]

5.4.1 Effect of Particle Concentration on Adsorption

Figure 5.12 shows typical dynamic interfacial tensions (γ_t) of the heptane-water interface in the presence of 3.2%BAPNIPAM microgels at different concentrations. The influence of microgel concentration on the adsorption kinetics was first investigated at 298 K, *i.e.* below the VPTT of the microgels. It can be clearly seen that γ_t decreases as a function of time, indicating the spontaneous adsorption of PNIPAM microgels to the heptane–water interface, which is in a good agreement with the recent results. Such an effective interfacial tension decrease is usually observed when proteins are absorbed onto the oil–water interface.[48–52] It is worth noting that conventional

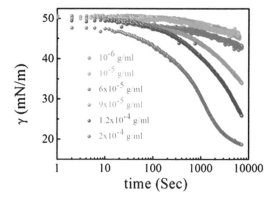

Figure 5.12   Microgel concentration dependence of γ_t at the heptane–water interface and 298 K, in which the used particles are 3.2%BA PNIPAM microgels. Reprinted with permission from reference 47. © 2013 The Royal Society of Chemistry.

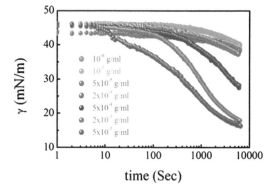

Figure 5.13   The effect of microgel concentrations on γ_t at a temperature above the VPTT (317 K), in which the used particles are 3.2%BA PNIPAM microgels. Reprinted with permission from reference 47. © 2013 The Royal Society of Chemistry.

Pickering stabilizers seldom decrease the interfacial tension, while organic colloidal particles usually lower the interfacial tension by about 10 mN m^{-1}.[53,54]

Figure 5.13 shows the evolution of interfacial tension with various microgel concentrations at a temperature of 317 K, well above the VPTT. Obviously, γ_t continuously decreases with time even at such a high temperature, indicating that microgels are still interfacially active, as these microgels can still reduce the interfacial tension even when they are in the collapsed state.

Figure 5.14 summarizes the meso-equilibrium interfacial tensions (γ_m is the dynamic interfacial tension value after 6000 seconds) as a function of microgel concentration at two above-mentioned temperatures, *i.e.*, below

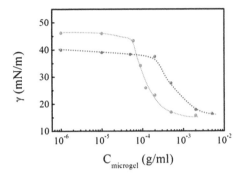

Figure 5.14 Meso-equilibrium interfacial tension γ_m as a function of 3.2%BA PNIPAM microgel concentrations (Cmicrogel) at 298 K (red points and dots) and 317 K (blue points and dots).
Reprinted with permission from reference 47. © 2013 The Royal Society of Chemistry.

and above the VPTT of the PNIPAM microgel. Similiar trends are observed at both 298 and 317 K: γ_m first decreases slightly, and then declines gradually, and ultimately reaches a plateau. Also, similar meso-equilibrium interfacial tensions are reached at these two distinctive temperatures at different microgel concentrations, indicating that the microgel has similar interfacial activity in both swollen and collapsed states.

A diffusion controlled process is commonly used to describe surfactant adsorption onto the interface.[55] Due to their large size, PNIPAM microgels are considered to be irreversibly adsorbed to the heptane–water interface. Thus a desorption process can be neglected in the diffusion controlled model derived by Ward and Tordai, leading to:

$$\Gamma_t = 2\sqrt{\frac{D}{\pi}} C_0 \sqrt{t} \tag{5.1}$$

where Γ_t is the interfacial PNIPAM microgel concentration, D is the diffusion coefficient and C_0 is the microgel bulk concentration.

Moreover, the interfacial concentration Γ_t can be correlated to the surface pressure Π_t *via*:

$$\Pi_t = \gamma_0 - \gamma_t = \Gamma_t RT \tag{5.2}$$

Combining Equations 5.1 and 5.2, Equation 5.3 can be obtained and expressed as:

$$\Pi_t = \gamma_0 - \gamma_t = \Gamma_t RT = 2RT\sqrt{\frac{D}{\pi}} C_0 \sqrt{t} \tag{5.3}$$

According to Equation 5.3, Π_t should be proportional to $\sqrt{t}$ for a purely diffusion controlled process and the slope should be proportional to the microgel bulk concentration C_0. Figure 5.15 shows Π_t *versus* $\sqrt{t}$ for microgels at 298 K at four different concentrations. At concentrations below

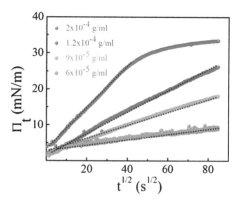

Figure 5.15   Influence of 3.2%BA PNIPAM microgel concentration on the evolution
of dynamic surface pressures (Π_t) *vs.* $\sqrt{t}$ at 298 K.
Reprinted with permission from reference 47. © 2013 The Royal Society
of Chemistry.

Table 5.1   The summary of k_i, C, R^2 Figure 15 by adopting Equation 5.4 at $T = 298$ K.
Reprinted with permission from reference 47. © 2013 The Royal Society
of Chemistry.

	6×10^{-5} g/ml	9×10^{-5} g/ml	1.2×10^{-4} g/ml	2×10^{-4} g/ml
k_i $(s^{-1/2} \cdot mN/m)$	0.0666	0.1873	0.2844	0.5861
C (mN/m)	3.524	2.469	2.472	2.922
R^2	0.9779	0.9975	0.9976	0.9993

2×10^{-4} g mL^{-1}, the curves can be fitted linearly very well. However, at
2×10^{-4} g mL^{-1} the curve can only be linearly fitted at the first 1600 seconds
when the surface pressure is lower than 30 mN m^{-1}.

The four dotted lines shown in Figure 5.15 were fitted with Equation 5.3:

$$\Pi_t = k_i\sqrt{t} + C \tag{5.4}$$

where C is a constant and the slope k_i should be:

$$k_i = 2RT\sqrt{\frac{D_{Interfacial}}{\pi}}C_0 \tag{5.5}$$

Table 5.1 summarizes the fitted slopes k_i and constant C from Figure 5.15
at four different microgel concentrations. The fitting parameters R^2 are
around 0.99, indicating the excellence of these fits.

At 317 K, microgel adsorption behaviours can also be well fitted by
Equation 5.3 at microgel concentrations less than 2×10^{-3} g mL^{-1}, and only
the initial stages can be fitted at microgel concentrations of 2×10^{-3} g mL^{-1}
and 5×10^{-3} g mL^{-1} as shown in Figure 5.16. Table 5.2 summarizes the fitted
results.

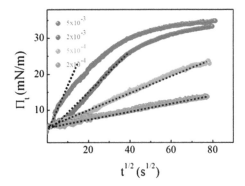

Figure 5.16 Influence of 3.2%BA PNIPAM microgel concentration on the correlation of dynamic surface pressures (Π_t) *vs.* $\sqrt{t}$ at 317 K.
Reprinted with permission from reference 47. © 2013 The Royal Society of Chemistry.

Table 5.2 The summary of k_i, C, R^2 Figure 16 by adopting Equation 5.4 at $T = 317$ K. Reprinted with permission from reference 47. © 2013 The Royal Society of Chemistry.

	2×10^{-4} g/ml	5×10^{-4} g/ml	2×10^{-3} g/ml	5×10^{-3} g/ml
k_i $(s^{-1/2} \cdot mN/m)$	0.1077	0.2422	0.6505	1.8169
C (mN/m)	5.457	5.050	1.193	1.442
R^2	0.9884	0.9937	0.9985	0.9504

Table 5.3 The comparison between $D_{\text{Interfacial}}$ obtained from the surface pressure *via* Equations 5.5 and 5.5 and bulk diffusion coefficient D_{Bulk} at two temperatures, 298 K and 317 K. Reprinted with permission from reference 47. © 2013 The Royal Society of Chemistry.

	$D_{\text{Interfacial}}{}^a$	$D_{\text{Bulk}}{}^b$
317 K	1 ± 0.02	1 ± 0.01
298 K	123.59 ± 0.02	0.19 ± 0.01

$^a D_{\text{Interfacial}}$ is extracted from the k_i in Table 1 and using Equation 5.5 and normalized with $D_{\text{Interfacial, 317K}}$, where the errors are derived from the fitting procedure in Figures 5.14 and 5.15.
$^b D_{\text{Bulk}}$ is determined from DLS, and D_{Bulk} is normalized with $D_{\text{Bulk, 317K}}$, where the errors are derived from the DLS measurements.

The normalized diffusion coefficients of these microgels at the interface with those in bulk at two above-mentioned temperatures are summarized in Table 5.3. Evidently, the diffusion in bulk becomes faster with increasing temperature. However, the diffusion coefficients calculated from the surface pressure are *smaller* at higher temperature. Figure 5.17 further shows the plots of the fitted k_i as a function of microgel concentrations at two distinct temperatures, 298 and 317 K. Note that for both cases, the linear fits are excellent (R^2 ca. 0.99), and that the slopes are strikingly different for these two temperatures.

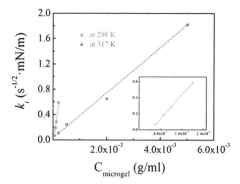

Figure 5.17   Plots of k_i vs. microgel concentrations $C_{microgel}$ at two distinct temperatures: 298 (red points and dots) and 317 K (blue points and dots). Inset highlights the plots at 298 K.
Reprinted with permission from reference 47. © 2013 The Royal Society of Chemistry.

It is worth pointing out that only the early adsorption stage shows a good linear fit with Equation 5.3, as in the late adsorption stage, the interactions between adsorbed microgels cannot be neglected.

Two conclusions can be drawn from the above data. First, similar to all other interfacially active materials, like small molecular surfactants, polymers,[56] proteins or other types of colloidal stabilizers,[49–52] microgels need to diffuse to the heptane–water interface. And this diffusion process is dependent on microgel concentration, as shown in Figures 5.12 and 5.13. Moreover, Figure 5.17 clearly demonstrates that the slope k_i is proportional to microgel concentration, validating the idea that microgel diffusion to the heptane–water interface is controlled by microgel concentration. And this is the first step for microgels to adsorb to the heptane–water interface. Second, microgel morphology dominates microgel diffusion behaviours at the heptane–water interface. As shown in Table 5.3, increasing the temperature from 298 to 317 K would accelerate the diffusion of microgels in bulk but largely decrease the diffusion at the interface (more than 100 times slower). Recent results demonstrated that the adsorption behaviours of PNIPAM microgels were dominated by NIPAM repeat units of microgels.[16] It was well documented that a conventional dense-core loose-shell (DCLS) structured microgel was prepared, as the reactivity between crosslinker BA and NIPAM was different. At 298 K (below the VPTT), microgels adopt a DCLS structure with large amounts of dangling PNIPAM chains swelling in the aqueous solution, and these dangling chains accelerate microgel diffusion at the interface, although they slow down microgel diffusion in bulk aqueous solution.[33] However, at 317 K (above the VPTT), microgels are collapsed and bear no dangling chains at the periphery; such distinctive structures slow down the diffusion behaviour of the microgels at the interface but accelerate the bulky diffusion.

5.4.2 Effect of Solution Temperature on Adsorption

Figure 5.18 shows γ_t of the heptane-water interface in the presence of fixed microgels concentration $(5\times10^{-3}$ g mL$^{-1})$ and at temperatures above 310 K. Note that all these temperatures are higher than the PNIPAM microgel VPTT but the hydrodynamic radius of PNIPAM microgels are the same (Figure 5.21). Obviously, the adsorption behaviours are slowed down with increasing temperatures. Similarly, we plotted the change of surface pressure as a function of $\sqrt{t}$, as shown in Figure 5.19. From the initial stages, four diffusion coefficients are derived from the adsorption isotherms. Table 5.4 summarizes the fitted slopes k_i at various microgel concentrations and constant C. And we further compared the diffusion coefficients of these microgels at the interface with those in bulk at various temperatures, as summarized in Table 5.5.

The data in Table 5.5 and Figure 5.18 show that even at $T>$VPTT, temperature has opposite effects on reduction rate of interfacial tension and diffusion coefficient in bulk: by increasing temperature from 310.4 to 329.2 K, $D_{\text{Interfacial}}$ decreases from 1 to 0.23, while D_{Bulk} increases from 1 to 1.54. Moreover, the microgel size does not change, at $T>310.4$ K (Figure 5.21). Thus, the observed different temperature effect on interface dynamics and on diffusion in bulk cannot be attributed to a change in particle morphology.

It is well known that the interfacial tension exerts strong forces on colloidal particles at the interface. Furst *et al.*[57] demonstrated that even polystyrene particles are deformed at the oil–water interface when heated above the glass transition temperature. Previous electron microscopy studies on microgels had already demonstrated that soft microgels are strongly deformed at the oil–water interface.[22,27] Microgels were found to be largely

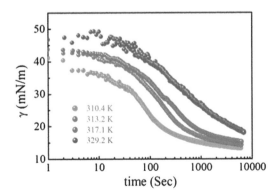

Figure 5.18 The effect of temperature ($T>$VPTT) on γ_t at the heptane–water interface, in which the used 3.2%BA PNIPAM microgel concentration is fixed at 5×10^{-3} g mL^{-1}.
Reprinted with permission from reference 47. © 2013 The Royal Society of Chemistry.

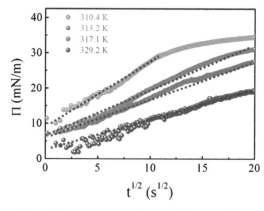

Figure 5.19   The effect of temperature $(T > \mathrm{VPTT})$ on surface pressure π at the heptane–water interface, in which the used 3.2%BA PNIPAM microgel concentration is fixed at 5×10^{-3} g mL^{-1}. Note that only the initial stages are enlarged, shown and fitted to extract the diffusion coefficient at the interface.
Reprinted with permission from reference 47. © 2013 The Royal Society of Chemistry.

Table 5.4   The summary of k_i, C, R^2 Figure 19 by adopting Equation 5.4. Reprinted with permission from reference 47. © 2013 The Royal Society of Chemistry.

	310.4 K	313.2 K	317.1 K	329.2 K
k_i (s$^{-1/2}$ · mN/m)	1.7809	1.4078	1.2269	0.9018
C (mN/m)	9.254	5.2815	4.9993	1.6494
R^2	0.9832	0.9831	0.9926	0.9802

Table 5.5   The comparison between interfacial diffusion coefficient $D_{\mathrm{Interfacial}}$ obtained from the surface pressure *via* Equations 5.4 and 5.5 and bulk diffusion coefficient D_{Bulk} at temperatures above VPTT. Reprinted with permission from reference 47. © 2013 The Royal Society of Chemistry.

	$D_{\mathrm{Interfacial}}$[a]	D_{Bulk}[b]
310.4 K	1 ± 0.02	1 ± 0.01
313.2 K	0.61 ± 0.01	1.06 ± 0.02
317.1 K	0.45 ± 0.02	1.21 ± 0.01
329.2 K	0.23 ± 0.01	1.54 ± 0.01

[a]The $D_{\mathrm{Interfacial}}$ is extracted from the k_i in Table 5.4 using Equation 5.5 and normalized with $D_{\mathrm{Interfacial,\,310.4K}}$, where the errors are derived from the fitting procedure in Figure 5.19.
[b]The D_{Bulk} is determined from DLS, the D_{Bulk} is normalized with $D_{\mathrm{Bulk,\,310.4K}}$, where the errors are derived from the DLS measurements.

deformed at the heptane–water interface, with a large proportion of the microgel residing in the aqueous phase and only small amounts protrude into the oil phase.[17,19]

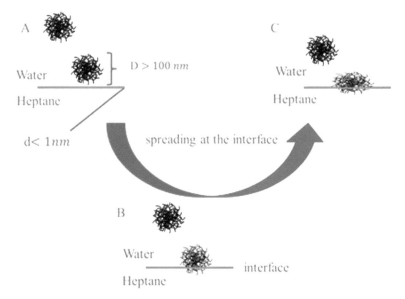

Figure 5.20 Schematic illustration of PNIPAM microgels spreading at the oil–water interface.
Reprinted with permission from reference 47. © 2013 The Royal Society of Chemistry.

Considering that the microgel we used (3.2%BA microgel) has a size greater than 100 nm and the heptane–water interface width is less than 1 nm, we propose that microgels undergo a spreading process at the heptane–water interface once they have diffused to the interface, as shown schematically in Figure 5.20. This model obviously relies on the deformability of the microgels. This, however, has been reported before and it was also demonstrated that softer microgels resulted in better emulsion stability.[22]

5.4.3 Effect of Particle Deformability on Adsorption

The normalized hydrodynamic radius of 3.2%BA and 12.8%BA microgels as a function of temperature is presented in Figure 5.21 and shows that the microgel with higher crosslinker content has a lower swelling ratio.

Figure 5.22 shows a comparison of the evolution of heptane–water interfacial tension in the presence of these two different PNIPAM microgels (3.2%BA and 12.8%BA) at 298 K at a microgel concentration of 5×10^{-3} g mL^{-1}. The induction period is much longer for the microgel with 12.8%BA content.

Figure 5.23 reveals the influence of temperature on the evolution of interfacial tension at constant microgel concentration for 12.8%BA PNIPAM microgel. Two observations should be noted: (i) the characteristic timescales below and above the VPTT are similar, in contrast to the behaviour of the less crosslinked microgel, where the process was much faster at

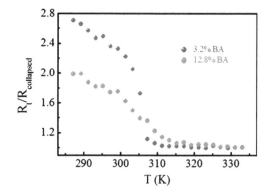

Figure 5.21 Normalized hydrodynamic radius of two pure PNIPAM microgels (3.2%BA and 12.8%BA) as a function of temperature.
Reprinted with permission from reference 47. © 2013 The Royal Society of Chemistry.

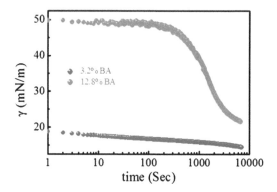

Figure 5.22 The effect of the softness of microgels (microgels of varied crosslink densities) on lowering heptane–water γ_t at 298 K.
Reprinted with permission from reference 47. © 2013 The Royal Society of Chemistry.

low temperatures. (ii) The process gets *faster* with increasing temperature, both below and above the VPTT, which is strikingly different from the behaviour of the less crosslinked microgel (3.2%BA microgel).

As further systems, we prepared two core-shell particles (S/N1 and S/N8) which consist of a rigid polystyrene core and a crosslinked PNIPAM shell. The two systems were prepared with different styrene/NIPAM (S/N) ratios to adjust the softness of the prepared particles. Figure 5.24 shows the normalized hydrodynamic radius of S/N1 and S/N8 microgels at various temperatures. S/N1 microgel shrinks upon heating, whereas S/N8 does not collapse. The particles were synthesized as described by Hellweg *et al.*[58] and their earlier report pointed out that PS-PNIPAM spheres with high styrene ratio (> 75%) did not collapse upon heating. Although both PNIPAM and polystyrene would randomly distribute within these microgels, PS abundant

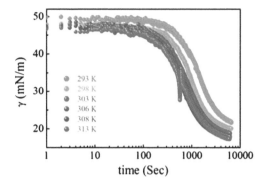

Figure 5.23 The effect of temperature on γ_t at the heptane–water interface, in which the 12.8%BA PNIPAM microgel concentration is fixed at 5×10^{-3} g mL^{-1}. Reprinted with permission from reference 47. © 2013 The Royal Society of Chemistry.

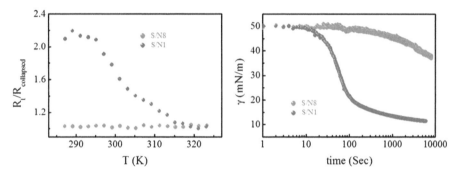

Figure 5.24 Properties of poly(styrene-*co*-NIPAM) particles (S/N1 and S/N8). Left: normalized hydrodynamic radius as a function of temperature. Right: evolution of interfacial tension at 298 K.
Reprinted with permission from reference 47. © 2013 The Royal Society of Chemistry.

core and PNIPAM abundant shell structured microgels were expected to form. At high temperatures, PNIPAM is still more hydrophilic than hydrophobic PS, and such PS-core PNIPAM-shell structures have been confirmed by small angle neutron scattering (SANS). Thus, these two microgels with varied S/N ratios will have similar surface properties. However, S/N1 can be regarded as a particle with a soft shell, resembling soft microgels, while S/N8 should be regarded as a rigid particle with a very thin layer of dangling PNIPAM chains at the surface.

Again the heptane–water γ_t was probed in the presence of these two different PNIPAM microgels (S/N1 and S/N8) at 298 K at a concentration of 5×10^{-3} g mL^{-1}. Two distinct γ_t curves were observed again, as shown in Figure 5.24. The S/N1 microgels are far more effective in lowering the interfacial tension as compared to the S/N8 particles. Note, however, that S/N8 particles reduce the interfacial tension slowly.

The different behaviours of the microgel with higher crosslink density (Figures 5.21 and 5.22) and of the core-shell particles with rigid core (Figure 5.24) are strong support for the model we proposed in Figure 5.20. These particles are less deformable, thus they reduce the heptane–water interfacial tension at a slower rate. This slower process is limited by the kinetic spreading process at the heptane–water interface, and not by diffusion from the bulk solution to the interface as sufficient microgel was supplied in these control experiments. This further shows how the adsorption behaviours of microgels at the heptane–water interface can be tuned by temperature, crosslinking density and morphology.[21]

The conventional Pickering stabilizers hardly lower the interfacial tension.[3,6,54] The observation that the interfacial tension is reduced even by the S/N8 particles indicates that the chemical structure of PNIPAM is relevant for the interfacial activity (Figure 5.24). The highly crosslinked microgel (12.8%BA) reduces the interfacial tension to similar values as the microgel with low crosslink density. The kinetics of the reduction in interfacial tension, however, becomes faster with increasing temperature in the entire temperature range, *i.e. below and above* the VPTT of the microgels, as shown in Figure 5.23. This observation is clearly different from those of microgels of low crosslink density (Figures 5.12–5.18). These experimental results clearly demonstrate the influence of microgel softness and thus the differences between microgels and conventional, rigid Pickering stabilizers.

In summary, we investigated the interfacial activity of PNIPAM-based microgels as a function of microgel concentration, temperature and crosslink density. PNIPAM microgels first diffuse to the heptane–water interface, in which the dominant parameter is microgel concentration, and then spread at the heptane–water interface to effectively lower the interfacial energy, where temperature and microgel softness are the controlling parameters.

5.5   Functional Materials Templating from Microgel-Stabilized Emulsions

5.5.1   Janus Microgel Particle Formation

Janus particles are a special class of colloidal particles with different chemical make-ups on their two hemispheres. They are named after the ancient Roman god Janus, who has two faces looking in opposite directions. Janus particles have received tremendous attention since de Gennes raised the concept in his Nobel Prize address.[59] Many theories and simulations show that Janus particles can assemble into novel structures, which facilitate the fabrication of new materials.[60]

Recently, Kawaguchi has shown that microgel particles at the oil–water interface are an excellent platform for the fabrication of Janus microgel particles (Figure 5.25).[61] PNIPAM-based microgel particles are assembled on the surfaces of oil droplets in water. The presence of the oil–water interface

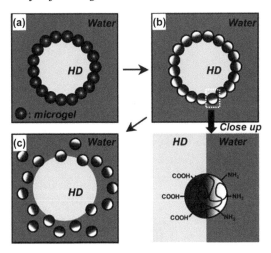

Figure 5.25 (a) Preparation of Pickering emulsion stabilized by microgel particles. (b) Hemispherical modification of amino groups into microgel particles attached at the oil–water interface. (c) Microgel collection by destabilizing emulsion.
Reprinted with permission from reference 61. © 2007 American Chemical Society.

of emulsions allows the selective introduction of a distributed amino group into the microgels on only one side (the side immersed in oil or the water phase), assuming that microgel particles did not strongly rotate at the interfaces. The resultant Janus microgel particles remain separate at pH 6, presumably due to both electrostatic repulsion and steric hindrance among particles. Interestingly, upon lowering the pH to 4, these Janus microgel particles self-assembled into string *via* electrostatics attractions (Figure 5.26).

5.5.2   Stimuli-Responsive Capsules

Colloidosomes, a term that was coined by Dinsmore *et al.*[62] by analogy with liposomes, are hollow and elastic shells composed of colloidal particles. With well-defined permeability and elasticity, colloidosomes are excellent carriers for encapsulation and controlled release and thus find many applications in drug delivery, pharmaceutical, cosmetic and other industrial fields. Stimuli-responsive colloidosomes, imitating the action of biological systems, are highly desirable for stimuli-triggered controlled release.[62,63]

Lawrence *et al.*[64] prepared the first temperature-sensitive semipermeable colloidosomes by the assembly of anionic poly(*N*-isopropylacrylamide-*co*-acrylic acid) PNIPAM-AA microgel particles into shells around W/O emulsion droplets, followed by locking the particles in the shell with the diblock copolymer poly(butadiene-*b*-*N*-methyl-4-vinyl pyridinium iodide) dispersed in oil. The colloidosomes were stably dispersed in aqueous solution and were

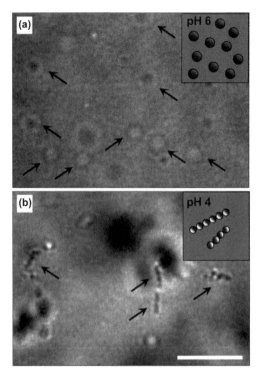

Figure 5.26 Optical microscope images of microgel particles dispersed in a pH 6 (a) and a pH 4 aqueous solution (b). The microgel particles indicated by the arrows are in almost the same focus. The scale bar is 10 μm.
Reprinted with permission from reference 61. © 2007 American Chemical Society.

responsive to temperature. Upon heating, the colloidosomes shrink reversibly by 13% or irreversibly by 40% in radius. They proposed that the increased van der Waals attraction among the microgel particles in a highly collapsed state may lead to the size change irreversibility.

Monodispersed stimuli-responsive colloidosomes, with reversible size-change behaviour similar to that of constituent poly(N-isopropylacrylamide-co-allylamine) PNIPAM-Ally microgel particles have been successfully produced with the help of microfluidic devices (Figure 5.27). In comparison to colloidosomes reported by Lawrence et al., Shah et al.[65] emulsified PNIPAM-Ally microgel dispersions in silicon oil containing a surfactant. Once emulsified, microgel particles self-assembled at the oil–water interfaces due to the presence of hydrophilic acrylamide groups and hydrophobic isopropyl groups. Glutaraldehyde was then added to crosslink the interfacial amine-functionalized microgel particles through an amine-aldehyde condensation reaction. Again, locking of the microgel particles at the oil–water interfaces is the key for the preparation of colloidosomes.

Pich et al.[66] have prepared responsive colloidosomes by employing poly(N-vinylcaprolactam) PVCL-based microgel particles as stabilizer.

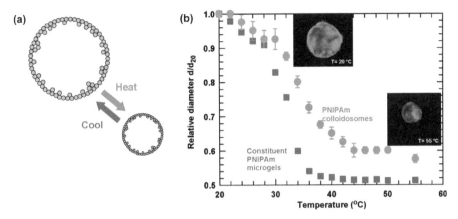

Figure 5.27 (a) Schematic representation of the thermoresponsive behaviour of a PNIPAM colloidosome. (b) Equilibrium size change of PNIPAM colloidosomes and the constituent PNIPAM microgel particles. Colloidosomes were dispersed in water and heated from 20 °C to 50 °C in fixed increments of 2 °C and then to 55 °C. Images were captured after allowing the sample to equilibrate for 30 min at each temperature. The sample was then cooled down to 20 °C using the same temperature steps. Dimensions of the colloidosomes were measured from the captured images using image-processing software. Size data of three different colloidosomes were averaged for better statistics. Size-change data of the constituent PNIPAM microgel particles over the same temperature range were collected using dynamic light scattering.
Reprinted with permission from reference 65. © 2010 American Chemical Society.

The PVCL microgel particles were assembled on the surfaces of chloroform droplets containing the biodegradable polymer poly(4-hydroxybutyrate-*co*-4-hydroxyvalerate) (PHBV). The evaporation of chloroform led to the formation of PHBV capsules with microgel particles decorated at the capsule's walls. As can be seen in Figure 5.28, these microgel particle integrated capsules exhibited strong temperature-dependent release behaviours. At temperatures below VPTT (25 °C), microgel particles in the capsule's walls are in the swollen state and this allows rapid diffusion of encapsulated FITC-dextran from the capsules to the aqueous solution. Upon increasing the temperature to 70 °C, the interfacial microgel particles are collapsed; consequently, the capsule's wall permeability decreases and results in a slow release profile.

5.5.3 High Internal Phase Emulsion (HIPE) and Porous Materials

High internal phase emulsions (HIPEs) are often defined as very concentrated emulsions where the volume fraction of the internal phase exceeds 0.74, the maximum packing density for monodisperse hard spheres.[2] Conventional HIPEs consisting of a continuous organic phase and an internal

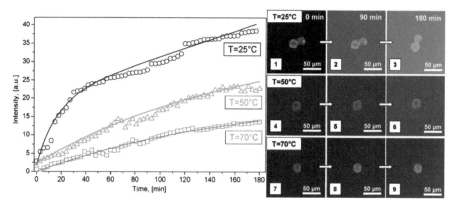

Figure 5.28 Fluorescence intensity *versus* time during release of FITC-labelled dextran from microgel-based capsules ($c_{microgel} = 0.083$ g L^{-1} and $c_{PHBV} = 30$ g L^{-1}) in water at different temperatures (inset shows fluorescence microscope images taken at different stages of release experiments).
Reprinted with permission from reference 66. © 2009 Wiley.

aqueous phase (water-in-oil, w/o emulsion) are commonly stabilized by large amounts of surfactants such as sorbitan monooleate (SPAN 80) or a mixture of non-ionic, anionic and cationic surfactants. If the continuous phase contains one or more monomeric species that are polymerizable, HIPEs can be used as templates for the fabrication of highly porous materials, which are proving to be very useful in a variety of applications including filtration membranes for molten metals and hot gases, bioreactors, catalyst carriers, and scaffolds for bone replacement and tissue engineering.[67–71]

Our group was the first to demonstrate the stabilization of hexane-in-water (O/W) HIPEs with volume fraction as high as 0.8 by using PNIPAM-MAA microgel particles.[26] Microgel particles are not only absorbed at the oil–water interface to hinder extensive droplet coalescence, excess microgel particles also simultaneously form a gel in the continuous phase to trap oil droplets in the gel matrix, which in turn inhibits the creaming of the oil droplets and enhances the emulsion stability (Figure 5.29). Moreover, rheological studies show that the microgel particles used can impart to the formed HIPEs their two-fold responsiveness to pH and temperature.

Following emulsification, the wet HIPEs were dried in air to remove oil and water. The relatively high vapour pressure of hexane and water (20.2 kPa and 2.34 kPa, respectively, at 25 °C) allowed for evaporation of the emulsion droplets at room temperature. The dried samples directly led to the macroporous material and the corresponding SEM images are depicted in Figure 5.30.[26] Note that no chemical reaction was involved before oil removal. The porous structures with average cavity size are comparable to the oil droplet diameter of the precursor emulsions (Figure 5.29), suggesting that these cavities resulted from a loss of the oil component. Therefore, it confirmed that the microgel particle-stabilized HIPEs used as initial

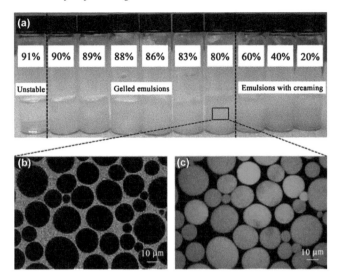

Figure 5.29 (a) Photograph of microgel-stabilized emulsions having an internal oil phase volume from 20 to 91% at room temperature, where the continuous phases of all mixtures consist of 2 wt% microgel particles in the initial dispersion with a solution pH value of approximately 6.0. (b, c) Confocal images of the emulsion with 80 vol% hexane oil stabilized by 2 wt% microgel particles excited by lasers with wavelengths of 408 and 543 nm, respectively.
Reprinted with permission from reference 26. © 2009 Wiley.

templates were sufficiently strong to withstand the high capillary stresses developed during drying, without distortion of their internal structure. Besides, the insets of Figure 5.30 (c) and (d) showed that the microstructure of resultant porous materials exhibited different degrees of pore interconnectivity between neighbouring droplets. For example, decreasing the microgel particle concentration in the initial dispersion to 0.05 wt% (Figure 5.30d) resulted in the formation of more open interconnected pores between touching droplets, which suggested that the microgel particle-stabilized HIPEs not only enabled us to tailor the pore size, but also simultaneously added flexibility to further tune the porosity between touching droplets by simply changing the microgel particle concentrations in the initial dispersion.

Using similar fabrication procedures, 3D hierarchical porous structures were obtained based on core-shell microgel particles (Figure 5.31).[72] More importantly, high-magnification images (Figure 5.31b) showed that the pore walls and surfaces of the resulting membranes were densely decorated with the core-shell microgel particles, which added a great flexibility to functionalize or enhance the surface roughness of the porous materials for a variety of applications in the future. So far the functionalization or modification of porous materials has often been achieved by the direct incorporation of second monomers or additives in the precursor emulsion, which in some

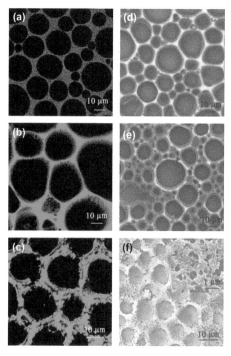

Figure 5.30 Porous materials obtained after dying the emulsions containing (a) 2 wt%, (b) 0.5 wt%, (c) 0.1 wt% and (d) 0.05 wt% microgel particles in the initial dispersion, respectively. The insets in (c) and (d) show the detailed appearance of the porous membrane at larger amplification.
Reprinted with permission from reference 26. © 2009 Wiley.

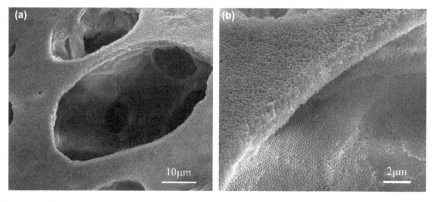

Figure 5.31 Microstructures of macroporous materials prepared from a hexane-in-water emulsion containing 80% internal phase volume of oil and 5 wt% PS-*co*-PNIPAM particles in the initial aqueous dispersion.
Reprinted with permission from reference 72. © 2010 American Chemical Society.

cases is followed by a post-modification step. This approach, however, has strict limitations because the stability of the HIPEs is a delicate hydrophobic/hydrophilic balance, *i.e.* for every change of monomer composition the process conditions have to be optimized. In particular, hydrophilic functional monomers are very difficult to incorporate as they destabilize the emulsion. Undoubtedly, microgel particle-stabilized HIPE templating provides a facile methodology to prepare functional pore materials because any additional properties from the particles can be directly imparted to the final composite materials.

Recently, we have extended our study to the production of HIPEs stabilized by two different kinds of particles.[73] Starting with an emulsion stabilized by microgels and silica nanoparticles, we were able, after drying and sintering, to obtain highly porous, three-dimensionally interconnected silica materials with ordering on three different scales. As shown in Figure 5.32, the direct emulsion droplets promoted macropores ($\sim$ 10–30 µm) and interconnecting windows ($\sim$ 3–5 µm) formation, while interfacial adsorbed microgel particles resulted in the formation of nanoscale porosity ($\sim$ 80 nm) throughout the macroporous walls. In principle, our developed method can be extended to prepare a range of materials by simply replacing the building

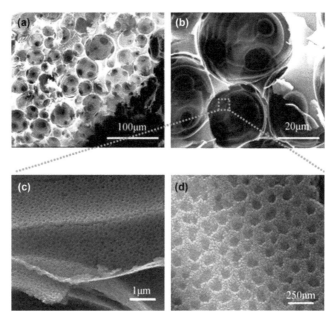

Figure 5.32 (a, b) SEM images of the hierarchical macropores and interconnected windows templating from HIPE droplets. (c, d) High magnification SEM images revealing the nanoscale pores throughout the cell walls due to the removal of microgel particles after calcination at 950 °C for 2 h. Reprinted with permission from reference 73. © 2010 The Royal Society of Chemistry.

blocks, silica nanoparticles, with other functional nanoparticles dispersed in aqueous solution.

5.5.4   Microgel-Stabilized Emulsions for Biocatalysis

Biocatalytic reactions are important to fabricate enatiopure substances, owing to enzymes' inherent enatioselectivity and mild reaction conditions. However, many substrates of interest are hardly soluble in water, while enzymes usually prefer aqueous environments. Thus, enzyme-catalysed biocatalytic reactions are often carried out at biphasic aqueous–organic reaction mixtures to achieve high substrate concentration and high productivity.[74,75] Moreover, enzyme structure and flexibility might be affected by the organic–aqueous interface of such biphasic reactions systems, therefore leading to reduced enzyme activity and selectivity.

To address the above-mentioned problem, Richtering *et al.*[14] presented a conceptually novel approach using microgel-stabilized smart emulsions for biocatalysis. As schematically illustrated in Figure 5.33, a two-phase system is emulsified wherein the organic (oil) phase contains dissolved substrate while the aqueous phase contains biocatalyst and microgel (Figure 5.33A). After emulsification the microgels are located at the droplet surface (as schematically shown in Figure 5.33B) and the enzyme converts the substrate to the reaction product. Finally, the emulsion is broken by increasing the temperature above the volume phase transition temperature (VPTT) of the

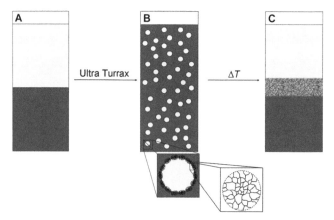

Figure 5.33   Schematic illustration of microgel-stabilized emulsions for biocatalysis: (A), emulsify it using an Ultra Turrax homogenizer, stabilize the emulsion droplets with responsive microgels (B), and afterwards use the sensitivity of the microgels for breaking the emulsion (C). In the beginning (A) the organic phase contains substrate and the aqueous phase enzyme and microgel. The enzyme reaction takes place in the emulsion (B). Finally the emulsion is broken (C). The organic phase contains the reaction product; the microgels flocculate in the upper region of the aqueous phase.
Reprinted with permission from reference 14. © 2009 Wiley.

microgels, thereby leading to macroscopic phase separation. The organic phase now contains the product of the enzyme reaction and can easily be separated while the aqueous phase still contains the biocatalyst and the now flocculated microgel (Figure 5.33C). Three different PNIPAM-based microgels are used to stabilize the biphasic emulsion system for reducing acetophenone to (R)-phenylethanol catalysed by alcohol dehydrogenase as model reaction. The choice of PNIPAM-based microgel systems is due to the fact that PNIPAM-based microgels are well known for their VPTT around 33 °C, thus retaining the reactivity of enzymes,[13,38] and at the same time that PNIPAM-based microgel stabilized emulsions can be broken efficiently under mild environmental conditions,[15,17] *e.g.* stirring for 15 min at 55 °C or stirring for 10 min at 50 °C.[14] Analysis of the organic as well as aqueous phase of the enzyme reaction system by gas chromatography revealed the feasibility of such reactions, indicating that microgels can be used to stabilize the enzyme against denaturation at the interface. Moreover, the flocculated microgels can be redissolved in the aqueous phase upon cooling, thus allowing for recycling of biocatalyst and emulsifiers.

5.6   Concluding Remarks

The adsorption of soft and deformable microgels to the oil–water interface provides a new model system for the study of Pickering emulsions. The hydrophobicity, charge, deformability and interaction force between particles, which can be tuned by varying temperature and solution pH, have been found key factors in determining the stabilizer efficiency. In addition, we illustrate the great potential of microgel-stabilized emulsions as a new enabling technology for the preparation of responsive capsules and porous materials. The remarkable stability of microgel-stabilized systems provides better control over the material's microstructures as compared to other approaches based on surfactant-stabilized emulsions. Although many recent works have been reported on the formulation of emulsions involving microgel particles, there are many fundamental challenges ahead. For example, our fundamental understanding of the connection between interfacial viscoelasticity and the structure and particle–particle interactions of these soft and deformable particles is still surprisingly poor, although this information is directly relevant in stabilizing interfacial systems such as foams and emulsions, which find wide use in food industry and personal care products. We hope this chapter can galvanize more researchers to get involved in this intriguing field.

Acknowledgements

The financial support of this work by the Hong Kong Special Administration Region (HKSAR) General Research Fund (CUHK403210, 2130237) is gratefully acknowledged.

References

1. A. W. Adamson and A. P. Gast, *Physical Chemistry of Surfaces*, 6th edn, Wiley-Interscience, Canada, 1997.
2. B. P. Binks, *Modern Aspects of Emulsion Science*, The Royal Society of Chemistry, Cambridge, 1998.
3. B. P. Binks and T. S. Horozov, *Colloidal Particles at Liquid Interfaces*, Cambridge University Press, Cambridge, 2006.
4. W. Ramsden, *Proc. R. Soc. London*, 1903, **72**, 156–164.
5. S. U. Pickering, *J. Chem. Soc., Trans.*, 1907, **91**, 2001–2021.
6. R. Aveyard, B. P. Binks and J. H. Clint, *Adv. Colloid Interface Sci.*, 2003, **100**, 503–546.
7. T. Ngai, S. H. Behrens and H. Auweter, *Chem. Commun.*, 2005, **3**, 331–333.
8. T. Ngai, H. Auweter and S. H. Behrens, *Macromolecules*, 2006, **39**, 8171–8177.
9. L. A. Lyon and A. Fernandez-Nieves, *Ann. Rev. Phys. Chem.*, 2012, **63**, 25–43.
10. M. Ballauff and Y. Lu, *Polymer*, 2007, **48**, 1815–1823.
11. M. Das, H. Zhang and E. Kumacheva, *Annu. Rev. Mater. Res.*, 2006, **36**, 117–142.
12. B. R. Saunders and B. Vincent, *Adv. Colloid Interface Sci.*, 1999, **80**, 1–25.
13. H. G. Schild, *Prog. Polym. Sci.*, 1992, **17**, 163–249.
14. S. Wiese, A. C. Spiess and W. Richtering, *Angew. Chem., Int. Ed.*, 2013, **52**, 576–579.
15. Z. Li and T. Ngai, *Nanoscale*, 2013, **5**, 1399–1410.
16. Y. Cohin, M. Fisson, K. Jourde, G. Fuller, N. Sanson, L. Talini and C. Monteux, *Rheol. Acta*, 2013, 1–10.
17. W. Richtering, *Langmuir*, 2012, **28**, 17218–17229.
18. T. T. Liu, S. Seiffert, J. Thiele, A. R. Abate, D. A. Weitz and W. Richtering, *Proc. Natl Acad. Sci. USA*, 2012, **109**, 384–389.
19. K. Geisel, L. Isa and W. Richtering, *Langmuir*, 2012, **28**, 15770–15776.
20. M. Destribats, V. Lapeyre, E. Sellier, F. Lea-Calderon, V. Ravaine and V. Schmitt, *Langmuir*, 2012, **28**, 3744–3755.
21. S. Schmidt, T. T. Liu, S. Rutten, K. H. Phan, M. Moller and W. Richtering, *Langmuir*, 2011, **27**, 9801–9806.
22. M. Destribats, V. Lapeyre, M. Wolfs, E. Sellier, F. Leal-Calderon, V. Ravaine and V. Schmitt, *Soft Matter*, 2011, **7**, 7689–7698.
23. M. Destribats, V. Lapeyre, E. Sellier, F. Leal-Calderon, V. Schmitt and V. Ravaine, *Langmuir*, 2011, **27**, 14096–14107.
24. C. Monteux, C. Marliere, P. Paris, N. Pantoustier, N. Sanson and P. Perrin, *Langmuir*, 2010, **26**, 13839–13846.
25. B. Brugger, J. Vermant and W. Richtering, *Phys. Chem. Chem. Phys.*, 2010, **12**, 14573–14578.
26. Z. F. Li, T. Ming, J. F. Wang and T. Ngai, *Angew. Chem., Int. Ed.*, 2009, **48**, 8490–8493.

27. B. Brugger, S. Rutten, K. H. Phan, M. Moller and W. Richtering, *Angew. Chem., Int. Ed.*, 2009, **48**, 3978–3981.
28. S. Tsuji and H. Kawaguchi, *Langmuir*, 2008, **24**, 3300–3305.
29. B. Brugger, B. A. Rosen and W. Richtering, *Langmuir*, 2008, **24**, 12202–12208.
30. B. Brugger and W. Richtering, *Langmuir*, 2008, **24**, 7769–7777.
31. B. Brugger and W. Richtering, *Adv. Mater.*, 2007, **19**, 2973–2978.
32. D. M. Kaz, R. McGorty, M. Mani, M. P. Brenner and V. N. Manoharan, *Nat. Mater.*, 2012, **11**, 138–142.
33. M. Stieger, W. Richtering, J. S. Pedersen and P. Lindner, *J. Chem. Phys.*, 2004, **120**, 6197–6206.
34. X. Wu, R. H. Pelton, A. E. Hamielec, D. R. Woods and W. McPhee, *Colloid Polym. Sci.*, 1994, **272**, 467–477.
35. M. H. Smith and L. A. Lyon, *Acc. Chem. Res.*, 2011, **45**, 985–993.
36. G. R. Hendrickson, M. H. Smith, A. B. South and L. A. Lyon, *Adv. Funct. Mater.*, 2010, **20**, 1697–1712.
37. L. A. Lyon, Z. Y. Meng, N. Singh, C. D. Sorrell and A. S. John, *Chem. Soc. Rev.*, 2009, **38**, 865–874.
38. S. Nayak and L. A. Lyon, *Angew. Chem., Int. Ed.*, 2005, **44**, 7686–7708.
39. A. Fernandez-Nieves, H. Wyss, J. Mattsson and D. A. Weitz, *Microgel Suspensions*, Wiley, Singapore, 2011.
40. R. Pelton, *Adv. Colloid Interface Sci.*, 2000, **85**, 1–33.
41. A. B. South and L. A. Lyon, *Angew. Chem., Int. Ed.*, 2010, **49**, 767–771.
42. R. H. Pelton and P. Chibante, *Colloids Surf.*, 1986, **20**, 247–256.
43. A. Pich and W. Richtering, *Adv. Polym. Sci.*, 2010, **234**, 1–37.
44. Z. Y. Meng, M. H. Smith and L. A. Lyon, *Colloid Polym. Sci.*, 2009, **287**, 277–285.
45. M. Kwok, Z. Li and T. Ngai, *Langmuir*, 2013, **29**, 9581–9591.
46. Z. F. Li, M. H. Kwok and T. Ngai, *Macromol. Rapid Commun.*, 2012, **33**, 419–425.
47. Z. Li, K. Geisel, W. Richtering and T. Ngai, *Soft Matter*, 2013, **9**, 9939–9946.
48. E. Dickinson, *Food Hydrocolloid*, 2003, **17**, 25–39.
49. E. Dickinson, *Colloid Surface B*, 2001, **20**, 197–210.
50. E. Dickinson, *Colloid Surface B*, 1999, **15**, 161–176.
51. E. Dickinson, *Trends Food Sci. Tech.*, 1998, **9**, 347–354.
52. E. Dickinson, *J. Chem. Soc. Faraday Trans.*, 1998, **94**, 1657–1669.
53. K. Du, E. Glogowski, T. Emrick, T. P. Russell and A. D. Dinsmore, *Langmuir*, 2010, **26**, 12518–12522.
54. E. Vignati, R. Piazza and T. P. Lockhart, *Langmuir*, 2003, **19**, 6650–6656.
55. A. F. H. Ward and L. Tordai, *J. Chem. Phys.*, 1946, **14**, 453–461.
56. R. Miller, P. Joos and V. B. Fainerman, *Adv. Colloid Interface Sci.*, 1994, **49**, 249–302.
57. B. J. Park and E. M. Furst, *Langmuir*, 2010, **26**, 10406–10410.
58. T. Hellweg, C. D. Dewhurst, W. Eimer and K. Kratz, *Langmuir*, 2004, **20**, 4330–4335.

59. P. G. Degennes, *Rev. Mod. Phys.*, 1992, **64**, 645–648.
60. S. Granick, S. Jiang and Q. Chen, *Phys. Today*, 2009, **62**, 68–69.
61. D. Suzuki, S. Tsuji and H. Kawaguchi, *J. Am. Chem. Soc.*, 2007, **129**, 8088–8089.
62. A. D. Dinsmore, M. F. Hsu, M. G. Nikolaides, M. Marquez, A. R. Bausch and D. A. Weitz, *Science*, 2002, **298**, 1006–1009.
63. D. Lee and D. A. Weitz, *Adv. Mater.*, 2008, **20**, 3498–3503.
64. D. B. Lawrence, T. Cai, Z. Hu, M. Marquez and A. D. Dinsmore, *Langmuir*, 2006, **23**, 395–398.
65. R. K. Shah, J. W. Kim and D. A. Weitz, *Langmuir*, 2010, **26**, 1561–1565.
66. S. Berger, H. P. Zhang and A. Pich, *Adv. Funct. Mater.*, 2009, **19**, 554–559.
67. N. R. Cameron and D. C. Sherrington, *Biopolymers Liquid Crystalline Polymers Phase Emulsion*, Advances in Polymer Science, Vol. 126, Springer, Berlin, 1996, pp. 163–214.
68. N. R. Cameron, D. C. Sherrington, L. Albiston and D. P. Gregory, *Colloid Polym. Sci.*, 1996, **274**, 592–595.
69. N. R. Cameron and D. C. Sherrington, *Macromolecules*, 1997, **30**, 5860–5869.
70. N. R. Cameron, *Polymer*, 2005, **46**, 1439–1449.
71. S. D. Kimmins and N. R. Cameron, *Adv. Funct. Mater.*, 2011, **21**, 211–225.
72. Z. F. Li and T. Ngai, *Langmuir*, 2010, **26**, 5088–5092.
73. Z. Li, X. Wei, T. Ming, J. Wang and T. Ngai, *Chem. Commun.*, 2010, **46**, 8767–8769.
74. J. E. Vick and C. Schmidt-Dannert, *Angew. Chem., Int. Ed.*, 2011, **50**, 7476–7478.
75. J. H. Schrittwieser, V. Resch, J. H. Sattler, W. D. Lienhart, K. Durchschein, A. Winkler, K. Gruber, P. Macheroux and W. Kroutil, *Angew. Chem., Int. Ed.*, 2011, **50**, 1068–1071.

CHAPTER 6

Bicontinuous Emulsions Stabilized by Colloidal Particles

JOE W. TAVACOLI,[a] JOB H. J. THIJSSEN[b] AND PAUL S. CLEGG*[b]

[a] Laboratoire de Physique de Solides, Bâtiment 510, Université Paris-Sud, F-91405 Orsay, France; [b] School of Physics and Astronomy, University of Edinburgh, Mayfield Road, Edinburgh EH9 3JZ, UK
*Email: paul.clegg@ed.ac.uk

6.1 Introduction

Bijels are a novel class of emulsion whose existence stems from a key difference between surfactant-stabilized and particle-stabilized interfaces. Emulsion droplets stabilized by low molecular weight surfactants have fluid interfaces and hence are always spherical; by contrast, interfaces densely coated by colloidal particles can behave like an elastic sheet. The final form of particle-stabilized droplets is dependent on their process history, potentially resulting in non-spherical droplets. Particle-stabilized emulsions were identified in the early 20th century by Ramsden and Pickering (they are often known as Pickering emulsions) and are now being exploited in several application areas.[1] The particles reside at the interface because they are trapped in an energy well whose depth is of the order of:

$$E = \pi a^2 \gamma \left(1 - |\cos \theta|\right)^2 \qquad (6.1)$$

RSC Soft Matter No. 3
Particle-Stabilized Emulsions and Colloids: Formation and Applications
Edited by To Ngai and Stefan A. F. Bon
© The Royal Society of Chemistry 2015
Published by the Royal Society of Chemistry, www.rsc.org

Here, a is the radius of the particle, γ is the interfacial tension and θ is the contact angle of a particle at the liquid–liquid interface. In the vast majority of cases, E in Equation 6.1 is orders of magnitude greater than k_BT and therefore particles tend to remain adhered to the interface. Note that Equation 6.1 ignores line tension effects, particle–particle interactions and assumes a flat interface. Whereas particle-stabilized droplets are an established technology, liquid bicontinuous architectures, maintained by interfacial particles, are only now being routinely fabricated in the laboratory. These structures are generally called bicontinuous interfacially jammed emulsion gels (bijels). Bijels consist of two tortuously entwined percolating liquid phases that are separated and stabilized by a layer of solid particles (Figure 6.1a, b) and can loosely be thought of as a non-equilibrium counterpart to the surfactant sponge phase. Unlike the highly dynamic sponge phase, the bijel's bicontinuous structure is locked in place because the particles are arranged into a jammed network at the liquid–liquid interface.

The bijel is a new class of soft solid with properties that are quite distinct from emulsion droplets because of its percolating sheet of particles. Bijels and/or their bicontinuous morphology have been suggested for applications in catalysis,[2,3] separation processes,[4,5] tissue engineering,[6,7] fuel cells,[8] solar cells,[9] barrier materials,[10] sensors[11] and co-continuous composites *via* nanoparticle infiltration of porous scaffolds.[12] They have also been envisaged as efficient microfluidic cross-flow chambers for chemical reactions or as vehicles for controlled release.[13–15] In Sections 6.4 and 6.5, the properties and potential applications of the bijel will be covered in greater detail.

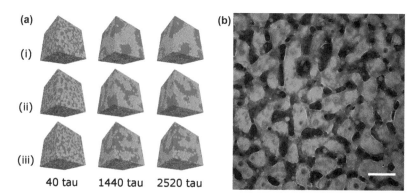

Figure 6.1 (a) Time sequence of snapshots from computer simulations of bijel formation. The two liquids are red and blue and the colloidal particles are green. (i), (ii) and (iii) correspond to particle volume fractions $\Phi = 20\%$, 25% and 30%, respectively. (b) Confocal image of an ethanediol–nitromethane (dark and red regions, respectively) bijel fabricated with silanized silica particles (green). The particles are FITC tagged silanized silica ($a = 453$ nm) and can be seen at the liquid–liquid interface. Scale bar 100 µm.
(a) Reprinted with permission from reference 20. © 2007, AIP Publishing LLC. (b) Adapted from reference 15.

In Section 6.3 we will consider how to tune the surface chemistry of the colloidal particles. Next, in Section 6.2, we outline some of the established and emerging protocols for fabricating bijels, starting with the most commonly used method: the spinodal decomposition of binary-liquid systems in the presence of particles that are neutrally wetted by the developing liquid phases.

6.2 Bijel Fabrication

6.2.1 Spinodal Decomposition

In order to fabricate bijels, two immiscible liquids must be induced to adopt a bicontinuous morphology, while the liquid–liquid interface is populated with colloidal particles. This rarely occurs with direct mixing of low molecular weight liquids: extended domains retract and/or break up into spherical droplets before full coverage of the interface by particles is reached.[16,17] This condition is met, however, during the phase separation of partially miscible liquids if the volumes of the two liquid phases are reasonably equal. Here, bicontinuous liquid domains evolve when the system is quenched into the spinodal region of its phase diagram (Figure 6.2). This kinetic pathway, spinodal decomposition, occurs when any fluctuation in sample composition is energetically favourable. Because significant fluctuations in composition take a long time to develop and small fluctuations create a large interfacial area, costing too much energy, an intermediate length scale of the fluctuations dominates, resulting in a 3D bicontinuous pattern of liquid domains which has a characteristic length scale that grows during the later stages of phase separation.[18] In a startling example of the power of computational physics, a study by Stratford *et al.*, employing lattice Boltzmann simulations, indicated that this 3D and bicontinuous pattern can be retained if nanoparticles that have a 90° three-phase contact angle are dispersed in a phase separating mixture.[13] In their simulations, the particles sequestered at the evolving interface, forming a monolayer which eventually jammed due to the reduction in interfacial area associated with coarsening of the liquid domains. By following the evolution of the characteristic domain length, Stratford *et al.* showed that on jamming, the phase separation was markedly curtailed; seemingly the particles lock in place the 3D bicontinuous architecture. Simulations applying the alternative approaches of dissipative particle dynamics and a different lattice Boltzmann model have subsequently corroborated the conclusions of the initial study,[19,20] as has a continuum model.[21]

Informed by the simulations of Stratford *et al.*, the first three-dimensional bijel was made in the laboratory by Herzig *et al.* using a water–lutidine binary-liquid system and Stöber silica particles (tagged with fluorescein isothiocyanate) which are neutrally wetted by the liquids.[22] The water–lutidine system has a lower critical solution temperature (LCST); spinodal decomposition was initiated by heating the three components from ambient temperature (single phase) to 40 °C (two phases) at a rate of 17 °C min^{-1},

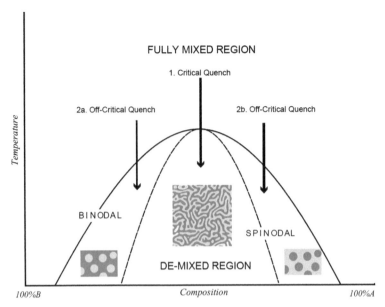

Figure 6.2   Schematic of a binary-liquid phase diagram for liquids *A* and *B*. A critical quench (pathway 1) leads to phase separation *via* spinodal decomposition. Here, any fluctuation in composition lowers the energy of the system leading to a bicontinuous pattern of liquid domains. Off-critical quenches (pathways 2a and 2b) into the binodal region lead to the nucleation of spherical droplets of *A* in *B* (pathway 2a) or *B* in *A* (pathway 2b).

directly through the critical point. Homogeneity of heating was essential and was achieved by inserting the glass cuvette containing the bijel premix into an aluminium block preheated to 40 °C. The quenching rate was also crucial: at a quench rate of 1 °C min^{-1} droplets were formed whereas at 0.1 °C min^{-1} evolving bijels collapsed, presumably because too much time was spent close to the critical point, where interfacial tension is very low.[14] Lee and Mohraz have since adapted the quenching protocol using microwave heating to warm the pre-mixture homogeneously and rapidly.[23] In the original experiments, the fluorescence of the silica particles allowed both visualization of the phase separation and 3D imaging of the final static structure, confirming that the particles were swept up by the newly formed interfaces and that the liquid domains were three-dimensional and bicontinuous. Significantly, as long as the bijel remained above the LCST of the water–lutidine system its morphology was retained. Herzig *et al.* confirmed, as previously proposed,[13] that the characteristic size of the bijel domains, *L*, follows:

$$L = \Lambda_{\mathrm{b}} \frac{d}{\Phi} \qquad (6.2)$$

where Λ_{b} is a pre-factor that depends on the packing of the interfacial colloids and the geometry of the interface, *d* is the particle diameter and Φ is the particle volume fraction.[22] The value of Λ_b extracted experimentally $(\Lambda_b \sim 1.24)$ is close to that of close packed particle-stabilized droplets $(\Lambda_d = 3^{-1/2}\pi \ (\sim 1.8))$ supporting the notion that a (disordered) monolayer of particles is jammed at the interface of the water–lutidine bijel. The d/Φ scaling permits the tuning of L from the nanoscale to many hundreds of microns, hence, the bijel can be tailored to address opportunities that require bicontinuity over this range of length scale.

There are some important differences between the experiments and the simulations performed by Stratford *et al.* The latter used nanoparticles with a purely repulsive potential dispersed in an idealised binary liquid system (*i.e.* one that was perfectly symmetrical and whose two liquid components had equal viscosity and density). The simulation units chosen map onto particles of size $a = 5$ nm trapped on an interface with $\gamma = 60$ mN m^{-1}. By contrast, the experiments employed a non-symmetrical system with liquid constituents of differing chemical and physical nature, together with silica particles around two orders of magnitude larger ($a = 290$ nm) with repulsive (electrostatic) and attractive (van der Waals (VdW)) interactions. Indeed, there is mounting evidence that the silica particles aggregate/flocculate at the interface whereas, because of their purely repulsive potential, the simulated particles remain separated at the liquid–liquid boundary.[14,24] The experimentally relevant interfacial tension is considerably lower $(\approx 0.2 \text{ mN m}^{-1})$ due to the proximity to a critical point. Although the simulations strongly implied retention of the bicontinuous liquid domains for long times, because of computational limits (the simulations did not extend greatly beyond the Brownian time of the colloids) the results were inconclusive.[13,14] Subsequent simulations, focusing on the intermediate evolution of the bijel,[25] point to a complicated energy landscape that lowers the barrier to particle detachment by a factor of order 10^{-2}. It is likely that the physical basis of this reduced barrier is the percolating nature of the particle monolayer that allows the (destabilizing) forces acting on particles to accumulate; the study did not rule out the possibility that the particles could be continually ejected from the interface. Consequently, whilst the experiments by Herzig *et al.* confirm the long-term stability of the bijel, the possibility remains that this is a consequence of attractive interactions between the particles. Additionally, particles larger than those employed in the simulations (*i.e.* those less influenced by k_BT) may be necessary; Jansen *et al.* reported convergence to a fixed domain size in their simulations where larger particles were utilized, although their computations were performed without considering thermal noise.[19]

Following the water–lutidine bijel, a nitromethane–ethanediol bijel has been developed using a similar scheme: a critical quench in the presence of particles that are neutrally wetted at the interface.[15,26,27] In addition to demonstrating the generality of the concept, the nitromethane–ethanediol bijel has proved to be more straightforward to fabricate and has facilitated a number of advances including studies of rheology (see Section 6.4.3) and the creation of bijel capsules (see Section 6.5.2).[15,26]

The shape of the underlying binary liquid phase diagram has an impact on bijel fabrication.[19,25] This relationship exists because the symmetry of the phase diagram dictates the relative volumes of the two liquid phases, which in turn has consequences for the stability of the spinodal pattern. The ratio of the volumes, V, of the two separated phases, α and β, is given by the Lever rule:

$$\frac{\varphi_i - \varphi_\alpha}{\varphi_\beta - \varphi_i} = \frac{V_\beta}{V_\alpha} \tag{6.3}$$

where i refers to the initial mixture and φ is the volume composition of each phase. For symmetric phase diagrams the volumes of the two phases are well matched whereas the opposite is true for unsymmetrical cases. Using simulations to study the crossover from bijels to particle-stabilized droplets, Jansen and Harting demonstrated that the volume ratio of the liquids is a controlling factor in determining whether a given set of particles can stabilize bicontinuous liquid domains.[19] Within the variation of particle contact angle explored (80 to 95°), bijels formed as long as the ratio of the volumes of the two separated phases did not fall below 0.65. Beyond this point, sensitivity to contact angle emerged, and at a phase volume ratio of 0.4 only particle-stabilized droplets remained after spinodal decomposition. These findings may account for why nitromethane and ethanediol work so well: the phase diagram for this system is unusually symmetrical, leading to two liquid domains of near-equal volume on separation.[15]

Because the interface separation of a bijel can be readily tuned, bijels serve as an ideal template from which to fashion other porous media. Lee and Mohraz, starting from the water–lutidine bijel, have assembled micron-scaled networks comprised of ceramic, copper, nickel and silver.[23,28] Here, post-processing of the bijel involves diffusing a soluble monomer and a photoinitiator into the lutidine or water-rich domain followed by UV polymerization. This process and its applications are described in greater detail in Section 6.5.1. We are now beginning to understand that post-processing like this relies on a specific feature of the water–lutidine/silica bijel system. Sanz *et al.* first showed that particles within water–lutidine bijels form a self-sustained network if the bijels are held above the LCST for some tens of minutes (compare Figures 6.3a and 6.3b).[24] At that point, structural integrity no longer depends on interfacial tension alone, *i.e.* the particle network remains even on complete remixing of the liquids (Figure 6.3b lower frame). The resulting particle scaffold was christened the 'monogel'. Computer simulations, also performed by Sanz *et al.*, suggested that an interplay between short-range attractive and long-range repulsive particle–particle interactions are responsible for the monogel's stability (Figure 6.3c). This is plausible because the silica particles that make up the monogel combine attractive VdW and repulsive electrostatic interactions. Specifically, Sanz *et al.* posited that because of capillary attractions, adjacent silica particles are able to overcome their mutual electrostatic repulsion, whereupon

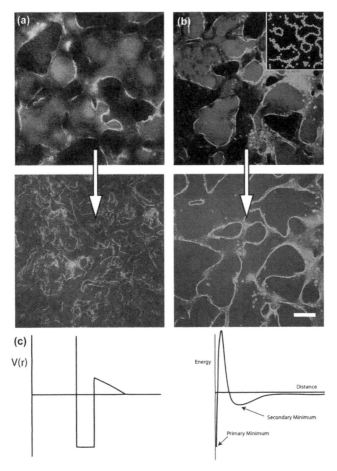

Figure 6.3 (a, b) Fluorescence confocal microscope images of water–lutidine bijels (upper images at 40 °C) responding to the liquid phases remixing as the system is cooled to room temperature, lower images. In (a) the liquids were remixed sooner than in (b); in the latter a monogel forms, *i.e.* the interface disappears but the particles remain in place. Scale bar is 100 μm. The inset to the LHS image in (b) is a computer simulation demonstrating that bicontinuous particle networks can be sustained without interfacial tension when the particles have a combination of long-range repulsive and short-range attractive potentials. (c) The particle potential used in the simulations: a square well short-range attraction combined with a long-range ramped repulsion (left). A schematic of a typical inter-particle potential for silica particles based on DLVO theory (right).
Adapted from reference 24.

they fall into their VdW minima, solidifying the particle network. In this picture, ageing is required because particles must first rearrange into a configuration where, for six nearest neighbours, the capillary attraction is of the same order of magnitude as the electrostatic repulsion.

Attempts to create a monogel from a nitromethane–ethanediol bijel stabilized by silanized silica particles proved less successful.[26] Lee *et al.* argued that the resistance to permanent network formation stemmed from the solvation of the particles with negative nitromethane ions. In this scenario, the ions increase the electrostatic repulsions between particles, impeding their access to VdW minima at the interface. A further possibility is that the stability of the monogel depends on the formation of covalent bonds between particles, an occurrence that is more likely in the water–lutidine case, where a condensation reaction between two silanol surface groups can readily take place to make a siloxane bridge, but less so in the nitromethane–ethanediol system because of the particle's coverage of tri-methyl groups (see Section 6.3.1).

Post-processing of the bijel is considerably eased if a monogel is formed first, because then the strength imparted to the structure is independent of interfacial tension. For this reason, the water–lutidine bijel is the system of choice for creating scaffolds from a bicontinuous template, as in the research led by Mohraz.

6.2.2   Polymer Systems

Polymer blends can have distinct and often superior properties compared to their individual parent components.[29] Furthermore, adding colloidal particles to polymers has long been used to produce composites with enhanced properties and can even promote the mixing of previously immiscible polymers.[30–32] A particular processing challenge associated with polymers, and polymer mixtures, lies in controlling and trapping a specific morphology that will lend some optimal properties to the material. An emerging method to achieve this is *via* the introduction of nanoparticles; a variety of particle/polymer combinations have been explored. Within this setting, then, perhaps it is not surprising that bijels have been independently developed. Just as with their low molecular weight counterparts, polymer-based bijels are likely to impact industry; controlling the periodicity of two polymer phases can lead to enhanced mechanical, electrical or textural properties, whilst the addition of conducting interfacial particles is a plausible route to flexible electronics.[31,33,34]

Before we describe the current polymer bijels, it is worth summarizing some key distinctions between polymers and low molecular weight liquids that are relevant to liquid phase separation accompanied by particles. Unlike low molecular weight liquids, which are always orders of magnitudes smaller, the radius of gyrations of polymers can be comparable to, or even greater in size, than colloidal particles. Therefore, entropic interactions that promote particle aggregation or that drive particles into defects present in the morphology can be present in polymer–particle mixtures. This process can be size selective and particles may segregate at a polymer–polymer interface even if the energy gains, associated with blocking the polymer–polymer contacts, are negligible.[31]

Additionally, the relative contribution of the entropy of mixing to the total free energy of mixing in polymers is far lower than in the low molecular weight liquid case.[18] Accordingly, even slightly unfavourable interactions between polymers renders them immiscible and therefore the accessing of spinodal decomposition, or alternative bicontinuous pathways, is rarely achieved through temperature variation alone.[29] Polymers also tend to be far more viscous than low molecular weight liquids. Whilst an increase in viscosity does not appear to alter the mechanism by which particles gather at the interface, it does slow down the kinetics of phase separation and often changes power law exponents that describe the dynamics of coarsening.[29,33] Finally, after annealing, many polymer systems do not require a jammed network of particles to impart solidity because they themselves are solids at ambient temperatures. Hence, although colloidal particles may segregate at the interface when polymers are molten, where they can regulate de-mixing, how they affect the long-term evolution of phase separation is often not known and sometimes is not even important for the practical application of the resultant composite material.

The first polymer bijel emerged from a series of studies led by Composto.[33–35] His group focused on regulating the phase separation of the polymers poly(methyl methacrylate) (PMMA), or its deuterated form (d-PMMA), and poly(styrene-ran-acrylonitrile) (SAN) using nanoparticles. In order to study this phenomenon, the polymers were mixed and prepared into a thin film by adding a volatile solvent and spin casting. Critical separation occurred in response to heating the film, under nitrogen, through its LCST of 165 °C to 195 °C. In the absence of particles, separation evolved in three stages.[36] Firstly, early in the process, a 3D bicontinuous structure developed simultaneously with upper and lower wetting layers of PMMA. Secondly, at intermediate times, these wetting layers grew to span the SAN mid-layer before forming 2D domains of PMMA and thirdly, at late stages, the film ruptured due to interfacial fluctuations. When the phase separation was repeated with particles that favourably partitioned into one of the polymer phases, as in the majority of cases, it was retarded due to the increased viscosity of the dispersion over the neat blend, but nevertheless continued *via* the outlined route.[33,34,37] However, on the addition of 10 wt% silica nanoparticles ($a = 9$ nm) grafted with chlorine-capped PMMA, film rupture was suppressed. On closer inspection with scanning electron microscopy, an internal 2D bicontinuous structure was revealed. The slow dynamics of the system permitted the domain size to be monitored which, in the presence of the PMMA-modified silica, plateaued. It seems, therefore, that the route to the bicontinuous architecture was the now established one, *i.e.* particle jamming at the interface during spinodal decomposition. Indeed, this mechanism was independently suggested by Composto *et al.*, who subsequently showed that the phase separation could be tuned through variation of the length and the end group of the PMMA chains.[37] Consistent with prior experiments on low and high molecular weight bijels, theory and dissipative dynamics simulations, further work by the Composto group on the

PMMA/SAN system revealed that bicontinuity is favoured by thicker films and higher volume fractions of particles.[19,20,36,37] Gam *et al.*[35] point out that in order to retain bicontinuity in the midplane of the film, the particles must, at a minimum, cover the interface between the bicontinuous network and the wetting layers of PMMA. Such a coverage requires a greater proportion of particles in thinner films and therefore explained the observed correlation.

Bicontinuity has also been maintained in immiscible polymer blends that incorporate colloidal particles of organoclay. Sokolov and Rafailovich demonstrated this with a number of systems, but focused strongly on both polycarbonate (PC)-SAN and polystyrene (PS)-PMMA.[38] Rather than thin films, these systems can ultimately take the form of large slabs that have enhanced mechanical properties and that have a potential application as fire retardants. Sokolov and Rafailovich introduced the clay particles to the polymer pairs during their melt-mixing to produce uniform dispersions which, after solidification with liquid nitrogen, were annealed at 190 °C for 14 hours. In this way they studied the evolution of the polymer systems over a range of composition. They discovered that an organoclay-PS-PMMA (10:27:63 (wt%)) system evolved into a bicontinuous network as did PC-SAN (70:30) samples loaded with particles above 5 wt%. Post inspection of these composites revealed that the particles resided at the polymer interface where they imparted a faceted character to the boundary. In the absence of particles, annealing produced droplets and/or larger-scale phase separation. The organoclay was shown to have an affinity for PC, SAN and PMMA (it dispersed easily in these polymers) but interacted unfavourably in PS (organoclay aggregated in PS). It is therefore at first surprising that the clay particles behaved broadly the same in both polymer systems. The researchers concluded that *in situ* grafting of polymer chains onto the organoclay during the melt-mixing was responsible for its seemingly ubiquitous interfacial activity and suggested this property can be used as a general method to prepare bicontinuous polymer blends. The starting characteristics of the clay particles are still crucial, however: additions of unmodified clay produced droplets rather than a bicontinuous morphology.

In 2009, Cheng used shear of partially miscible polymers to fabricate tortuous domains separated by interfacial particles.[39] One possible explanation for this phenomena could be that mechanical agitation breaks down droplets to the molecular scale, creating an effective shift in the critical temperature. His system was comprised of polyisoprene ($M_w = 29$ k), polyisobutylene ($M_w = 1.3$ k) and iron hydroxide colloids; static liquid domains of both positive and negative mean curvature resulted simply from the hand mixing of the three components with a spatula. For the particles to collect at the polymer boundary it was necessary for them to be pre-dispersed in the polyisoprene-rich phase. We note that Cheng's system comes close to satisfying the condition for inversion under shear:

$$\frac{\varphi_1}{\eta_1} \sim \frac{\varphi_2}{\eta_2} \tag{6.4}$$

where subscripts 1 and 2 denote the two liquids in the system, φ is the phase volume fraction and η is the phase viscosity.[40] It therefore cannot be discounted that the miscibility gap remained constant during the mixing and separation process. This research is poised at a tantalising point: further characterization could test whether these polymer blends constitute the first bijel formed *via* mechanical agitation.

6.2.3   Nucleation and Bijels

So far we have dealt with critical or critical-like phase separations where bicontinuous liquid patterns can be directly accessed. For non-critical mixtures, spinodal decomposition only occurs after first crossing the binodal (Figure 6.2); hence, the sample will be susceptible to de-mixing *via* droplet nucleation. Homogeneous nucleation is driven by thermal fluctuations in the local composition which must exceed a critical domain size for further coarsening to become energetically favourable. In the presence of impurities that act as nucleation sites (for example colloidal particles that have an affinity for the droplet phase), this process is called heterogeneous nucleation.

By quenching rapidly across the binodal and the spinodal lines, nucleation can be avoided however, and an advantage of using non-critical quenches is that the volume ratio of the formed phases can be chosen as required (see Equation 6.2). This is often necessary with unsymmetrical phase diagrams where critical quenches lead to a vast predominance of one phase, and therefore a loss of bicontinuity.

Indeed, non-critical quenches were employed in the initial attempts by Clegg *et al.* to fabricate bijels from alcohol–alkane systems.[41] Here, the concentrations of the pre-mixtures were chosen as halfway between a critical quench and a 1:1 phase volume ratio. Two binary liquid systems were used: dodecane–ethanol and hexane–methanol, both in combination with silanized Stöber silica. These systems have an upper critical solution temperature and on their de-mixing with the silica particles, a range of structures formed depending on the rate and depth of the temperature quench. Bicontinuity began to emerge with rapid quenches to an intermediate depth (-78 °C, with dry ice) and was widespread in deep and relatively slow quenches (-198 °C, using liquid nitrogen). Clegg *et al.* suggested that the quick quenches resulted in limited bicontinuity because, as noticed in their samples, they promote the formation of tubular liquid domains that 'compete' with bicontinuous ones.[42] They also hypothesized that the inertia and viscous drag of the particles hinders their capture by the rapidly moving interfaces associated with fast changes in temperature. However, even after quenching slowly and deeply, the alcohol–alkane bijels were quasi-2D formations, spanning one domain length in thickness at best. Attempts to form a three-dimensional bijel resulted in a tortuous structure which reverted to being spherical droplets in the centre of the sample.[41]

By combining the pathways of nucleation and spinodal decomposition, bijels with increased mechanical stability can be constructed.[27] Such bijels, christened bridged bijels, have one of their continuous phases packed full of zipped droplets and were first fabricated by Witt *et al.* using the nitromethane–ethanediol binary liquid pair.[27] To form bridged bijels, Witt *et al.* induced spinodal decomposition in the presence of particles with a wetting angle distribution centred on $\theta \sim 60°$. On the onset of de-mixing, some particles are swept up by the newly formed liquid–liquid interface while the remainder partition into one of the two liquid domains. The division of particles resulted from the low interfacial tension at this juncture: only particles with near neutral wetting were permanently attached to the interface. However, as the quench deepened, droplets developed within the bicontinuous domains, entrapping the remaining particles at their interfaces. The droplets continued to grow and were eventually 'zipped' together by particles that straddled two of their boundaries. Because of the presence of a thin film of the continuous phase, the particles were able to maintain an equilibrium contact angle at each droplet.

Bridged bijels are able to access domain sizes almost an order of magnitude larger than standard bijels, *i.e.* millimetres, which is a useful scale of porosity for tissue engineering (Section 6.5.1). Interestingly, despite their mixed morphology, the bridged bijels retain the rheological signature of bijels with channels that are free of droplets.

6.3 Particles and Wettability

The choice of particles has a profound effect on the structure and stability of the bijel.[22,24,26] Specifically, it appears that particles will only support bicontinuous liquid domains if they have a near 90° contact angle at the liquid–liquid interface.[22] The tuning of particles for the latter condition is demanding, and generally the controlling step in bijel fabrication.[15,43,44] Even with appropriate particle wettability, bijel formation can still be undermined by strong electrostatic repulsions between particles, an accumulated body force due to external fields or a compressional stress acting on the interface.[15,45,46]

During the development of the bijel, colloidal silica was employed to establish the bicontinuous liquid networks and silica particles remain integral to the development of this material.[15,23,27] However, increasingly more exotic particles are now being investigated as alternatives. These include anisotropic and magnetic particles as well as graphene oxide sheets.[47–50] Using other types of particles within bijels can result in structures that appear to gain their stability in quite different ways.

6.3.1 Silica Particles

Almost without exception, bijels fabricated from low molecular weight liquids rely on silica particles, or other particles with a surface layer of silica. It is easy to see why: silica particles are readily synthesized in the laboratory, where they can be produced in a large range of sizes and with low polydispersity, whilst their surfaces can be readily functionalized.[51] The latter feature is particularly suited to the needs of bijel fabrication, where a prerequisite is near neutral wetting of the particles by the two liquid phases. Often the appropriate wetting conditions for the silica surfaces are only found *via* trial and error. Nevertheless, we routinely adapt silica for both the water–lutidine and nitromethane–ethanediol bijel systems.[15,22]

The most frequently used route to silica particles for bijel preparation is the Stöber synthesis. Four chemical ingredients are required: water, a short-chained alcohol (typically ethanol), ammonia and tetraethylorthosilicalate (TEOS).[52,53] In this mixture, the ammonia acts as a catalyst and, after the hydrolysis of TEOS, silica is produced from the condensation of silanol groups. The overall reaction is:

$$Si(OC_2H_5)_4 + 2H_2O \rightarrow SiO_2 + 4C_2H_5OH$$

Silanol groups dominate the surfaces of native Stöber silica, but there are also appreciable numbers of siloxane bridges, ethoxy groups and surface-bound water. Whereas silanol groups and adsorbed water render the surface hydrophilic, siloxane bridges and ethoxy groups lend it a more hydrophobic character. In the majority of cases, the silanol moiety is the key to chemical grafting onto silica.[54]

In the first forays into fabricating three-dimensional bijels, the surface properties of Stöber silica were tuned for the water–lutidine bijel by incorporating 3-(aminopropyl)triethoxysilane (APS) into the silica matrix.[22,43,44] This molecule is used to bind the fluorescent dye, fluorescein isothiocyanate (FITC), to silica. Through a series of systematic experiments, White *et al.* discovered a positive correlation between free APS content (APS bound to silica but free of FITC) and the hydrophilicity of the silica.[43,44] To explain this somewhat counter-intuitive relation, White *et al.* placed it in the context of the water–lutidine binary system where the crucial factor in determining the contact angle of APS-modified silica particles (APS-silica) is the extent to which lutidine can adsorb onto their surface prior to, and during, phase separation.[55] Indeed, during demixing/remixing cycles within the binary liquid, the character of silica particles can shift from hydrophilic to hydrophobic as a surface layer of lutidine develops.[44] White *et al.* suggested that the presence of APS molecules impedes this process because they replace deprotonated silanol and siloxane groups that can both form π-bonds with lutidine molecules. In this picture, particles containing large amounts of APS maintain an affinity for the water, whereas those with low proportions prefer the lutidine-rich phase, but only after progressively adsorbing lutidine. Practically, the concentration of APS provides a handle with which to adjust the contact angle of the particles, and White *et al.* were able to create bijels when employing silica that contained intermediate levels of dye/APS.

In a further study, White *et al.* tuned the wettability of APS-silica by removing its surface-bound water in a controlled manner by drying the

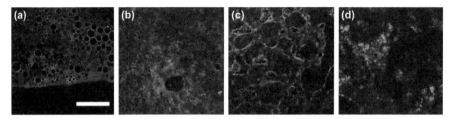

Figure 6.4 Fluorescence confocal microscopy images of emulsions formed from the critical demixing of water and lutidine in the presence of APS-modified silica. Green areas of the image correspond to the particles, red areas to the lutidine-rich phase and black areas to the water-rich phase. The scale bar is 100 μm. As we proceed from (a) to (d) more APS is added to the silica particles and their affinity for lutidine progressively decreases. This is reflected by the nature of the emulsions formed: (a,b) The particles prefer lutidine and hence water-in-lutidine droplets are stabilized, (c) the particles have near equal affinity for water and lutidine and a bijel is maintained, and (d) the particles prefer water resulting in lutidine-in-water emulsions.
Reproduced from reference 43.

particles at temperatures close to 170 °C.[44] Through a simple variation of drying time, three different types of particle-stabilized emulsions were formed from the critical demixing of water and lutidine: lutidine-in-water emulsions (short drying times, the particles favoured the water-rich phase), bijels (intermediate drying times, the particles had near-neutral wetting at the interface) and water-in-lutidine emulsions (long drying times, the particles favoured the lutidine-rich phase) (Figure 6.4). White *et al.* suggested that a competition between water and lutidine for silanol groups can explain this behaviour. They argued that when water is removed it vacates silanol groups, freeing these otherwise hindered moieties to π-bond with lutidine molecules during the process of phase separation. Therefore, by progressively varying their drying time, the APS-silica can be tailored to hold a range of water–lutidine ratios during demixing, book-ended by a near fully physisorbed layer of water (minimal drying) and an adsorbed lutidine layer (complete drying). Evidently, at some intermediate point between the two the particle surfaces support bijel production.

Two further points are important to mention on the subject of silica and the water–lutidine system. Firstly, drying of the APS-silica must be performed at temperatures below 180 °C. Above that, silanol groups condense to siloxane bridges and the nature of the surface is irreversibly (from a practical viewpoint) altered. Secondly, White *et al.* were unable to sustain water–lutidine bijels using native silica particles (particles lacking APS) regardless of their drying protocol. They found that 'wet' native silica remained in the water-rich phase, having no affinity for the interface, whereas 'dry' particles proceeded to partition into the lutidine-rich phase. It seems that the latter behaviour is a consequence of the dry particles rapidly obtaining a wetting layer of lutidine.[44]

Whilst the behaviour of silica particles in the water–lutidine system is becoming more understood, it is system specific, and much of the wisdom gained is difficult to extend to other binary liquids. In developing alternative bijel systems, it is worth using a more generic approach to tune the particle wettability: one that can be more easily adopted for a broader range of binary liquid pairs. Promising candidates are silanization reactions,[54,56–60] in which a silanol group is directly replaced by the hydrophobic group(s) of choice. The degree of substitution can be moderated as required, with higher proportions associated with increased hydrophobicity. Studies on silica-stabilized emulsions, formed by direct mixing, have highlighted the power of this approach where either oil-in-water or water-in-oil types formed subject to the degree of silanization of the particles.[56] A common silanization agent is dichlorodimethylsilane (DCDMS) and silica particles treated with this chemical have successfully supported bicontinuous domains in some instances. For example, 2D alcohol–alkane bijels (see Section 6.2.3) have been prepared with silica particles treated in this manner.[41]

It is commonly believed that chlorosilanes modify the surfaces of silica *via* a two-step reaction.[61] The first reaction step is the hydrolysis of the chlorosilane by water present at the silica surface:

$$2H_2O + Cl_2SiO(CH_3)_2 \rightarrow (OH)_2Si(CH_3)_2 + 2HCl$$

The second step is a condensation reaction between the silonal groups on the silica and the alkylsilanols (produced by the first reaction) to form siloxane bonds:

$$Si_sOH + (OH)_2Si(CH_3)_2 \rightarrow Si_sOSi(OH)(CH_3)_2 + H_2O$$

The subscript S denotes the silica surface.[59,61,62] Intermolecular siloxane bonds may also form between alkylsilanes. In doing so multiple layers of the silanizing agents can assemble around the silica particles.[61]

The reaction is effective in imparting a hydrophobic character to silica but it is tricky to control and to implement reproducibly.[15,62] Predominately, such difficulties arise because the reaction is highly sensitive to the amount of water present; too much water and widespread polymerization of DCDMS occurs, whereas in its complete absence the reaction does not proceed. Further, the treatment generally requires a hydrocarbon reaction medium within which it is difficult to fully disperse native silica and its flocculation is therefore often unavoidable. Indeed, Clegg *et al.* commented on the presence of a fluffy and ill-defined particle layer of DCDMS-modified silica at the interfaces of their bijels.[41] It is tempting to link their observation to the prior instability of the silica colloid during its silanization. Because of these difficulties, an alternative silanization protocol has recently been introduced to modify silica particles for bijels, one that employs the molecule hexamethyldisilazane (HMDS).[15,26,27]

In this process, trimethyl groups replace the silanol groups on silica. The significant advantage here is that the reaction can be performed under conditions similar to those used for the Stöber synthesis and therefore large-scale treatments can be carried out on fully dispersed native Stöber silica. Moreover, the treatment is less sensitive to changing proportions of water. These features improve the reproducibility of the process and the hydrophobicity of silica particles can be readily tuned as a function of re-action time/HMDS concentration.

The reaction in solution is, however, subtle. Si MAS NMR studies indicate it proceeds *via* two mechanisms,[54] a fast one where HMDS directly reacts with silica:

$$(CH_3)_3Si\text{-}NH\text{-}Si(CH_3)_3 + 2Si_sOH \rightarrow 2Si_sOSi(CH_3)_3 + NH_3$$

and a slow one where HMDS is first transformed into ethoxytrimethylsilane and then condenses with surface silanols:

$$(CH_3)_3Si\text{-}NH\text{-}Si(CH_3)_3 + 2CH_3CH_2OH \rightarrow 2CH_3CH_2OSi(CH_3)_3 + NH_3$$
$$CH_3CH_2OSi(CH_3)_3 + Si_sOH \rightarrow 2Si_sOSi(CH_3)_3 + CH_3CH_2OH$$

Because the HMDS reaction with a silica surface proceeds only *via* the silica's silanol groups, increasing the quantity of these groups permits a higher degree of silanization. This can be achieved by adding increasing proportions of water to the reaction media, because water converts ethoxy groups, present on the silica surface, to silanol groups.

Following these procedures, surface coverage of the trialkyl group can extend to ~35%. For water droplets on HMDS-treated silica surfaces, this gives a contact angle of ~135° as opposed to ~15° for a hydrophilic silica surface (estimated from ref. 54). HMDS-treated silica particles are now ha-bitually used within nitromethane–ethanediol bijels and the relatively facile preparation of the particles has facilitated the advances associated with this bijel system, such as bijel capsules and bridged bijels.[15,27]

6.3.2 Future Prospects: Anisotropic and Field Responsive Particles

At a liquid interface, anisotropic particles can have quite different inter-actions to those of their spherical equivalents. For instance, whereas iso-tropic smooth spheres have a straight contact line, in order to satisfy the Young relationship, anisotropic particles have more varied contact lines that can result in long-range, shape-dependent capillary interactions, producing forces and torques that drive their assembly towards favoured conform-ations.[60–63] In addition, 2D packing fractions and percolation thresholds of anisotropic particles vary from those of the isotropic case.[64,65] Unsurprisingly then, the character of solid-stabilized emulsions can be substantially influenced by the shape of the particle constituents. For ex-ample, it has been shown that particle-stabilized emulsions have increased robustness when ellipsoid particles are employed.[66] This enhancement may derive from attractive capillary forces; the emulsification performance of the ellipsoids improves as their aspect ratio is increased. Extrapolated to bijels, it is therefore likely that the inclusion of non-spherical particles will augment their structural stability and could be used to ease the post-processing of the nitromethane–ethanediol bijel, for instance. Anisotropic particles may also have some influence over the large-scale architecture of the bijel and therefore they could be employed as a handle to adjust it. From another perspective, the bijel itself offers a unique platform from which to study the effect of interface curvature on particle assembly.[67] To explore these avenues, some systematic work is required and Hijnen and Clegg have begun this process.[48] They added hollow silica rods, with aspect ratios of up to 10 (Figure 6.5a,b), to the spinodal decomposition of the water–lutidine system. This resulted in static liquid domains, of both positive and negative mean curvature, that were held by an interfacial layer of the rods (Figure 6.5c).

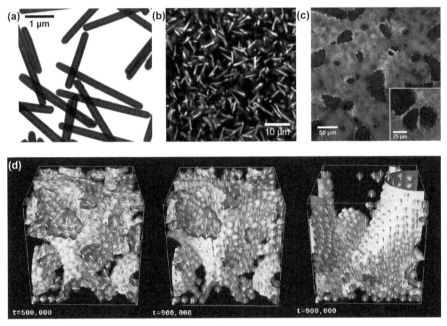

Figure 6.5 (a) Hollow silica colloidal rods obtained after dissolving the akaganeite cores of core-shell particles. (b) A fluorescent confocal microscopy snapshot of hollow FITC-labelled colloidal rods dispersed in an index matching solvent. (c) A fluorescent confocal microscopy image of a continuous emulsion containing water (black) and lutidine-rich (red) liquid domains and stabilised by colloidal rods with a surface of silica (green). Inset image: a capillary bridge between domains containing and maintained by rods. (d) Snapshots of lattice Boltzmann simulations of bijels containing magnetic particles subjected to a vertical field gradient. Left: the pre-field configuration of the bijel. Centre and right: the bijel subjected to a field gradient corresponding to $\xi_o = 0.01$ and $\xi_o = 0.1$, respectively (see text for details). The gradient was applied for 4×10^5 time steps.
(a) to (c) Adapted from reference 48; (d) reprinted with permission from reference 79. © 2010 American Chemical Society.

Bulk bicontinuity has yet to be demonstrated, but nevertheless some intriguing features, relevant to bijels, can be extracted from this work. For example, rather than being smooth, the interface was a little 'porcupine'-like, indicating that the long axes of the rods were not always lying flat. Because isolated rods spontaneously adsorb with their long axis in the plane of the interface, Hijnen and Clegg suggested that the rods had 'flipped out' of the interfacial plane as they were mutually compressed during domain coarsening. Such 'flipping' has been observed previously in compressed 2D networks of ellipsoids on planar liquid interfaces.[64] Further, the interface of the rod-stabilized bijel seems more faceted than in the regular case; similar to polymer-based bijels where lamellar clay particles are used (see Section 6.2.2). Thin capillary bridges stabilized by, and with similar dimensions to, the rods were also noticed between domains of the tortuous structure. Such bridges have been reported in simulations on bijels containing ellipsoid particles and are absent in standard bijels.[68] The bridges may add further structural integrity to bijels secured with colloidal rods.

Simulations by Günther *et al.* on bijels containing ellipsoidal particles have demonstrated that fewer ellipsoidal particles are required to sustain a bijel in comparison to their spherical analogues, a reflection of the ellipsoid's larger surface area to volume ratio.[69] Here, the ellipsoids remained adsorbed along their length and it is interesting to note that more ellipsoids will be required than spheres if the ellipsoids flip out of the interfacial plane.

In contrast to these rods and ellipsoids, graphene is shaped like a platelet. Chemically, it is a honeycomb 2D lattice of σ-bonded carbon atoms which each contribute a π-orbital to a delocalized cloud of electrons. It is highly conductive to heat and electricity, almost optically transparent and mechanically strong. A great deal of the excitement around this material originates from its potential to be integrated with others, such as polymers, to form composites with superior properties.[70] To retain conductivity within a composite, graphene must form a percolated network and the liquid interfaces of the bijel offer an interesting template with which to guide this process. To explore the behaviour of graphene at liquid interfaces it is more convenient to employ the graphene derivative/precursor, graphene oxide (GO), which has additional alcohol, epoxide and carboxylic acid groups.[46,47,49,71,72] Significantly, through oxidation and reduction reactions, the surface of GO can be rendered more hydrophobic and hydrophilic, respectively, and can hence be adapted to exhibit partial wettability at liquid interfaces.[46,49,50] Using aromatic solvents and water, Kim *et al.* made particle-stabilized emulsions with sheets of GO and other groups have since followed suit.[47,49,50] In almost all cases, oil-in-water emulsions are formed whose stability can be enhanced by adding salt, decreasing pH and by partially reducing the GO sheets.

Some researchers have highlighted the considerable robustness of monolayers of GO at liquid interfaces.[46,73] Imperiali *et al.*, for instance, explored in depth the mechanical properties of GO sheets (lateral dimension ~ 1 μm, thickness ~ 0.8 nm) at a planar air–water boundary.[46] In this setting, they thrice compressed/expanded, biaxially, monolayer assemblies of GO using a Langmuir trough. From analysis of the resultant surface pressure (P) versus surface area (A) curves, they concluded that the first cycle of compression and expansion rearranged the GO monolayer into a denser configuration. However, even at high compressional stresses (> 50 mN m^{-1}) the sheets were resistant to overlap. Rather, they buckled out of plane to form periodic wrinkles that were spaced with a characteristic wavelength, λ. The onset of wrinkling corresponded to a plateau in the P–A curves and such buckling has been reported before in compressed monolayers of spherical colloids.[74] The fact that wrinkling, rather than particle ejection, was observed highlights the strong trapping of the sheets at the interface. λ can be related to the monolayer bending stiffness, b, *via*:

$$b = \frac{\Delta \rho g \lambda^4}{16\pi^4} \qquad (6.5)$$

where $\Delta \rho$ is the density difference between two fluids across the interface and g is the acceleration due to gravity.[75] After the extraction of λ, *via* image analysis, a value of b similar to that of regular nanoparticles was calculated by Imperiali *et al.*, a surprisingly low value given that the graphene sheets have dimensions on the micron scale. Imperiali *et al.* attributed the GO monolayer's relatively low bending stiffness to connections between sheets that acted as hinges to support their vertical displacement and to the inherent flexibility of individual sheets themselves.[46] In addition, interfacial creep and recovery tests revealed that the GO monolayer had a surface modulus value similar to that of an aggregated surface layer of microparticles and that it could support far greater strains before yielding than typical particle-laden interfaces. Interestingly, the structural integrity of GO monolayers held even in the absence of a liquid interface; centimetre-scaled free-standing films were consolidated after compression.

The results of the above studies suggest that GO sheets can be successfully incorporated into bijels and, just recently, this has been implemented (Figure 6.6) by Imperiali *et al.* employing a water–lutidine mixture containing GO particles to create a GO-stabilized bijel structure.[76] The 2D nature of GO sheets, combined with an appropriate surface chemistry simply controlled by their carbon to oxygen ratio, causes them to collect and assemble at the interfaces and create highly elastic layers that arrest the phase separation process. Soft solid GO frameworks are created which have sufficient mechanical stability to allow removal of the solvents by freeze drying. The advantage of the method is that the structure and mesh size of the frameworks can be easily controlled, providing direct access to the GO surfaces while creating a mechanically robust and percolating network of GO tiles. This could be used, for example, as electrodes for supercapacitors or fuel cells (see Section 6.5).

An interesting development in emulsion technology is the use of field-responsive particles as interfacial stabilizers. Droplets holding such particles

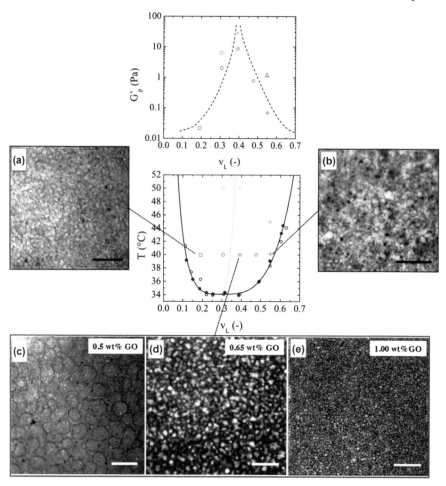

Figure 6.6 Phase behaviour and mechanical properties: water–lutidine mixtures with GO sheets at different lutidine volume fraction v_L, GO concentration and temperature. (a,b) Optical micrographs of droplet matrix structures obtained at 0.8 wt% GO, $T = 40$ °C and at v_L indicated on the phase diagram. Scale bar: 100 μm. (c, d, e) Microscopic observations of the arrested spinodal decomposition for mixtures at $v_L = 0.39$ and different GO concentrations (as indicated). Scale bar: 300 μm. The phase diagram is shown in the middle. The full black line is the binodal derived from optical observations; the dotted line is the geometric locus where the volume ratio of the coexisting phases is 50/50 and the maximum degree of co-continuity is expected. The bulk plateau storage modulus G_p' is plotted on the upper graph for mixtures with a fixed GO content of 0.8 wt%, at v_L and temperatures indicated on the phase diagram (same legend). The maximum values of G_p' are measured for the co-continuous morphologies (the line is a guide to the eye).
Reproduced from reference 76.

can be remotely translated, rotated, distorted and also forced to coalesce or break down in response to external fields.[45,77,78] Kim *et al.* have explored this theme computationally on bijels stabilized by magnetic particles.[79] Here, the bijel was subjected to both uniform and gradient fields. On applications of the former, an increase in the mean spacing between interfaces was observed along the field direction, compensated for by its decrease in the perpendicular directions. In other words, the homogeneous field introduced structural anisotropy to the magnetic bijels. This anisotropy was accentuated by application of the field earlier on in the phase separation process and hints at an intriguing prospect of a bicontinuous to lamellar switch in internal architecture. Kim *et al.* pointed out that such a switch would convert the structure from one that was isotropically permeable to one that only permits flow in a single direction.

The magnetic bijel's response to field gradients was quite different. At sufficiently high gradients particles were pulled off the interface, permitting coarsening of domains and presumably, had the simulations run for longer, breakdown of the bijel's bicontinuous structure (Figure 6.5d). Kim *et al.* characterized this behaviour with the two relevant forces acting on the bijel's particles; a magnetic (field gradient) and a capillary one. If directed perpendicular to an interface, the former 'pulls' a particle away from a boundary whereas the latter resists this motion. The ratio of the two forces is given by ξ_0:

$$\xi_0 = \frac{m|\nabla B|}{\gamma \pi a} \tag{6.6}$$

where m is the dipole moment of the magnetic particle and B is the field strength. Drastic destabilization of the bijel was witnessed at $\xi_0 = 0.1$ (*i.e.* capillary force > magnetic force) and Kim *et al.* reasoned that contacts between particles allow the magnetic forces to accumulate along the bijel. That is to say, in order to define the stability of magnetic bijels, Equation 6.6 should include a pre-factor that scales in some way with the number of magnetic particles contained within a bijel. Such accumulation of forces through contact has been hinted at in particle-stabilized droplets holding superparamagnetic colloids and demonstrated explicitly in huge droplets maintained by very large and dense beads.[45,80,81] Indeed, because of the small scale of the simulations, structural breakdown was witnessed at practically incredible values of B. Nevertheless, Kim *et al.* suggested that realistic field gradients applied to real bijels should reproduce this behaviour because such bijels can contain many orders of magnitude more particles than can be included in simulation studies. The theme of field-responsive bijels still requires addressing experimentally.

6.4 Morphology and Scaling Behaviour

In this section we compare the morphology and aging behaviour of the bijel to those of phase-separating polymer blends; we then consider how this morphology is likely to influence the properties of the composite. We finish by outlining the predicted and observed scaling behaviour of the bijel rheology. In the following section we consider technological problems to which either the bijel or materials derived from the bijel might be applied.

6.4.1 Morphology and Spinodal Decomposition

The domain morphology generated by spinodal decomposition develops so quickly for low molecular weight fluids that it is difficult to study using slow, high-resolution imaging techniques. However, phase separation can become slow or even completely stop for polymer blends and for bijels. This situation was first exploited for the polymer case where three-dimensional stacks of high-resolution images were captured using laser scanning confocal microscopy.[82] Doping with anthracene rendered one of the polymers fluorescent. This direct-space visualization provides an immediate representation of the shape of the interfaces that is not readily available from the structure factors from scattering experiments.

Characterization techniques have been developed for use with these stacks of images which yield values for the mean ($H = 1/R_1 + 1/R_2$) and Gaussian ($K = 1/R_1R_2$) curvatures, the volume averaged surface area and the more traditional characteristic length scale. Rather than average values, this approach yields curvature values for all points on the interface, allowing a direct and local visualization of the aging processes.

The studies of phase-separating polymer blends showed that the spinodal structure is hyperbolic (the mean curvature $\langle H \rangle = 0$ and the Gaussian curvature $\langle K \rangle < 0$.[83] Data was captured for blends aged for a range of times and the statistical distribution of curvatures was determined for all cases. As spinodal decomposition progressed the distribution of mean and Gaussian curvatures became more sharply peaked. The quantity: interfacial area per unit volume has units of 1/length. The reciprocal of this value was taken as a length scale characteristic of the phase separation process; it was employed to convincingly demonstrate that the statistical distributions of curvatures exhibited dynamical scaling.

The spread in values of the mean curvature is quite significant and can be contrasted with the situation for periodic bicontinuous structures where it can be zero everywhere.[84] The broad distribution of H is indicative of a pressure difference across the interface, as described by the Laplace equation: $\Delta p = 2\gamma H$. Such pressure differences, which vary from place to place, imply that there must be stresses in the plane of the interface. These stresses are resolved as spinodal decomposition proceeds, *i.e.* the distribution of curvatures changes, favouring smoother flatter interfaces. The image analysis approach makes it possible to spatially resolve the curvature values, indicating how elliptical regions ($\langle H \rangle \neq 0$ and $\langle K \rangle = 0$) tend to become flatter with time; while hyperbolic regions with significantly negative Gaussian curvature values tend to pinch off as the structure ages.[85]

Recently, the techniques developed for polymer blend phase separation have been applied to characterize the static structure of the bijel.[23] Again, stacks of high-resolution images were captured using laser scanning confocal microscopy. These were analysed using commercial realizations of the techniques developed previously. Distributions of the mean and Gaussian curvatures have been presented for the bijel structure as initially formed and for the structure once one domain has been filled with a polymer (Figure 6.7a–d). An arrested bijel shows a distribution of curvatures in good agreement with that shown for phase separating polymer blends (the mean curvature $\langle H \rangle = 0$ and the Gaussian curvature $\langle K \rangle < 0$. During bijel processing the Gaussian curvature becomes more sharply peaked, which is similar to what happens during an aging process. This might suggest that some of the more extreme necks may pinch off during filling with monomer and polymerizing. More ambiguous changes to the mean curvature were

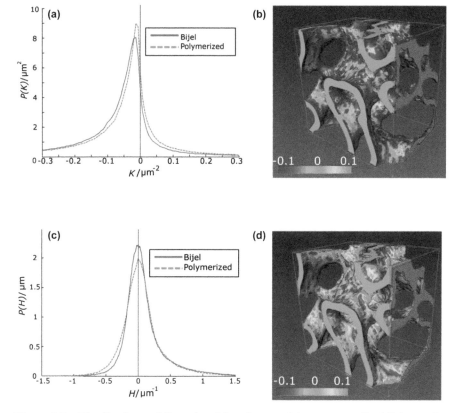

Figure 6.7 Distributions of Gaussian (a) and mean (c) curvatures for bijel samples (blue) and polymerized bijel samples (red, dashed) and a corresponding surface reconstruction of a bijel for each curvature type (b, d). The size of the box in the renderings corresponds to $33 \times 33 \times 26$ μm^3.
Reproduced from reference 23 with permission from Wiley. © 2010 WILEY-VCH Verlag GmbH & Co.

observed: the distribution of values on the negative side grew slightly while the positive side of the distribution remained fixed. Analysis of a larger sample of data may be required to confirm this trend.

In most research to date bijels have been prepared with similar sized particles ($d \approx 400$ nm) and interfacial separations ($L \approx 50 \mu$m), hence the mechanical properties have not varied greatly. In the future, bijels may be prepared with significantly smaller particles and a wide range of interface separations. This will lead to a very much broader range of structures, some of which may exhibit relatively soft interfaces. The ability of the interfaces to support stresses due to the variation in the mean curvature may vary considerably along with this variation in structure and it will be very interesting to see whether we observe a concomitant change in the distribution of interfacial curvatures.

6.4.2 Morphology and Properties

In order to, for example, understand building materials and to predict ideal porous media for tissue scaffolds, computational and theoretical research has focused on determining the relationship between structure and properties for composite solid materials.[86–88] Pertinent here: a material that has elastic properties due to interfacial layers of particles is described as a sheet solid, to contrast it with node-bonded/network solids that resemble a network of struts. A percolating sheet of particles, as observed in the monogel structure,[24] can be generated computationally *via* a modification of Cahn's Gaussian Random Field (GRF) model. The GRF is an isotropic pattern of overlapping sine waves of random amplitudes designed to reproduce the domain pattern that forms due to spinodal decomposition.[89] It can be divided into two domains by imposing a threshold amplitude as the dividing line. This procedure has been extended to a pair of positive and negative thresholds to yield a random bicontinuous structure with a defined interfacial layer.[90] Adjusting the original pattern of sine waves yields control over the volume fraction and the interfacial roughness.

Calculations of the mechanical properties and simulations of transport properties show that sheet-like solids have superior conductivity and mechanical strength compared to materials which owe their strength to junctions between domains.[86] Within the spectrum of sheet-like materials it was shown that roughening the interface undermines performance. The roughness wastes structural material and adds unnecessary tortuosity for transport phenomena. Clearly, the monogel/bijel is an intriguing model system for mechanical investigations, perhaps in comparison to high internal-phase Pickering emulsions.

Structure/property relationships have also been considered for more exotic morphologies. Periodic sheet and network solids based on triply periodic minimal surfaces have been compared specifically as candidate materials for tissue scaffolds.[88] Due to the periodicity of the structures, statistical techniques are not required to generate the simulated materials. A vast array of different periodic structures have been explored: in general, the network solids tend to have struts of varying thicknesses leading to weak points. By contrast, the sheet solids are defined by a layer of uniform thickness, a very efficient use of ingredients, hence mechanical properties are greatly enhanced. However, the transport properties of the network solids tend to be better because of the concentration of material into thick struts or channels. Experimentally, creating rigorously periodic materials derived from emulsions is extremely challenging!

6.4.3   Rheology and Scaling

Stratford *et al.* predicted that the bijel would undergo shear melting above a critical strain, required to unjam the particle layer. Beyond this point they anticipated large-scale rearrangement of the bijel, with the rearrangements locked in place on removal of the shear stress. *In situ* visualization of a nitromethane–ethanediol bijel subjected to a simple shear stress seems to confirm this picture.[15] Here, the bijel's domains were observed to break, make new connections and align themselves with the shear. On removal of the shear the bijel remained in an anisotropic arrangement.[15]

Stratford *et al.* also predicted that the static modulus, G_0, of the bijel would scale with its interfacial energy density:

$$G_0 \sim \frac{\gamma}{L} \tag{6.7}$$

because L can be tuned so in turn can G_0.[13] The first mechanical tests on the water–lutidine bijel revealed a lower bound yield stress two orders of magnitude greater than anticipated by this scaling. This discrepancy may well arise from the difficulty of making bijels that are devoid of particle–particle interactions and it is suspected that attractive VdW interactions between particles may, at least partly, explain the water–lutidine's higher than predicted yield stress.[22,24] In support of this idea, Lee *et al.* have measured storage moduli consistent with Equation 6.7 for bijel structures prepared from the nitromethane/ethanediol liquid pair using different quench temperatures to vary γ.[26] Here, attractive interactions between particles are likely to be far less important. More recently, Witt *et al.* found exponential scaling behaviour in bridged bijels (discussed further below) which appears to be unique although its physical origins are as yet still unknown.[27]

6.5   Post-Processing and Applications

As described above, the bijel (should) have unusual materials properties. Ever since the concept was first proposed *via* lattice Boltzmann simulations in 2005,[13] it has been recognized that these could be relevant to various applications.[14]

- The bijel provides a uniform domain size and a percolating path in 3D for both constituent materials. Materials with such a bicontinuous morphology are valued in a broad array of applications for their large and tunable interfacial area,[4,91,92] efficient transport characteristics,[93,94] enhanced structural integrity[95,96] and percolating material properties.[97,98]

- If the layer of particles on the liquid–liquid interface truly solidifies, the bijel itself should become a 3D solid gel with a finite elastic modulus and a yield stress. This is different from (uncompressed) particle-stabilized emulsions, which only have one continuous phase, and very different from microemulsions, which have a fluid interface and are therefore not much more viscous than their component fluids. The finite elastic modulus and yield stress have been confirmed in experiments.[15,22,26]
- If the particles are purely repulsive, the mechanical properties of bijels are dominated by interfacial tension. This could potentially lead to transient fluidization, *i.e.* a large enough mechanical strain could unjam the interfacial particles, allowing fluid-like rearrangements before rejamming in a partially relaxed configuration. Both experiments and simulations have shown transient fluidization in the form of (self-)healing[15,22] and orientational training.[15,79]
- If the particles become attractive once on the liquid–liquid interface, the spinodal particle structure can remain stable even upon remixing of the liquids (*i.e.* the interfacial tension is removed), forming a so-called 'monogel'.[24] Simulations have shown that particles with a short-ranged attraction plus a repulsive barrier can indeed stabilize monogels.[24] Experiments have confirmed that interparticle interactions play a pivotal role in monogel formation, which in turn facilitates bijel post-processing.[26]
- Both the typical pore size L and the yield stress τ_y can be controlled by varying the radius a and/or the volume fraction Φ of the particles. Specifically, assuming all the particles are on the liquid–liquid interface, L scales as a/Φ, while τ_y scales as γ/L with γ the interfacial tension between the liquids. As a can be varied from nanometres to microns and Φ from 0.01 to 0.1, this provides about three decades of tuning range for both L and τ_y. Over a limited range, experiments have confirmed the scaling for L[15,22] but suggest an exponential dependence for τ_y.[27]

Below, we focus on several applications in (i) porous materials, (ii) encapsulation and delivery and (iii) microfluidics. We will also discuss future challenges for developing real-life applications based on bijels.

6.5.1 Porous Materials

In a 2010 paper, Lee and Mohraz showed that bijels can be used as soft templates to synthesize (hierarchically) porous composites, typical pore size ~25 μm, with flexible chemistry and co-continuous morphology.[23]

Essentially, the post-processing protocol involves exposing a 3D bijel to a reservoir of liquid monomer that partitions into one of the two liquid phases. After curing *via* UV photopolymerization, the remaining liquid phases are drained, resulting in a bicontinuous material consisting of a particle-coated polymer channel and an air channel (Figure 6.8a). This polymerized bijel can then be used as a template for a variety of bicontinuous materials, including macroporous ceramics and spinodal metal shells (Figure 6.8b). In 2011, Lee and Mohraz specifically extended this work using a common nanocasting procedure to include bicontinuous metal monoliths with hierarchical pore morphology across a wide range of length scales (Figure 6.8c).[28]

The crucial step in the process described above is the infiltration of the bijel with a liquid monomer – if the bijel collapses during this step, the bicontinuous morphology will not be copied into a polymer composite. As elucidated in a collaboration with the Edinburgh group in 2013, this infiltration step is particularly robust if a monogel is formed, *i.e.* bijels stabilized

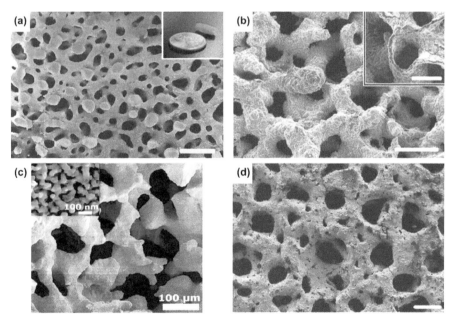

Figure 6.8 Scanning electron microscopy (SEM) images. (a) Bijel after selective polymerization with bicontinuous microstructure present throughout the macroscopic sample pictured next to a US penny (photo inset). (b) A spinodal nickel shell (the inset shows a sectioned specimen). (c) Silver monolith with continuous pores on two widely separated length scales. (d) A copper-coated macroporous polymer. Scale bars: (a) 150 μm, (b, d) 50 μm and (c) 100 μm. Inset scale bars: (b) 20 μm and (c) 100 nm.
Panels (a, b, d) reproduced from reference 23 with permission from Wiley. © 2010 WILEY-VCH Verlag GmbH & Co. Panel (c) is reprinted with permission from reference 28. © 2011 American Chemical Society.

by interfacial tension alone are more easily destroyed upon monomer loading, presumably because steep gradients in chemical composition induce interfacial tension gradients and concomitant Marangoni stresses along the liquid–liquid interface, leading to rupturing of the liquid domains. Using bijels as soft templates for porous composites retains their advantageous morphology in the final materials, but their unique rheological properties are lost along the way. All the same, the bijel architecture itself could lead to significant improvements in the design of materials in various applications, some of which we discuss below.

6.5.1.1   *Tissue Engineering*

Due to structural similarities, bicontinuous macroporous composites had already been considered as tissue scaffolds, *e.g.* using calcareous materials for bone repair.[6] Moreover, interconnected hydrogels made of biocompatible materials such as poly(ethylene glycol) (PEG) could be valuable in tissue engineering where cell penetration and nutrient diffusion are necessary for tissue regeneration.[7] As bijels can be copied into PEG,[26,28] they have great potential as templates for bicontinuous biocompatible hydrogels, as the length scale of the uniform pore morphology can be tuned from 10 to 100 μm. Unfortunately, the bijel domain size appears to have an upper limit of 100 μm, which is too small for some applications in tissue engineering.[99] This upper limit could be due to gravitational breakdown of the liquid domains[27] and/or a preferred curvature of the liquid–liquid interface that undermines emulsion connectivity.[22] As described in Section 6.2.3, Witt *et al.* reported bridged bijels in which one of the fluid channels is reinforced by a high internal phase solid-stabilized emulsion. This hybrid morphology is the result of a combined spinodal decomposition and secondary nucleation and growth mechanism in binary liquid mixtures containing interfacially active colloids with partial affinity for one fluid phase. The largest reported domain size in these bridged bijels is 450 μm, thus significantly expanding the potential technological applications of bijel-based materials.[27]

6.5.1.2   *Catalysis and Sensors*

Materials with a bicontinuous structure and a bimodal pore size distribution have great potential for applications in catalysis,[100] gas storage[101] and osmotic power ('blue energy').[102] The idea is that combining interconnected pores at the nanometre and micrometre scales allows simultaneous optimization of active surface area and mass transport.[103] In other words, the small pores enhance the active surface area for adsorption and heterogeneous reactions, while the large pores facilitate the rapid flow of fluid reactants or macromolecular solutions. Such materials could be obtained from bijels by using the polymerized version as a template for nanocasting, generating micropores down to 10 nm (Figure 6.8c).[28]

Such bicontinuous bimodal porous materials could also be excellent chemical sensors by using a responsive polymer either during or after the post-processing step. As demonstrated by Lee and Mohraz,[23] a polymerized bijel can be coated with a thin layer of conductive material (*e.g.* copper), while retaining the softness of the polymer (Figure 6.8d). In such a hybrid sample, a responsive polymer will swell or contract upon contact with a chemical stimulus, thereby changing the density of electrically conductive pathways. In this case, the small pores provide enhanced surface area for sensitivity, whereas the spinodal morphology ensures that the entire internal surface area is readily accessible to the chemical to be detected. The latter feature in particular could offer superior performance over other materials based on soft templates, such as polymer foams[104] and colloidal crystals,[105] which typically have one continuous phase only.

6.5.2 Encapsulation and Delivery

The prospect of a responsive composite holding two contrasting materials in intimate contact prior to use makes bijels an interesting alternative for encapsulation and (targeted) delivery applications. For example, in a haircare product, one would like to encapsulate a dye and a bleach separately but release them simultaneously. One could argue that Pickering multiple emulsions can be used instead, but bijel-based systems have several advantages, including two rather than one continuous phases (so more or less equal phase volumes) and stability vs gravity (whereas internal droplets may sediment). Note that neither Pickering emulsions nor bijels need suffer from coarsening *via* Ostwald ripening as domain shrinkage requires particles to be expelled from the liquid–liquid interface (which is energetically expensive).

Unfortunately, in any application of particle-stabilized liquid–liquid composites involving foreign components in the liquid phase, the interfacial tension gradients induced by the payload may cause severe problems for stability and function.[26] Here again, as in the case of bijel post-processing, the formation of a monogel (without actually remixing the liquids) could greatly enhance the ability of the bijel to cope with steep gradients in chemical composition. Hence, the ability to tune interactions between particles at a liquid–liquid interface will be pivotal in developing bijels for encapsulation and delivery applications.

6.5.2.1 Bijel Capsules

The potential for bijels as delivery vehicles was demonstrated in 2011 by Tavacoli *et al.*[15] They developed bijel capsules *via* spinodal decomposition of a binary liquid within Pickering drops in a continuous oil phase (Figure 6.9a), where the capsule and the encapsulated spinodal domain sizes can be controlled independently through variations in the concentrations of droplet and bijel-stabilizing particles. This independent control of two

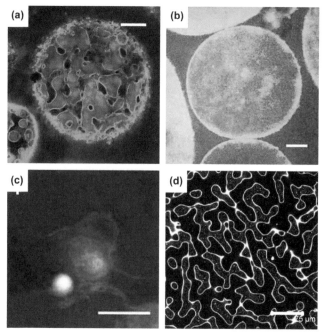

Figure 6.9 (a) Confocal fluorescence micrograph of a bijel capsule. The green regions correspond to fluorescently labelled particles on the interface (silanized silica within the capsule and silica stabilized with poly-(12-hydroxystearic acid) chains at the capsule periphery). The ethanediol (dark) and nitromethane (red) rich phase reside within the capsule which sits in a continuous dodecane phase (dark). (b) A similar capsule after internal remixing due to a temperature trigger (*i.e.* heating to 46.3° C). (c) After collapse due to addition of ethanol to the continuous oil phase; note the release of the encapsulated materials into the continuous phase. (d) Confocal reflection micrograph of a phase-separated mixed biopolymer system containing 0.7 wt% polystyrene latex particles. Scale bars: (a–c) 100 μm and (d) 75 μm.
Panels (a–c) adapted from reference 15. Panel (d) adapted with permission from reference 108. © 2009 American Chemical Society.

different length scales is reminiscent of the bijels with bimodal pore size distributions as developed by Lee and Mohraz,[28] although the capsule geometry seems more suitable for delivery applications.

Tavacoli *et al.* also envisaged three ways in which release or mixing could be achieved.[15] Firstly, just as for vesicle-based delivery systems,[106] the application of shear will break the capsule; a stress of order 50 Pa is required to break open the outer particle shell of a 500 μm bijel capsule. Secondly, a change in temperature can push the bijel system into the single-phase region of its phase diagram, thereby causing the constituent liquids to mix internally, *i.e.* without release into the continuous phase. Hence, this temperature trigger could be used to mix previously compartmentalized reactants close to the targeted point of delivery for the reaction product(s). Finally, Tavacoli *et al.* added small amounts of a trigger species (ethanol) to the continuous phase, inducing partial mixing of all four liquids, allowing for the release of a payload into the continuous phase. Other triggers could be (i) salt for internal mixing (as salt can decrease/increase the LCST/UCST of binary liquids) and/or (ii) pH for release into the continuous phase by using pH-sensitive particles to initially stabilize the Pickering drop.[107]

6.5.2.2 Edible Bijels

From the point of view of applications in encapsulation and delivery, it is unfortunate that most bijel systems discussed so far in this applications section are fabricated from at least one component that is non-edible or even toxic (2,6-lutidine, ethanediol and nitromethane). However, in 2009, Firoozmand *et al.* reported bicontinuous structures in a water-in-water system based on a phase-separating aqueous solution of gelatin + starch (Figure 6.9d).[108] In this mixture of incompatible biopolymers, as the spinodal microstructure develops, the added 300 nm diameter polystyrene particles effectively flocculate at the developing biopolymer-depleted liquid–liquid boundaries. The authors suggest that the main thermodynamic driving force for this process is an osmotic repulsion between the particles and the non-adsorbed biopolymer molecules.

At least in this particular investigation,[108] the particles accumulated at the liquid–liquid interface and reduced the rate of coarsening, but they could not prevent the growth of the bicontinuous phase-separated biopolymer structure. One suggested explanation for the hindered coarsening is the development of elastic character within the large-scale particle-rich interfacial regions, as inferred from jagged and irregular particle-laden interfaces in the microscopy images. These observations may well agree with simulations by Aland *et al.*, which seem to suggest that coarsening can only be halted by particles jamming at the interface.[21] In other words, the high concentration of polystyrene particles in the biopolymer-depleted region can endow the liquid–liquid interface with elasticity that can slow down the coarsening, but if the particles do not actually jam then they cannot prevent coarsening. Fortunately, if required, the spinodal structure can be frozen in by gelation of the gelatin-rich phase.

Although the authors themselves state that "it still remains to be seen whether these structures have practical application in the fabrication of nanoscale biopolymer systems for use in delivery and encapsulation technologies",[108] we would like to emphasize here that their work can be considered in the broader context of liquid–liquid composites consisting of and/or stabilized by food-grade constituents, in particular edible particles[109] or emulsion droplets.[110,111] As such, some of the unique mechanical properties of bijels could be used to tune the rheology of food products, which have been reported to play a major role in determining mouthfeel.[112]

6.5.3 Microfluidics

In the initial publication on bijels, Stratford *et al.* speculated that bijels might have potential as cross-flow microreactors, in which two liquids are made to flow in opposite directions through the two separate channels (Figure 6.10a), allowing intimate contact between mutually insoluble reagents.[13,14] These reagents can meet at the interstices between the interfacial particles, which take up at most $\sim$ 90% of the liquid–liquid interface, so the particles pose a significant barrier for reagents or products only if they are of comparable size (a feature which may even be used for size selection). Notably, the set-up also purifies the reaction, as any reaction product that is soluble in either phase can be swept out continuously. This is in stark contrast to microreactor designs based on droplet emulsions,[113] which have to run as batch processes as reagents and/or products can only be injected into/extracted from the dispersed phase by breaking the emulsion. Intriguingly, a bijel device should 'plug-and-play' when brought into contact with adjacent bulk phases of the two liquid phases (*e.g.* containing the reagents), as either phase should connect automatically to the corresponding bijel channel only.

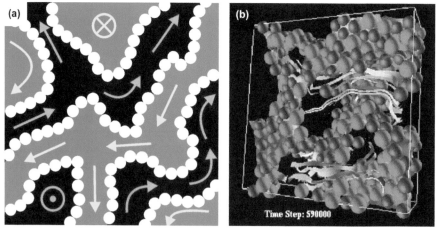

Figure 6.10 (a) Schematic geometry of a cross-flow microreactor based on the bijel geometry. The bicontinuous morphology of the bijel should allow two immiscible fluids to be passed through the material in opposite directions *via* a continuous process. Black: aqueous domain, green: oil domain, white: particles. Note that fluid in neighbouring domains flows in opposite directions (arrows). (b) Section of a near-arrested bicontinuous structure from computer simulations, showing velocity streamlines for the two component fluids under cross-flow forcing with velocity $\approx 0.01\gamma/\eta$, with γ the liquid–liquid interfacial tension and η the liquid viscosity. There is no discernible motion of the interfacial structure at this flow rate.

Panel (b) from reference 13. Reprinted with permission from AAAS.

As before, the mechanical properties of bijels are crucial for their application, in this case as microfluidic devices. Comparing the yield stress with the viscous stresses generated locally in a crossflow geometry, a bijel should be able to support a velocity difference Δv between the two fluids of order $0.01\ \gamma/\eta$ without disruption. For binary pairs of low molecular weight liquids, the interfacial tension $\gamma \sim 1\ \text{mN m}^{-1}$ and the viscosity $\eta \sim 1\ \text{mPa}$, so Δv is of order centimetres per second, independent of the pore size L.[14] Note that this is a remarkably high value – even for $L \sim 25\ \mu\text{m}$, local shear rates are of order $10^3\ \text{s}^{-1}$. As such, bijels may also find applications as 3D microfluidic mixers, where the convoluted void space promotes chaotic advection of the passing fluids.[23] Although simulations have suggested the potential of bijels as microfluidic devices (Figure 6.10b), only preliminary experiments have been carried out so far.[22] Note that the simulations did not take into account any potential reagents/reaction products in the liquid phases. The associated gradients in chemical composition across the particle-laden interface may lead to domain rupturing *via* Marangoni stresses,[26] although monogel formation *via* attractive interactions between interfacial particles may again reinforce the bijel.

6.5.4 Application Challenges

Looking ahead, we identify below several challenges that may have to be addressed in order for bijel applications to come to fruition, although we do not claim this list is exhaustive.

- Scaling: For some applications at least, bijel fabrication needs to be scaled up. At the moment typical sample volumes are of the order of a millilitre. This is mainly because the required spinodal phase separation is triggered *via* a temperature quench, so large samples tend to suffer from temperature inhomogeneities, leading to nucleation and growth in some parts of the sample. Approaches that employ direct mixing at the molecular level might alleviate this problem.
- Cost: Some bijel components currently used, *e.g.* Stöber silica particles, are relatively expensive. For bulk manufacturing, it would be necessary to switch to particles that are produced in bulk, like fumed silica. Controlling the wetting of such fractal-like particles may not be trivial, although alternative approaches using non-spherical particles or carefully balanced mixtures of hydrophilic and hydrophobic particles may provide an appropriate alternative.
- Health and sustainability: At the moment, most binary liquids used for bijels have at least one component that is not benign to humans or the environment. The use of biocompatible phase-separating polymers is an interesting development, but the challenge here is to develop biocompatible/edible particles with the same exquisite control over their contact angle as for example in the case of silica.

- Monogels: Quite a few bijel applications depend on the formation of a monogel, whether or not the liquids are actually allowed to remix. Although significant progress has been made (Section 6.2.1) in explaining why some systems do readily form monogels and others not, designing a system to form a monogel seems to require a daunting level of control of the interactions between interfacial particles, especially considering that the particle contact angle needs to be tuned simultaneously for neutral wetting.

Although the challenges identified above may seem substantial, we would like to emphasize here that the field of bijels is not even 10 years old yet. Given the encouraging amount of progress since 2005, through a combined effort of simulators and experimentalists, we are optimistic about bijels eventually enhancing the performance of real-life materials and devices.

6.6   Two Percolating Solid Domains in a Liquid – the Bigel

Many of the applications just described make use of the continuous pathways through the bijel for charge conduction, delivery or flow. The open pathways originate from the two percolating liquid domains; they are separated by a particle-coated interface with the character of a solid sheet.[22] This architecture relates to the physical process of liquid–liquid phase separation *via* spinodal decomposition in a conceptually simple way. It is also interesting to consider the idea of creating a structure which has two co-continuous solid domains separated by a single liquid domain. While colloidal gas–liquid phase separation is well known (and can result in gel formation), phase separation into two separate colloidal liquids, both of which go on to form stable gels, had not previously been considered. This problem has now been approached by two simulation groups and one experimental group.[114–116] Each has succeeded in creating interpenetrating colloidal gels but they have all tackled the challenge in different ways.

Verrato *et al.* explore conventional phase separation for an unconventional population of colloids theoretically and *via* molecular dynamics simulations.[115] They have two populations of colloidal particles with attractive interactions between particles of the same kind and purely repulsive interactions between particles of different kinds. The attractions are strong and short range; a situation which is known to result in gel formation during phase separation. They explore the parameter space of volume fraction, concentration of each type of particle and temperature. They identify the regions for which gas–liquid phase separation could be anticipated and where liquid–liquid phase separation is going to occur. During the simulations the system falls out of equilibrium, necessitating repeat runs for ensemble averaging. In this impressive piece of work, the authors manage to identify the regions of parameter space for which interpenetrating gels can be found and they distinguish percolating and non-percolating networks *via* consideration of the Gaussian curvature and the Euler characteristic. They christen the structures, with two percolating colloidal gels, bigels.

Earlier, Goyal *et al.* used discontinuous molecular dynamics to simulate two different populations of particles interacting *via* forces with dipolar symmetry.[114,117] In this work, they explore the cases of pairs of particles with different sizes and the same strength of dipole, the same sizes and different dipoles and finally, with an important modification to the interaction strength, different sizes and different dipoles. For all these cases they were able to scan out a composition–temperature phase diagram extending from a fluid phase at low concentration and high temperatures to various crystal phases at high concentrations and low temperatures. The most exciting aspects are the gel 'phases' which form as fluid phases were very slowly cooled. Due to the dipolar symmetry these gels are comprised of chains of particles. In the case where the dipoles had different strengths but the particles had the same sizes interpenetrating gels are found to form. The formation process was quite different to the simulations of particles with isotropic interactions described above. Here, the two populations form gels sequentially as slow cooling is carried out. As the temperature falls, the thermal energy is insufficient to keep the particles with larger dipoles apart; as these gel, the particles with smaller dipoles continue to exist as fluid phases in between the gel strands. On further cooling the smaller dipoles form a gel as well. Slow cooling is crucial. In the third set of simulations, very small particles with very small dipoles were modified with the addition of an isotropic attraction (this was to mimic van der Waals attractions). In this case the second population of particles form a 'shell' around the gel strands that formed first.

Most recently, Di Michele *et al.* have successfully made bigels experimentally (see Figure 6.11).[116] Conceptually their approach shares some features of both of the simulation studies described above: as with Varrato *et al.* they have isotropic attractions between 'like' particles alone,[115] as with Goyal *et al.* they rely on slow cooling and interactions that become important sequentially.[114] In practice the interactions are created by meticulously coating the colloidal particles with DNA strands with appropriate base-pair sequences.[118] The sequences are chosen so that there are two separate populations which can only interact amongst themselves with strengths tailored to give the desired sequential gelation. Hence, on cooling, one population of particles first becomes 'sticky' and gels, and subsequently the other follows suit. Ultimately, this gives the same structures as those observed by Verrato *et al.* even if the kinetic pathway is different (Figure 6.11). It is a ground-breaking achievement. Additionally, by modifying the DNA on the second population of particles such that it will begin to stick to the first particle population to gel, Di Michele *et al.* were also able to make a core-shell network. Both the bigel and the core-shell networks could have a wide variety of applications as composite materials with connected morphologies.[116]

Figure 6.11 Confocal fluorescence images of three arrested bigel samples with $\varphi_a = 5\%$ and, from left to right, $\varphi_b = 5$, 7.5 and 10% ('a' component in red, 'b' component in green). Scale bar, 50 μm.
Reprinted by permission from reference 116. © 2013 Macmillan Publishers Ltd.

In addition to the different architecture, the conceptual approach also provides an important contrast. The bijel is prepared by a physical route: arrested phase separation. This approach has then been applied to simple liquids, polymer blends and water–water systems. Colloidal spheres and rods, clay platelets and graphene sheets have all been employed using essentially the same method. By contrast, the bigel is currently prepared by carefully tuning the particle–particle interactions. At the moment this is a chemically specific strategy although perhaps it might ultimately be possible to prepare bigels *via* a physical route.

6.7   Conclusions

We have reviewed here the development of the bijel from computational discovery through to realization in the laboratory. We have presented the original experimental approach together with more recent experiments and simulations using a greater range of liquids, polymers and particles. Because it is a key step, we have described the treatment of the particle's surfaces in some detail. The morphology, properties and potential applications of the bijel have been described together with the challenges that need to be overcome before this composite material can be fully exploited. Finally, we have considered different but related work where specific interactions between particles are used to create co-continuous solid domains surrounded by a liquid. Both the underlying approach and the final structure provide an illuminating contrast to the bijel.

References

1. B. P. Binks and T. S. Horozov (eds), *Colloidal Particles at Liquid Interfaces*, Cambridge University Press, Cambridge, 2006.
2. I. K. Sung, C. M. Mitchell, D. P. Kim and P. J. A. Kenis, *Adv. Funct. Mater.*, 2005, **15**, 1336.

3. A. N. Pestryakov, V. V. Lunin, A. N. Devochkin, L. A. Petrov, N. E. Bogdanchikova and V. P. Petranovskii, *Appl. Catal., A*, 2002, **227**, 125.

4. N. Tanaka, H. Kobayashi, N. Ishizuka, H. Minakuchi, K. Nakanishi, K. Hosoya and T. Ikegami, *J. Chromatogr., A*, 2002, **965**, 35.

5. M. Motokawa, H. Kobayashi, N. Ishizuka, H. Minakuchi, K. Nakanishi, H. Jinnai, K. Hosoya, T. Ikegami and N. Tanaka, *J. Chromatogr., A*, 2002, **961**, 53.

6. M. Martina, G. Subramanyam, J. C. Weaver, D. W. Hutmacher, D. E. Morse and S. Valiyaveettil, *Biomaterials*, 2005, **26**, 5609.

7. S. G. Levesque, R. M. Lim and M. S. Shoichet, *Biomaterials*, 2005, **26**, 7436.

8. J. R. Wilson, J. S. Cronin, A. T. Duong, S. Rukes, H.-Y. Chen, K. Thornton, D. R. Mumm and S. Barnett, *J. Power Sources*, 2010, **195**, 1829.

9. W. L. Ma, C. Y. Yang, X. Gong, K. Lee and A. J. Heeger, *Adv. Funct. Mater.*, 2005, **15**, 1617.

10. M. Ulbricht, *Polymer*, 2006, **47**, 2217.

11. C. A. L. Colard, R. A. Cave, N. Grossiord, J. A. Covington and S. A. F. Bon, *Adv. Mater.*, 2009, **21**, 2894.

12. S. Polarz and M. Antonietti, *Chem. Commun.*, 2002, **22**, 2593.

13. K. Stratford, R. Adhikari, I. Pagonabarraga, J. C. Desplat and M. E. Cates, *Science*, 2005, **309**, 2198.

14. M. E. Cates and P. S. Clegg, *Soft Matter*, 2008, **4**, 2132.

15. J. W. Tavacoli, J. H. J. Thijssen, A. B. Schofield and P. S. Clegg, *Adv. Funct. Mater.*, 2011, **21**, 2020.

16. H. P. Grace, *Chem. Eng. Commun.*, 1982, **14**, 225.

17. M. E. Cates, Complex Fluids: The Physics of Emulsions, *arXiv:1209.2290*, 2012.

18. R. A. L. Jones, *Soft Condensed Matter*; OUP, Oxford, 2002.

19. F. Jansen and J. Harting, *Phys. Rev. E*, 2011, **83**, 046707.

20. M. J. A. Hore and M. Laradji, *J. Chem. Phys.*, 2007, **126**, 244903.

21. S. Aland, J. Lowengrub and A. Voigt, *Phys. Fluids*, 2011, **23**, 062103.

22. E. M. Herzig, K. A. White, A. B. Schofield, W. C. K. Poon and P. S. Clegg, *Nat. Mater.*, 2007, **6**, 966.

23. M. N. Lee and A. Mohraz, *Adv. Mater.*, 2010, **22**, 4836.

24. E. Sanz, K. A. White, P. S. Clegg and M. E. Cates, *Phys. Rev. Lett.*, 2009, **103**, 255502.

25. E. Kim, K. Stratford, R. Adhikari and M. E. Cates, *Langmuir*, 2008, **24**, 6549.

26. M. N. Lee, J. H. J. Thijssen, J. A. Witt, P. S. Clegg and A. Mohraz, *Adv. Funct. Mater.*, 2013, **23**, 417.

27. J. A. Witt, D. R. Mumm and A. Mohraz, *Soft Matter*, 2013, **9**, 6773.

28. M. N. Lee and A. Mohraz, *J. Am. Chem. Soc.*, 2011, **133**, 6945.

29. P. S. Clegg, *J. Phys.: Condens. Matter*, 2008, **20**, 113101.

30. A. Bansal, H. Yang, C. Li, B. C. Benicewicz, S. K. Kumar and L. S. Schadler, *J. Polym. Sci., Part B: Polym. Phys.*, 2006, **44**, 2944.

31. A. C. Balazs, T. Emrick and T. P. Russell, *Science*, 2006, **314**, 1107.

32. M. Si, M. Goldman, G. Rudomen, M. Y. Gelfer, J. C. Sokolov and M. H. Rafailovich, *Macromol. Mater. Eng.*, 2006, **291**, 602.
33. H.-J. Chung, A. Taubert, R. D. Deshmukh and R. J. Composto, *Europhys. Lett.*, 2004, **68**, 219.
34. H. Chung, K. Ohno, T. Fukuda and R. J. Composto, *Nano Lett.*, 2005, **5**, 1878.
35. S. Gam, A. Corlu, H. Chung, K. Ohno, J. A. Hore and R. J. Composto, *Soft Matter*, 2011, 7, 7262.
36. H. Wang and R. J. Composto, *J. Chem. Phys.*, 2000, **113**, 10386.
37. H. Chung, J. Kim, K. Ohno and R. J. Composto, *ACS Macro Lett.*, 2012, **1**, 252.
38. M. Si, T. Araki, H. Ade, A. L. D. Kilcoyne, R. Fisher, J. C. Sokolov and M. H. Rafailovich, *Macromolecules*, 2006, **39**, 4793.
39. H.-L. Cheng, *Spreading and Jamming Phenomena of Particle-Laden Interfaces*, University of Pittsburgh, 2009.
40. A. Onuki, *Phase Transition Dynamics*, CUP, Cambridge, 2002.
41. P. S. Clegg, E. M. Herzig, A. B. Schofield, S. U. Egelhaaf, T. S. Horozov, B. P. Binks, M. E. Cates and W. C. K. Poon, *Langmuir*, 2007, **23**, 5984.
42. H. Tanaka, *Phys. Rev. E.*, 1996, **54**, 1709.
43. K. A. White, A. B. Schofield, B. P. Binks and P. S. Clegg, *J. Phys.: Condens. Matter*, 2008, **20**, 494223.
44. K. A. White, A. B. Schofield, P. Wormald, J. W. Tavacoli, B. P. Binks and P. S. Clegg, *J. Colloid Interface Sci.*, 2011, **359**, 126.
45. S. Melle, M. Lask and G. G. Fuller, *Langmuir*, 2005, **21**, 2158.
46. L. Imperiali, K.-H. Liao, C. Clasen, J. Fransaer, C. W. Macosko and J. Vermant, *Langmuir*, 2012, **28**, 7990.
47. J. Kim, L. J. Cote, F. Kim, W. Yuan, K. R. Shull and J. Huang, *J. Am. Chem. Soc.*, 2010, **132**, 8180.
48. N. Hijnen and P. S. Clegg, *Chem. Mater.*, 2012, **24**, 3449.
49. Y. He, F. Wu, X. Sun, R. Li, Y. Guo, C. Li, L. Zhang, F. Xing, W. Wang and J. Gao, *ACS Appl. Mater. Interfaces*, 2013, **5**, 4843.
50. C. Cheng and D. Li, *Adv. Mater.*, 2013, **25**, 13.
51. R. K. Iler, *The Chemistry of Silica: Solubility, Polymerisation, Colloid and Surface Properties, and Biochemistry*, Wiley-Interscience, New York, 1979.
52. W. Stöber, *J. Colloid Interface Sci.*, 1968, **69**, 62.
53. I. A. M. Ibrahim, A. A. F. Zikry and M. A. Sharaf, *J. Am. Sci.*, 2010, **6**, 985.
54. T. I. Suratwala, M. L. Hanna, E. L. Miller, P. K. Whitman, I. M. Thomas, P. R. Ehrmann, R. S. Maxwell and A. K. Burnham, *J. Non-Cryst. Solids*, 2003, **316**, 349.
55. M. Privat, M. Amara, A. Hamraou, H. Sellami and A. M. Mear, *Ber. Bunsenges. Phys. Chem.*, 1994, **98**, 626.
56. B. P. Binks, *Adv. Mater.*, 2002, **14**, 1824.
57. B. P. Binks and J. H. Clint, *Langmuir*, 2002, **18**, 1270.
58. B. P. Binks and R. Murakami, *Nat. Mater.*, 2006, **5**, 865.

59. P. D. I. Fletcher and B. L. Holt, *Langmuir*, 2011, **27**, 12869.
60. N. D. Hegde and A. Venkateswara Rao, *Appl. Surf. Sci.*, 2006, **253**, 1566.
61. W. Hertl and M. L. Hair, *J. Phys. Chem.*, 1969, **76**, 3.
62. P. Silberzan, L. Léger, D. Ausserré and J. J. Benattar, *Langmuir*, 1991, **7**, 1647.
63. K. J. Stebe, E. Lewandowski and M. Ghosh, *Science*, 2009, **325**, 159.
64. M. G. Basavaraj, G. G. Fuller, J. Fransaer and J. Vermant, *Langmuir*, 2006, **22**, 6605.
65. T. Schilling, S. Jungblut and M. Miller, *Phys. Rev. Lett.*, 2007, **98**, 108303.
66. B. Madivala, S. Vandebril, J. Fransaer and J. Vermant, *Soft Matter*, 2009, **5**, 1717.
67. M. Cavallaro, L. Botto, E. P. Lewandowski, M. Wang and K. J. Stebe, *Proc. Natl. Acad. Sci. U. S. A.*, 2011, **108**, 20923.
68. T.-L. Cheng and Y. U. Wang, *J. Colloid Interface Sci.*, 2013, **402**, 267.
69. F. Günther, F. Janoschek, S. Frijters and J. Harting, *Comput. Fluids*, 2013, **80**, 184.
70. H. Kim, A. A. Abdala and C. W. Macosko, *Macromolecules*, 2010, **43**, 6515.
71. Y. Zhu, S. Murali, W. Cai, X. Li, J. W. Suk, J. R. Potts and R. S. Ruoff, *Adv. Mater.*, 2010, **22**, 3906.
72. X. Zhang, H. Sun and S. Yang, *J. Phys. Chem. C*, 2012, **116**, 19018.
73. L. J. Cote, F. Kim and J. Huang, *J. Am. Chem. Soc.*, 2009, **131**, 1043.
74. R. Aveyard, J. H. Clint, D. Nees and V. N. Paunov, *Langmuir*, 2000, **16**, 1969.
75. K. D. Danov, P. A. Kralchevsky and S. D. Stoyanov, *Langmuir*, 2010, **26**, 143.
76. L. Imperiali, C. Clasen, J. Fransaer, C. W. Macosko and J. Vermant, *Mater. Horiz.*, 2014, **1**, 139–145.
77. K. Hwang, P. Singh and N. Aubry, *Electrophoresis*, 2010, **31**, 850.
78. P. Dommersnes, Z. Rozynek, A. Mikkelsen, R. Castberg, K. Kjerstad, K. Hersvik and J. Otto Fossum, *Nat. Commun.*, 2013, **4**, 1.
79. E. Kim, K. Stratford and M. E. Cates, *Langmuir*, 2010, **26**, 7928.
80. J. W. Tavacoli, J. H. J. Thijssen and P. S. Clegg, *Soft Matter*, 2011, **7**, 7969.
81. J. W. Tavacoli, G. Katgert, E. G. Kim, M. E. Cates and P. S. Clegg, *Phys. Rev. Lett.*, 2012, **108**, 268306.
82. H. Jinnai, T. Koga, Y. Nishikawa, T. Hashimoto and S. T. Hyde, *Phys. Rev. Lett.*, 1997, **78**, 2248.
83. H. Jinnai, Y. Nishikawa and T. Hashimoto, *Phys. Rev. E*, 1999, **59**, R2554.
84. Y. Nishikawa, T. Koga, T. Hashimoto and H. Jinnai, *Langmuir*, 2001, **17**, 3254.
85. H. Jinnai, Y. Nishikawa, H. Morimoto, T. Koga and T. Hashimoto, *Langmuir*, 2000, **16**, 4380.
86. A. P. Roberts and M. A. Knackstedt, *Phys. Rev. E*, 1996, **54**, 2313.
87. A. P. Roberts and E. J. Garboczi, *J. Mech. Phys. Solids*, 2002, **50**, 33.
88. A. C. Kapfer, S. T. Hyde, K. Mecke, C. H. Arns and G. E. Schröder-Turk, *Biomaterials*, 2011, **32**, 6875.

89. J. W. Cahn, *J. Chem. Phys.*, 1965, **42**, 93.
90. N. F. Berk, *Phys. Rev. Lett.*, 1987, **58**, 2718.
91. Y. Ding, M. W. Chen and J. Erlebacher, *J. Am. Chem. Soc.*, 2004, **126**, 6876.
92. B. H. Jones and T. P. Lodge, *J. Am. Chem. Soc.*, 2009, **131**, 1676.
93. H.-Y. Chen, Y. Kwon and K. Thornton, *Scripta Materialia*, 2009, **61**, 52.
94. S. Torquato, S. Hyun and A. Donev, *Phys. Rev. Lett.*, 2002, **89**, 266601.
95. M. Lai, A. N. Kulak, D. Law, Z. Zhang, F. C. Meldrum and D. J. Riley, *Chem. Commun.*, 2007, **34**, 3547.
96. L. Wang, J. Lau, E. L. Thomas and M. C. Boyce, *Adv. Mater.*, 2011, **23**, 1524.
97. B. G. Soares, F. Gubbels, R. Jerome, P. Teyssie, E. Vanlathem and R. Deltour, *Polym. Bull.*, 1995, **35**, 223.
98. F. Gubbels, R. Jerome, E. Vanlathem, R. Deltour, S. Blacher and F. Brouers, *Chem. Mater.*, 1998, **10**, 1227.
99. T. P. Kraehenbuehl, R. Langer and L. S. Ferreira, *Nat. Methods*, 2011, **8**, 731.
100. G. S. Chai, S. B. Yoon, J. S. Yu, J. H. Choi and Y. E. Sung, *J. Phys. Chem. B*, 2004, **108**, 7074.
101. S. Sepehri, Y. Liu and G. Cao, *Adv. Mater. Res. (Durnten-Zurich, Switz.)*, 2010, **132**, 1.
102. N. Boon and R. van Roij, *Mol. Phys.*, 2011, **109**, 1229.
103. Z. Y. Yuan and B. L. Su, *J. Mater. Chem.*, 2006, **16**, 663.
104. Y. B. Wang, G. A. Sotzing and R. A. Weiss, *Chem. Mater.*, 2003, **15**, 375.
105. J. H. Holtz and S. A. Asher, *Nature*, 1997, **389**, 829.
106. H. N. Yow and A. F. Routh, *Soft Matter*, 2006, **2**, 940.
107. B. Brugger, B. A. Rosen and W. Richtering, *Langmuir*, 2008, **24**, 12202.
108. H. Firoozmand, B. S. Murray and E. Dickinson, *Langmuir*, 2009, **25**, 1300.
109. E. Dickinson, *Curr. Opin. Colloid Interface Sci.*, 2010, **15**, 40.
110. T. Hanazawa and B. S. Murray, *Langmuir*, 2012, **29**, 9841.
111. T. Hanazawa and B. S. Murray, *Food Hydrocolloids*, 2013, **34**, 128.
112. T. van Vliet, G. A. van Aken, H. H. J. de Jongh and R. J. Hamer, *Adv. Colloid Interface Sci.*, 2009, **150**, 27.
113. D. G. Shchukin, E. A. Ustinovich, G. B. Sukhorukov, H. Mohwald and D. V. Sviridov, *Adv. Mater.*, 2005, **17**, 468.
114. A. Goyal, C. K. Hall and O. D. Velev, *Soft Matter*, 2010, **6**, 480.
115. F. Varrato, L. Di Michele, M. Belushkin, N. Dorsaz, S. H. Nathan, E. Eiser and G. Foffi, *Proc. Natl. Acad. Sci. USA*, 2012, **109**, 19155.
116. L. Di Michele, F. Varrato, J. K. Otar, S. H. Nathan, G. Foffi and E. Eiser, *Nat. Commun.*, 2013, **4**, 2007.
117. B. J. Alder and T. E. Wainwright, *J. Chem. Phys.*, 1959, **31**, 459.
118. L. Di Michele and E. Eiser, *Phys. Chem. Chem. Phys.*, 2013, **15**, 3115.

CHAPTER 7
Complex Pickering Emulsions

YOSHIMUNE NONOMURA

Department of Biochemical Engineering, Graduate School of Science and Engineering, Yamagata University, Japan
Email: nonoy@yz.yamagata-u.ac.jp

7.1 Introduction

Solid particles exhibiting suitable wettability are adsorbed at liquid–liquid or air–liquid interfaces, and they stabilize emulsions and foams.[1–3] This phenomenon has been extensively studied and followed closely by many researchers, because some materials including colloidosomes, dry water and particle films were prepared by particles adsorbed at interfaces.[4–7] Many of these materials were simple spherical emulsions because surface tension gives liquid droplets a spherical shape, minimizing the surface area for a given volume.[8] On the other hand, supracolloidal systems with complex structures have been produced to create new materials such as network structures,[9,10] bicontinuous structures,[11,12] squeezing air bubbles,[13] and anisotropic structures formed by the arrested coalescence of the droplets.[14]

Such particle behaviours have been discussed on the basis of energy changes with adsorption at interfaces.[15] Mbamala *et al.* showed that there is often a barrier that prevents spontaneous adsorption of particles when the particles have positive or negative charges.[16] However, once this barrier is overcome, the particles are trapped at the interface because of capillary forces. The dipole–dipole interactions organize the solid particles into a two-dimensional triangular lattice and the above-mentioned supracolloidal systems.[17]

RSC Soft Matter No. 3
Particle-Stabilized Emulsions and Colloids: Formation and Applications
Edited by To Ngai and Stefan A. F. Bon
© The Royal Society of Chemistry 2015
Published by the Royal Society of Chemistry, www.rsc.org

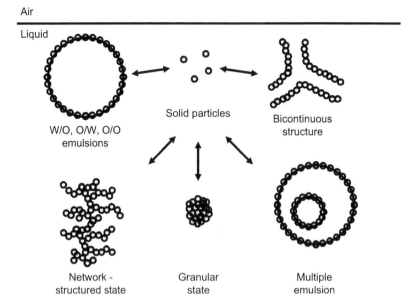

Figure 7.1   Conceptual scheme of complex Pickering emulsions.

Facile preparation methods are required for such complex structures in the pharmaceutical, food and cosmetics industries.[18] Here, I show some complex Pickering emulsions such as oil-in-oil emulsions (O/O emulsions), multiple emulsions, non-spherical emulsions, high internal phase emulsions and bicontinuous structures (Figure 7.1).

7.2   Emulsions and Related Systems Stabilized by Solid Particles

7.2.1   O/W Emulsions and W/O Emulsions

In ternary systems containing solid fine particles, water and oil, hydrophilic particles are dispersed in water, whereas lipophilic particles are dispersed in oil. In reality, some amphiphilic solid particles, such as hydrophobic silica, talc and carbon black, are adsorbed at a liquid–liquid interface and form stable emulsions. Figures 7.2 and 7.3 show the state diagram and microscopic images of ternary systems consisting of hydrophobic mica, water and silicone oil.[19] The water-in-oil emulsion (W/O emulsion), in which spherical water droplets were dispersed in silicone oil, was obtained when water was added to a hydrophobic mica–silicone oil dispersion. The emulsion state was maintained for more than 30 days; however, flocculation of water droplets and separation of the excess oil phase were observed in the emulsion system just after it was prepared. These spherical emulsions are stabilized by numerous other solid particles, such as polymer latexes, metal oxide particles, clay minerals and biological materials.[20,21]

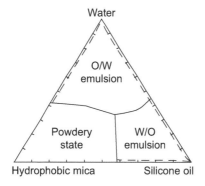

Figure 7.2 A mixed state diagram of hydrophobic mica/silicone oil/water mixtures.

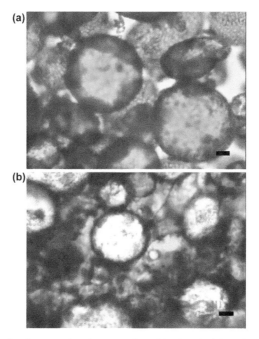

Figure 7.3 Optical microscopic photographs of hydrophobic mica/silicone oil/water mixtures: (a) W/O emulsion; (b) O/W emulsion. Scale bar: 50 μm.

The addition of water induces a change from a W/O emulsion to an oil-in-water emulsion (O/W emulsion) when the amount of the oil phase exceeds the limiting amount of oil absorbed by the solid particles. This type of phase inversion has been reported for some emulsion systems stabilized by surfactants or solid particles as follows.[22,23] Similar phase inversion has been induced by the change in the wettability of solid particles: the W/O emulsions stabilized by hydrophobic silica changed to O/W emulsions when hydrophilic silica was added.

7.2.2   O/O Emulsions

Pickering emulsions are not limited to O/W and W/O emulsions consisting of solid particles, water and oil. We found some O/O emulsions stabilized by solid particles.[9] Figures 7.4 and 7.5 show a mixed-state diagram and some microscopic images of the fluorinated silicone resin particles–fluorinated oil–silicone oil ternary system. The silicone resin particles were treated with a fluorinated agent, the diethanolamine salt of perfluoroalkyl phosphate. If the system contained a suitable amount of silicone oil, silicone oil droplets covered with solid particles were dispersed in the fluorinated oil phase. However, when the silicone oil concentration was insufficient, the mixture was in the network-structured state. In this state, solid particles formed a network structure in the fluorinated oil. When the silicone oil concentration was greater than that in the O/O emulsion state, the mixture was in the separation state consisting of the O/O emulsion and two oil phases. The silicone oil phase separated by coalescence, whereas the fluorinated oil phase separated by creaming. When the ternary system contained more solid particles, the mixture formed a granular state in which the solid particles were granulated in the silicone oil. Fluorinated oil was absorbed by solid particles and acted as a binding material.

We also found the previously described characteristic states, including O/O emulsions, a network-structured state and a granular state, in the ternary system that contained fluorinated plate-shaped organic crystalline particles, fluorinated oil and silicone oil.[10] Particle shape is one of the important factors that changes adsorption behaviour at liquid–liquid interfaces.[24–27] The plate-shaped particles were wettable with both fluorinated oil and silicone oil because their surfaces were treated with the diethanolamine salt of perfluoroalkyl phosphate. The separated state, in which a transparent silicone oil phase was separated from a milky white phase containing the solid particles and fluorinated oil, was observed.

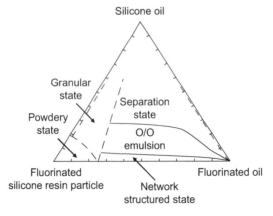

Figure 7.4   A mixed state diagram of fluorinated silicone resin particles/silicone oil/fluorinated oil mixtures.

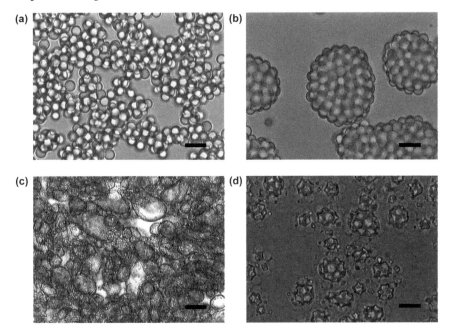

Figure 7.5 Optical microscopic photographs of fluorinated silicone resin particles/ silicone oil/fluorinated oil mixtures: (a) network structured state; (b) O/O emulsion; (c) separation state; (d) granular state. Scale bar: 10 μm (a, b, d) and 40 μm (c).

The shear-stress curve and contact angle measurements, as well as theoretical analysis, indicated that these interesting observations originated from the plate-type shape of the organic crystalline particles and their characteristic network in the fluorinated oil.

7.2.3 Non-Spherical Emulsions

As previously mentioned, most of the fluid droplets in emulsions and foams are simple spheres because surface tension gives liquid droplets a spherical shape.[8] However, some non-spherical Pickering emulsions and foams have recently been reported. In 2005, researchers achieved fusion of the solid-stabilized bubbles by squeezing these bubbles between two glass plates.[13] The fused armoured bubble maintained a stable ellipsoidal shape even after the side plates were removed; they were unable to relax to a spherical shape because of the jamming of the particles on the closed interface. The armoured bubbles could be remodelled into various stable anisotropic shapes because the interfacial composite material is able to undergo extensive particle-scale rearrangements to accommodate external inhomogeneous stresses. These shape changes occur with apparently no hysteresis and at relatively low forces, which is equivalent to perfect plasticity in continuum mechanics.

Bon *et al.* reported a route for the production of stable non-spherical emulsion droplets by pushing millimetre-sized liquid droplets stabilized by an excess amount of solid particles through a narrow capillary.[28] The excess amount of solid particles allows for full coverage of the newly created droplet interface during deformation. Schmitt-Rozières found a transition in the shape of emulsion droplets from spherical to polymorphous.[29] In emulsions stabilized by hexadecyltrimethylammonium bromide and silica nanoparticles, a transition occurs in their geometry depending on the concentration of the stabilizer. Pawar *et al.* studied the coalescence of structured droplets containing a network of anisotropic colloids, whose internal elasticity provides a resistance to full shape relaxation and interfacial energy minimization during coalescence.[30]

7.2.4 Pickering High Internal Phase Emulsions

High internal phase emulsions, which are characterized by a minimum internal phase volume ratio of 0.74, are important for a wide range of applications in the food, cosmetics, pharmaceutical and petroleum industries. They can be used as templates for the synthesis of highly porous polymers with potential applications as low-weight structures or scaffolds in tissue engineering. Bismarck *et al.* and Bon *et al.* reported the stabilization of Pickering W/O emulsions with high internal phases using silica nanoparticles or microgel particles.[31–33] Pickering high internal phase emulsions have been used to prepare some novel materials, such as emulsions and porous foams containing titania nanoparticles[34] and poly(tetrafluoroethylene).[35]

7.2.5 Emulsions with Wrinkled Interfaces

When emulsion-type formulations are prepared for external medications or cosmetics, they should be designed with consideration of their collapse processes because of external forces or drying. In a previous study, we focused on the collapse of Pickering emulsions in drying processes.[36] Spherical silicone resin particles were adsorbed at the oil–water interfaces and stabilized W/O emulsions, which had an average diameter of several tens of micrometres. These emulsions maintained their dispersed states in a glass tube for more than a month. However, the droplet shape drastically changed when the emulsion was deposited onto a glass slide; a number of wrinkles occurred and expanded gradually, as shown in Figure 7.6. Several tens of minutes after the emulsion was deposited onto the glass slide, the droplet surfaces wrinkled similar to a dry raisin. Interestingly, the emulsion droplets suddenly disappeared, and the red marks of pigment in the water phase remained where the emulsion droplets previously existed. Therefore, even if the surface area of the emulsion droplets decreases after drying, the solid particles are adsorbed at oil–water interfaces and some wrinkles are formed at the interfaces to maintain the interfacial area. These wrinkles form

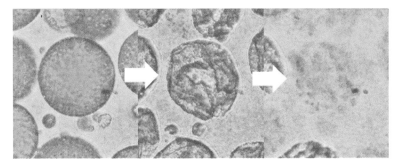

Figure 7.6 Microscopic images of W/O emulsions of spherical silicone resin particle *n*-dodecane/water mixture. The images are water droplets at 0 min (left), 10 min (middle) and 40 min (right) after application. Scale bar: 50 μm.

because the solid particles are not desorbed from the oil–water interfaces as the size of the particles is several tens of thousands times larger than that of normal surfactant molecules.[37] Similar deformation phenomena have been reported in some Pickeirng systems.[38–43]

7.2.6 Multiple Emulsions

Stable multiple emulsions have been prepared by two-step emulsification methods.[44–46] In the case of the preparation of water-in-oil-in-water (W/O/W) emulsions, fine water droplets are dispersed in an oil phase by a low-hydrophilic lipophilic balance (low-HLB) surfactant. In the subsequent process, the obtained W/O emulsion is emulsified in a water phase by a high-HLB surfactant. We discovered that microbowls, which are silicone particles with holes on their particle surfaces, were adsorbed at liquid–liquid interfaces and exhibited anomalous emulsification behaviours.[47] For example, multiple emulsions, W/O/W and oil-in-water-in-oil (O/W/O) emulsions, were formed by simply mixing microbowls with water and oil. The W/O/W emulsions are dispersions of oil droplets containing fine water droplets, whereas O/W/O emulsions are emulsions of water droplets containing fine oil droplets (Figures 7.7 and 7.8).

Why do microbowls form multiple emulsions only by mixing? In solid particles–oil–water ternary systems, multiple emulsions were formed when the particles showed sufficient contact-angle hysteresis as well as intermediate hydrophobicity. In general, the contact angle of solid particles at an oil–water interface is unique and determined by Young's law. However, at the defects of the solid surfaces, the contact angle is not unique. We postulated that microbowls demonstrated large contact angle hysteresis because of their distinctive particle shape. In addition, we hypothesized that the microbowls performed a double role as both a high- and low-HLB surfactant because of the contact angle hysteresis while stabilizing both the outer and inner drop surfaces of multiple emulsions. The same particles exhibited a different contact angle with the oil–water interface because the

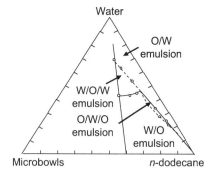

Figure 7.7 A mixed state diagram of microbowl/*n*-dodecane/water mixtures.

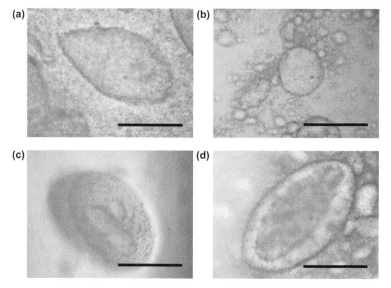

Figure 7.8 Optical microscopic photographs of microbowl/*n*-dodecane/water mixtures: (a) W/O emulsion; (b) O/W emulsion; (c) O/W/O emulsion; (d) W/O/W emulsion. Scale bar: 100 μm.

wettability of solid particles changes drastically with liquid in the holes; they are hydrophilic when the holes are suffused by water and lipophilic when they are suffused by oil or air.

7.2.7   Bicontinuous Structures

Bicontinuous interfacially jammed emulsion gels (bijels) are soft materials in which interpenetrating, continuous domains of two immiscible fluids are maintained in a rigid state by a jammed layer of colloidal particles at their interface. They are of interest to many physical and material scientists because bijels are three-dimensional solid gels with a finite elastic modulus, a yield stress and non-linear mechanical properties unlike those of the common Pickering emulsions.[48] This state was proposed in 2005 as a hypothetical new class of soft materials. Stratford *et al.* reported large-scale computer simulations of the demixing of a binary solvent containing solid particles.[11] The jammed colloidal layer appeared to enter a glassy state, creating a multiply connected, solid-like film in three dimensions. The creation of bijels in the laboratory is difficult because fast quench rates are necessary to bypass nucleation and even a small degree of unequal wettability of the particles by the two liquids can lead to the formation of lumpy interfacial layers and, therefore, irreproducible material properties. Some years after the computer simulations of Stratford *et al.*, a bijel was prepared in a real ternary system containing solid particles, oil and water: Herzig *et al.* reported a reproducible protocol for the preparation of three-dimensional samples of bijels consisting of fluorescent silica colloids, water and 2,6-lutidine.[12] Recently, bijels were applied to some functional materials, such as soft-matter templates for materials synthesis and hierarchically porous silver monoliths.[49,50]

7.3 Conclusions

Here we have discussed some complex Pickering emulsions such as O/O emulsions, multiple emulsions, non-spherical emulsions, high internal phase emulsions and bicontinuous structures. Solid particles can be adsorbed at liquid–liquid interfaces and stabilize complex structures as well as surfactant molecules. These complex materials have been applied in food, cosmetic and medical products.

References

1. W. Ramsden, *Proc. R. Soc.*, 1903, **72**, 156.
2. S. U. Pickering, *J. Chem. Soc.*, 1907, **91**, 2001.
3. S. I. Kam and W. R. Rossen, *J. Colloid Interface Sci.*, 1999, **213**, 329.
4. A. D. Dinsmore, M. F. Hsu, M. G. Nikolaides, M. Marquez, A. R. Bausch and D. A. Weitz, *Science*, 2002, **298**, 1006.
5. P. Noble, O. J. Cayre, R. G. Alargova, O. D. Velev and V. N. Paunov, *J. Am. Chem. Soc.*, 2004, **126**, 8092.
6. P. Aussilous and D. Quéré, *Nature*, 2001, **411**, 924.
7. B. P. Binks and R. Murakami, *Nat. Mater.*, 2006, **5**, 865.
8. P.-G. de Gennes, F. Brochard-Wyart and D. Quèrè, *Capillarity and Wetting Phenomena: Drops, Bubbles, Pearls, Waves*, Springer, Berlin, 2003, p. 9.
9. Y. Nonomura, T. Sugawara, A. Kashimoto, K. Fukuda, H. Hotta and K. Tsujii, *Langmuir*, 2002, **18**, 10163.
10. Y. Nonomura, K. Fukuda, S. Komura and K. Tsujii, *Langmuir*, 2003, **19**, 10152.
11. K. Stratford, R. Adhikari, I. Pagonabarraga, J.-C. Desplat and M. E. Cates, *Science*, 2005, **309**, 2198.

12. E. M. Herzig, K. A. White, A. B. Schofield, W. C. K. Poon and P. S. Clegg, *Nat. Mater.*, 2007, **6**, 966.

13. A. B. Subramaniam, M. Abkarian, L. Mahadevan and H. A. Stone, *Nature*, 2005, **438**, 930.

14. A. R. Studart, H. C. Shum and D. A. Weitz, *J. Phys. Chem. B*, 2009, **113**, 3914.

15. S. Levine, B. D. Bowen and S. J. Partridge, *Colloids Surf.*, 1989, **38**, 325.

16. E. C. Mbamala and H. H. von Grunberg, *J. Phys.: Condens. Matter*, 2002, **14**, 4881.

17. P. Pieranski, *Phys. Rev. Lett.*, 1980, **45**, 569.

18. A. S. Utada, E. Lorenceau, D. R. Link, P. D. Kaplan, H. A. Stone and D. A. Weitz, *Science*, 2005, **308**, 537.

19. Y. Nonomura and N. Kobayashi, *J. Colloid Interface Sci.*, 2009, **330**, 463.

20. B. P. Binks, *Curr. Opinion Colloid Interface Sci.*, 2002, 7, 21.

21. E. Dickinson, *Curr. Opinion Colloid Interface Sci.*, 2010, **15**, 40.

22. B. P. Binks and S. O. Lumsdon, *Langmuir*, 2000, **16**, 2539.

23. B. P. Binks and S. O. Lumsdon, *Langmuir*, 2000, **16**, 3748.

24. Y. Nonomura, S. Komura and K. Tsujii, *Langmuir*, 2004, **20**, 11821.

25. Y. Nonomura, S. Komura and K. Tsujii, *J. Oleo Sci*, 2004, **53**, 607.

26. Y. Nonomura, S. Komura and K. Tsujii, *J. Phys. Chem. B*, 2006, **110**, 13124.

27. Y. Nonomura and S. Komura, *J. Colloid Interface Sci.*, 2008, **317**, 501.

28. S. A. F. Bon, S. D. Mookhoek, P. J. Colver, H. R. Fischer and S. van der Zwaag, *Eur. Polymer J.*, 2007, **43**, 4839.

29. M. Schmitt-Rozières, J. Krägel, D. O. Grigoriev, L. Liggieri, R. Miller, S. Vincent-Bonnieu and M. Antoni, *Langmuir*, 2009, **25**, 4266.

30. A. B. Pawar, M. Caggioni, R. W. Hartel and P. T. Spicer, *Faraday Discuss.*, 2012, **158**, 341.

31. A. Menner and A. Bismarck, *Macromol. Symp.*, 2006, **242**, 19.

32. V. O. Ikem, A. Menner and A. Bismarck, *Angew. Chem., Int. Ed.*, 2008, **47**, 8277.

33. P. J. Colver and S. A. F. Bon, *Chem. Mater.*, 2007, **19**, 1537.

34. A. Menner, V. Ikem, M. Salgueiro, M. S. P. Shaffer and A. Bismarck, *Chem. Commun.*, 2007, **19**, 4274.

35. J. C. H. Wong, E. Tervoort, S. Busato, U. T. Gonzenbach, A. R. Studart, P. Ermanni and L. J. Gauckler, *J. Mater. Chem.*, 2010, **20**, 5628.

36. N. Nakagawa and Y. Nonomura, *Chem. Lett.*, 2011, **40**, 818.

37. B. P. Binks and T. S. Horozov (eds) *Colloidal Particles at Liquid Interfaces*, Cambridge University Press, Cambridge, 2006.

38. S. O. Asekomhe, R. Chiang, J. H. Masliyah and J. A. W. Elliott, *Ind. Eng. Chem. Res.*, 2005, **44**, 1241.

39. H. Xu, S. Melle, K. Golemanov and G. Fuller, *Langmuir*, 2005, **21**, 10016.

40. M. Abkarian, A. B. Subramaniam, S.-H. Kim, R. J. Larsen, S.-M. Yang and H. A. Stone, *Phys. Rev. Lett.*, 2007, **99**, 188301.

41. S. Fujii, A. Aichi, M. Muraoka, N. Kishimoto, K. Iwahori, Y. Nakamura and I. Yamashita, *J. Colloid Interface Sci.*, 2009, **338**, 222.

42. S. S. Datta, H. C. Shum and D. A. Weitz, *Langmuir*, 2010, **26**, 18612.
43. M. Dandan and H. Y. Erbil, *Langmuir*, 2009, **25**, 8362.
44. N. Garti, A. Aserin, I. Tiunova and H. Binyamin, *J. Am. Oil Chem. Soc.*, 1999, **76**, 383.
45. T. Sekine, K. Yoshida, F. Matsuzaki, T. Yanaki and M. Yamaguchi, *J. Surf. Deterg.*, 1999, **2**, 309.
46. S. Arditty, V. Schmitt, J. Giermanska-Kahn and F. Leal-Calderon, *J. Colloid Interface Sci.*, 2004, **275**, 659.
47. Y. Nonomura, N. Kobayashi and N. Nakagawa, *Langmuir*, 2011, **27**, 4557.
48. M. E. Cates and P. S. Clegg, *Soft Matter*, 2008, **4**, 2132.
49. J. A. Witt, D. R. Mumma and A. Mohraz, *Soft Matter*, 2013, **9**, 6773.
50. M. N. Lee and A. Mohraz, *J. Am. Chem. Soc.*, 2011, **133**, 6945.

CHAPTER 8

Multiple Pickering Emulsions for Functional Materials

YU YANG,[†] YIN NING,[†] ZHEN TONG AND CHAOYANG WANG*

Research Institute of Materials Science, South China University of Technology, Guangzhou 510641, China
*Email: zhywang@scut.edu.cn

8.1 Introduction

Conventional double emulsions, which are emulsions with smaller droplets of a third fluid within the larger drops, are generally stabilized by surfactants. Seifriz *et al.* first presented the double emulsions in 1925,[1] but detailed studies on multiple emulsions were reported only in the last two decades. It is well known that the main drawbacks associated with traditional emulsions lie in low stability against demulsification, toxicity brought from surfactants and the tough post-processing. However, double emulsions still play critical roles in many areas, such as controlled release,[2–4] separation,[5] and encapsulation of nutrients and flavours for food additives.[6–8] Hence, alternative multiple emulsions that can overcome the shortcomings but retain the advantages of conventional multiple emulsions are urgently needed.

Multiple Pickering emulsions are desired candidates who can conquer the disadvantages of traditional multiple emulsions. Single Pickering emulsions are intensely studied and well-established systems which are stabilized by solid particles instead of surfactants, offering more stability compared with

[†]These authors are thought to have equal contributions.

RSC Soft Matter No. 3
Particle-Stabilized Emulsions and Colloids: Formation and Applications
Edited by To Ngai and Stefan A. F. Bon
Published by the Royal Society of Chemistry, www.rsc.org

surfactant-stabilized emulsions.[9–27] This is because the attachment energy of solid particles onto oil–water interfaces is much higher than that of surfactants, making the solid particles strongly adsorb in the liquid–liquid interface, giving rise to extremely stable emulsion droplets.[28–43] Moreover, double Pickering emulsions provide a convenient approach to fabricate organic/inorganic composite materials by simply using functional inorganic particles to stable emulsions, endowing the products with specific performances like magnetic, optical and catalytic properties. Even though multiple Pickering emulsions are much like conventional multiple emulsions in term of multiple phase structure, the research on multiple Pickering emulsions is just in the initial stages. Here, multiple Pickering emulsions will be systematically described by focusing on their preparation and applications in the following sections.

8.2 Preparation of Multiple Pickering Emulsions

In general, the solid double Pickering emulsion stabilizers are not much different to single Pickering emulsion stabilizers. To be specific, multiple Pickering emulsions are stabilized by two kinds of solid particles with opposite wettability, of which one is hydrophilic for the O/W interface and the other is hydrophobic for the W/O interface. So far as we know, there are three preparation methods for multiple Pickering emulsions using solid particles as stabilizers.

8.2.1 One-Step Method

Figure 8.1 (top row) shows the one-step method for the preparation of double Pickering emulsions. First, two kinds of solid particles with different amphipathicity are separately dispersed in two phases; then an ultrasonic is employed to facilitate the arrangement of solid particles onto liquid–liquid interfaces. Double Pickering emulsions can be obtained under proper operation. This method is time-saving, but it involves more complex parameters, leading to instability of the resulting double emulsion droplets because ultrasonic not only enables adsorption of solid particles onto the interface, but also destabilizes the double Pickering emulsions formed. Many factors, including proper solid particle types, oil to water volume ratio, emulsification time, *etc.* should be taken into consideration to obtain a stable double Pickering emulsion by this method. Hanson *et al.*[44] were the first to show that W/O/W multiple emulsions could be prepared in a direct process using single-component, synthetic amphiphilic diblock copolypeptide surfactants with polydimethylsiloxane as oil phase. However, these reported multiple emulsions need surfactant blends and are usually described as transitory or temporary systems. Hong *et al.*[45] presented a one-step phase inversion process to produce water-in-oil-in-water (W/O/W) multiple emulsions stabilized solely by a synthetic diblock copolymer (Figure 8.2). Unlike the use of small molecule surfactant combinations, block copolymer

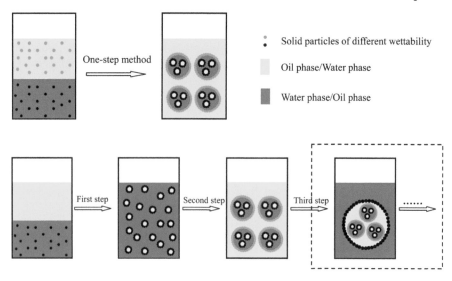

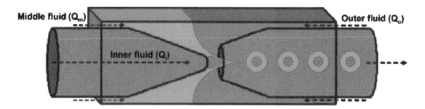

Figure 8.1 Schematic representation of one-step method (top row), two-step method (middle row) and microcapillary device (bottle row) for double emulsion generation.
Taken from reference 72 with permission of Wiley-VCH.

stabilized multiple emulsions are remarkably stable and show the ability to separately encapsulate both polar and non-polar cargos. Very recently, we had successfully prepared W/O/W double Pickering emulsions solely stabilized by solid particles using one-step methods (Figure 8.3). Firstly, PLGA was dissolved in CH_2Cl_2, and hydrophilic SiO_2 nanoparticles were dispersed in water. Then the oil and water phases were mixed with a volume ratio of 1 to 10, forming a water-in-CH_2Cl_2-in-water double Pickering emulsion in one step. The mechanism and the factors of this novel one-step fabrication system are under-researched.

8.2.2   Two-Step Method

The two-step method is the most common approach to preparing double Pickering emulsions.[46–53] The first stage involves the preparation of a single emulsion (W/O or O/W). Commonly, solid particles are pre-dispersed in the

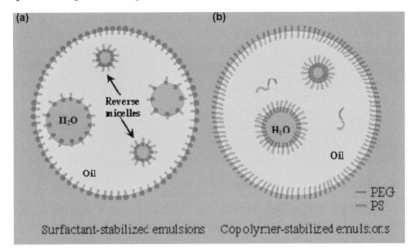

Figure 8.2 Schematic illustration of the influence of surfactants and block co-polymers on emulsion stability control: L (a) W/O/W multiple emulsions stabilized by small molecule surfactants and (b) W/O/W emulsion stabilized by PEG45-*b*-PS66 block copolymers.
Taken from reference 45 with the permission of the American Chemical Society.

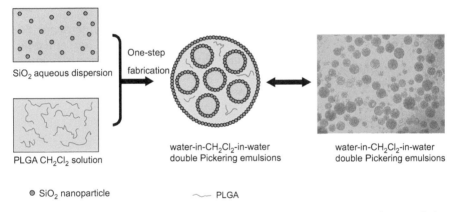

Figure 8.3 Schematic illustration of one-step fabrication of water-in-CH₂Cl₂-in-water double Pickering emulsions.

continuous phase; single Pickering emulsions are then generated by addition of disperse phase. In the second step, the as-prepared single Pickering emulsions are further emulsified into the outmost phase containing solid particles by homogenizer or hand shaking. It is worth noting that here ultrasonic should be avoided because in many cases it will destroy the system stability. Similarly, to get triple Pickering emulsions, another re-emulsification of double Pickering emulsion droplets into the fourth phase is needed, as shown in the dotted rectangle of Figure 8.1 (middle row).

Compared with the one-step method, the two-step method is a much more feasible way of producing multiple Pickering emulsions because the sizes of the inner and outer droplets can be easily tuned in the first stage. At the same time, two or more kinds of solid particles are introduced into the system, and these functional particles can give different specific properties to the double Pickering emulsion droplets, such as magnetic, optical and catalytic performance. More importantly, the stability is remarkably improved compared with the one-step fabrication of double Pickering emulsions.

8.2.3 Microcapillary Device

A microcapillary device (see Figure 8.1 (middle row)), also known as the microfluidic technique, plays a significant role in multiple emulsion preparation because it can precisely control the monodispersity and the number of inner droplets.[54–71] In 2008, Weitz's group presented a monodisperse water-in-oil-in-water (W/O/W) double Pickering emulsion produced by using glass capillary microfluidic devices.[72] The reported microfluidic device combines a flow focusing and co-flowing geometry, which results in hydrodynamic flow focusing on three different fluid streams at the orifice of the collection tube and leads to the formation of double emulsion droplets. The double Pickering emulsions are stabilized by hydrophobic silica nanoparticles, which are pre-dispersed in the middle oil phase.

In 2012, Russell's group[73] exploited the simultaneous interfacial adsorption of independently functionalized semiconductor and metallic nanoparticles to give stable W/O/W and O/W/O double-emulsion structures based on a glass microcapillary fluidic device. To be specific, they used tetra(ethylene glycol) (TEG) functionalized Au NPs that were known to stabilize O/W droplets, and CdSe QDs that were known to stabilize W/O droplets when the QDs were functionalized with their native alkyl ligands (Figure 8.4). Then, a glass capillary device combining both co-flow and flow focusing was used to form W/O/W double-emulsion droplets, as shown in Figure 8.5a. The innermost aqueous phase was deionized water, while the middle oil phase consisted of 2 mg mL^{-1} CdSe QDs in TCB (with 1 vol% Coumarin 153 added to facilitate characterization by fluorescence microscopy), and the outer phase consisted of 5 mg mL^{-1} Au NPs in water. The size of the outer droplets is controlled by the size of the exit capillary orifice, with larger openings yielding larger droplets, and the relative flow rates and viscosities of the outermost and middle fluid phases. The number and sizes of internal droplets are determined by similar considerations of the inlet capillary nozzle and the relative rates and viscosities of the middle and inner phases. Typical resulting W/O/W emulsion droplets are shown in the optical and fluorescence images of Figure 8.5. In this experiment, the outer droplet diameter is seen to range from 17 to 40 μm, representing a substantial improvement in droplet uniformity relative to the double emulsions formed simply by shaking.[74–89] Moreover, the flow-focusing technique produced

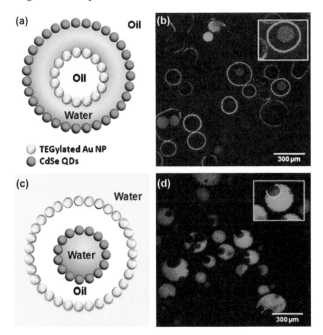

Figure 8.4 Schematic representations and fluorescence confocal microscopy images of NP-stabilized double emulsions: (a, b) O/W/O double-emulsion droplets, and (c, d) W/O/W double-emulsion droplets. The red fluorescence is from the CdSe QDs, and the green fluorescence is from Coumarin 153. Taken from reference 73 with the permission of Wiley-VCH.

samples in which nearly every oil droplet contained a single internal water droplet.

It is obvious that this method can offer easy control over the size distribution of emulsion droplets, but unfortunately it has an inherent drawback in that its output is limited, which inevitably precludes its wide application. More efforts should be devoted to increasing the production rate based on this excellent device.

8.3 Multiple Pickering Emulsions as Templates for Functional Materials

8.3.1 Multihollow Nanocomposite Microspheres

It is obvious that multiple Pickering emulsions can act as a facile template for the preparation of multihollow nanocomposite microspheres because they are stabilized by solid nanoparticles and these nanoparticles will be retained in the inner and outer surface of the resultant microspheres; multihollow structures will be feasibly obtained after the removal of the inner phase.

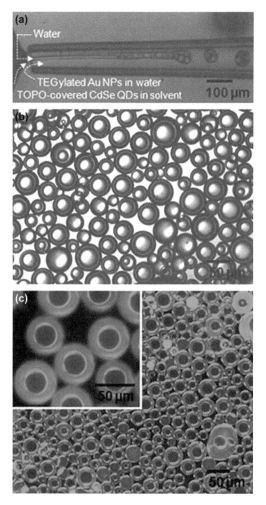

Figure 8.5   (a) Formation of NP/QD-stabilized double emulsions in a microcapillary
device with a flow-focusing geometry. (b) Optical and (c) fluorescence
microscopy images of W/O/W double emulsions stabilized by Au NPs
and CdSe QDs.
Taken from reference 73 with the permission of Wiley-VCH.

Ge's group[90] successfully fabricated hybrid hollow microspheres from
double emulsions stabilized by single-component amphiphilic silica nanoparticles, which were formed and partially modified *in situ* at the interface
of water-in-oil emulsions. Figure 8.6 shows the schematic procedure for
preparation of hybrid hollow microspheres using double Pickering emulsions as templates. The water-in-oil (W/O) emulsions were first prepared
using the mixture of styrene (St), tetraethoxysilane (TEOS), hexadecane
and g-(trimethoxy-silyl) propyl methacrylate (MPS) as the oil phase,
and aqueous triethylamine (TEA) solution as the inner water phase,

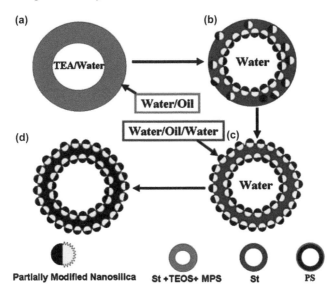

Figure 8.6 Schematic illustration of the preparation of hybrid hollow microspheres using double Pickering emulsions as templates. (a) W/O emulsions with the monomers/initiator as the oil phase and TEA/H$_2$O as the water phase are obtained by using a magnetic stirrer. (b) A silica-stabilized inverse Pickering emulsion is obtained after silica particles are formed under basic conditions. (c) W/O/W emulsions are formed by using a homogenizer to apply a shear force. (d) Polymerization is carried out for 12 h and hybrid hollow spheres are obtained.
Taken from reference 90 with the permission of The Royal Society of Chemistry.

respectively (Figure 8.6a). Silica particles were formed and modified by MPS at the O/W interface based on the hydrolysis–condensation of TEOS (Figure 8.6b). W/O/W emulsions were fabricated by adding water as the outer phase, and the partially modified silica nanoparticles stabilized both inner and outer droplets of the double emulsions (Figure 8.6c). Finally, hybrid hollow microspheres were obtained after the further polymerization of the interlayer phase (Figure 8.6d).

Figure 8.7 shows TEM images of hybrid hollow microspheres, demonstrating their hollow structures. By increasing the amount of TEOS, the content of silica particles in the medium increased accordingly, which provided more Pickering emulsifiers to stabilize the double emulsions and led to the formation of monomer droplets of smaller size. Therefore, when the TEOS amount increased from 4.0 g to 8.0 g, the average size of hollow microspheres decreased from 800 to 400 nm (Figure 8.7a and b). Moreover, unstable W/O/W emulsions were observed with less MPS in the formulation because silica particles had not been fully modified; 30 wt% or more MPS relative to TEOS was needed. With 15 wt% MPS relative to TEOS in the formulation, hollow microspheres thus obtained were

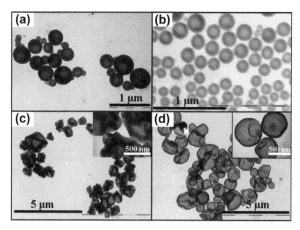

Figure 8.7   TEM images of hybrid hollow microspheres prepared with volume ratio
of oil/water phase = 10/15 and different ratios of St/TEOS/MPS: (a) St/
TEOS/MPS = 16/4/1.2, (b) St/TEOS/MPS = 12/8/2.4, (c) St/TEOS/MPS = 12/
8/1.2, (d) St/heptane/TEOS/MPS = 10/6/4/1.2.
Taken from reference 90 with the permission of The Royal Society of
Chemistry.

deformable (Figure 8.7c). When part of the polymerizable St was replaced
with volatile heptane in the formulation, the wall of the hollow microspheres
was thin and fragile under sonification (Figure 8.7d). Generally, it was shown
that the double Pickering emulsions which were prepared by using the *in situ*
formed and modified silica particles as Pickering emulsifiers were ideal
templates to fabricate inorganic/organic hollow spheres.

Moreover, Marda *et al.* have demonstrated that multihollow HAp/PLLA
nanocomposite microspheres can be fabricated by evaporating CH_2Cl_2 from
a water-in-(CH_2Cl_2 solution of poly(L-lactic acid)(PLLA))-in-water multiple
emulsion template, which is stabilized by hydroxyapatite nanoparticles and
PLLA molecules as wettability modifiers.[91] The interaction between PLLA
and HAp nanoparticles at the oil–water interfaces plays a crucial role in the
preparation of stable multiple emulsion and multihollow microspheres.
Figure 8.8a shows optical micrographs of the W/O/W emulsion stabilized
with SHAp and PLLA-RHAp treated at 200 °C. After evaporation of CH_2Cl_2,
the number-average diameter of the droplets decreased from 14.9 ± 12.8 μm
to 9.1 ± 8.8 μm (Figure 8.8b). The morphologies of microspheres were nearly
spherical, with some roughness on the surfaces (Figure 8.8c). Free SHAp
were not observed, indicating that almost all SHAp were adsorbed on the
microsphere surfaces. Thanks to two kinds of HAp nanoparticles with different morphologies, it is easy to observe where the nanoparticles locate in
the multihollow microspheres in electron microscopy studies. As shown in
Figure 8.8e, the microspheres had hollows inside the PLLA matrix. The
hollows were filled with epoxy resin. The TEM image also shows that RHAp
nanoparticles, which were not adsorbed at the oil–water interface, were
dispersed in the PLLA matrix. At the same time, the magnified images shown

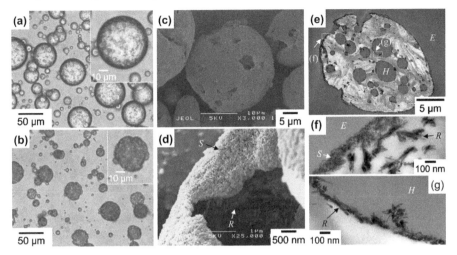

Figure 8.8 (a, b) Optical micrographs of the W/O/W emulsion stabilized with SHAp and PLLA-RHAp treated at 200 °C (a) before and (b) after evaporation of CH₂Cl₂. The insets in micrographs (a) and (b) show magnified images. (c, d) SEM images of the multihollow microspheres. Image (d) is a magnified image. (e, g) TEM images of ultrathin cross-sections of the microspheres. Images (f, g) show magnified images of the area shown in image (e). The symbols E, H, S and R indicate the exterior of the microsphere (epoxy resin), hollow inside the microsphere, SHAp and RHAp, respectively. Taken from reference 91 with the permission of the American Chemical Society.

in Figure 8.8f and g reveal that the outer surfaces of the microspheres were coated only with hydrophilic SHAp, and the inner domain surfaces were coated only with hydrophobic PLLA-RHAp. In the magnified SEM image shown in Figure 8.8d, the outer surfaces of the microspheres were uniformly coated with SHAp, and the surfaces of the dents were covered with RHAp, which indicates that the dents on the surface were derived from the water domains stabilized with PLLA-HAp. These results demonstrated that these novel organic/inorganic multihollow nanocomposite microspheres could be successfully prepared based on a water-in-(CH₂Cl₂ solution of PLLA)-in-water double Pickering emulsion template.

In our group, robust multihollow microspheres are successfully prepared using double Pickering emulsions as templates.[92,93] In these cases, two kinds of stabilizers are used. One is a hydrophilic nanoparticle (Fe₂O₃, Fe₃O₄, or clay) that can stabilize inner O/W interfaces while the other is a hydrophobic nanoparticle (hydrophobically modified silica, Fe₂O₃ or Fe₃O₄, *etc.*) for outer W/O interfaces. Multiple W/O/W Pickering emulsions with polymerizable monomers such as styrene and divinylbenzene as the oil phase or O/W/O Pickering emulsions with acrylamide (Am) in the middle aqueous phase are prepared in a similar two-step process. Subsequently, polymeric microspheres with dual nanocomposite multihollow structure or aqueous multi-cores are obtained after the polymerization of the middle phase.

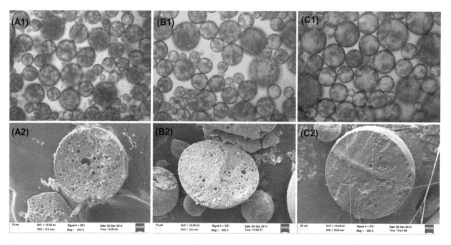

Figure 8.9 Optical photographs of W/O/W double Pickering emulsions (top row); SEM images of resultant polymeric multihollow microspheres (bottom row). W/O volume ratios of the primary emulsions for A, B and C are 1:4, 1:6 and 1:8, respectively.

More importantly, the pore content within the microsphere can be easily adjusted by changing the two-phase volume ratio of primary emulsion. The pore content will increase accordingly as the two-phase volume ratio ((W/O)/ W) increases (as shown in Figure 8.9). The pore size can also be controlled by varying the usage of single emulsion stabilizer and volume ratio of the primary emulsion. In general, regardless of the volume ratio of the primary emulsion and emulsification time, the more the usage is, the smaller the pore size gets.

From Figure 8.10A, we can clearly distinguish the water globules, each of them containing many internal oil droplets. These dual hybrid multihollow microgels are prepared *via* suspension polymerization of Am dissolved in the water globules, which is carried out in an ice-water bath under 30 mW cm^{-2} UV radiation for 5 min. Hybrid PAm microspheres are labelled with fluorescein isothiocyanate (FITC) for confocal laser scanning microscopy (CLSM) observation. The fluorescence image of microgels redispersed in water is presented in Figure 8.10B. The hybrid microgels have a good dispersibility in water and the PAm microgels with multiple compartment structure are clearly observed in Figure 8.10B.

As the above discussion shows, the multihollow nanocomposite microspheres fabricated based on double Pickering emulsions have a common disadvantage: the inner pores did not connect to each other. It is a common understanding that the inner pore structure was determined by the primary emulsion: a low inner phase fraction emulsion leads to a separated pore structure, but with the increase in the inner phase fraction of the primary emulsion, the space between the neighbouring pores becomes much smaller; if the primary emulsion is a high internal phase emulsion,

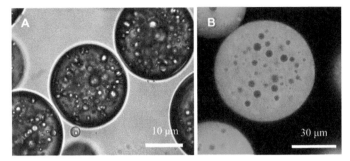

Figure 8.10 (A) Light microscopy image of multiple O/W/O emulsion stabilized by clay and silica nanoparticles and (B) fluorescence image of nanocomposite multihollow PAm microgels redispersed in water.
Taken from reference 92 with the permission of The Royal Society of Chemistry.

the space between the neighbouring pores can be ignored, resulting in multihollow nanocomposite microspheres with interconnected pore structure. Based on this viewpoint, we successfully prepared micrometre-sized dual nanocomposite polymer microspheres with tunable pore structures even in the interconnected pore structures using Pickering double emulsions.[94]

First, a primary water-in-styrene (oil) emulsion (W_1/O) was prepared using the hydrophobic silica nanoparticles as a particulate emulsifier without any molecular surfactants. Then, a water-in-styrene-in-water (W_1/O/W_2) Pickering emulsion was produced by emulsification of the primary W_1/O emulsion into water using Fe_3O_4 nanoparticles as external emulsifier. Nanocomposite polymer microspheres were formed by polymerizing the styrene in the middle phase of the double emulsions. The pore structure of the microspheres could be tuned by adjusting the volume ratio of the internal water phase to the medium oil phase (W_1:O) of the primary emulsions. More importantly, with increasing W_1:O from 1:8 to 4:1, the amount of pores in one microsphere increased gradually and the pore structures changed from close to interconnected (Figure 8.11).

Some typical optical microscopy images of the emulsions are shown in Figure 8.12. For low and medium ratios of W_1:O from 1:8 to 1:1, the emulsion droplets of the primary W_1/O emulsions (Figure 8.12a–c) and the double W_1/O/W_2 emulsions (Figure 8.12e–g) are spherical and polydisperse. The mean sizes of the primary emulsions are about 7 μm, 15 μm and 20 μm for W_1:O of 1:8, 1:4 and 1:1, respectively. The droplet size increases with increasing the volume ratio of W_1:O. This observation is consistent with those reported previously.[95–105] It is worth pointing out that the number of small water droplets in the big oil droplets increases with increasing volume ratios of W_1:O. Therefore, the inner structure of the double emulsion can be easily tuned by changing the volume ratio of W_1:O. The primary emulsion and corresponding double emulsion are shown in Figure 8.12d and h.

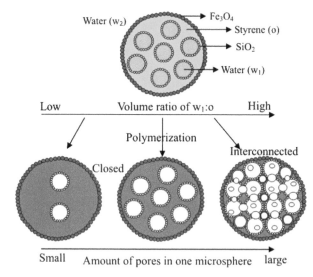

Figure 8.11   Schematic representation of the preparation of multihollow microspheres with tunable pore structures by double Pickering emulsion
templating.
Taken from reference 94 with the permission of Elsevier.

The pore morphology of the whole and sectioned microspheres
was observed using SEM and the images are shown in Figure 8.13a–c.
From the SEM images, it can be clearly observed that nanocomposite
PS microspheres consist of a porous structure and that each inner pore
is closed. Note that the size and number of the inner pores increase
with the increasing ratio of W_1:O. More importantly, when the ratio of
W_1:O reaches 4:1, the pore structure of the resulting microspheres
become interconnected as shown in Figure 8.14. This is attributed to sufficiently thin films between neighbouring water droplets, regarded as
the probable mechanism of pore throat formation. In other words, the
multihollow microspheres with tunable inner pore structures from closed
and fewer pores to interconnected and more pores were obtained via
increasing the ratio of W_1:O. Moreover, the energy dispersive spectroscopy
(EDS) spectra of the different locations of the nanocomposite PS microspheres (Figure 8.13d–f) show that the inorganic particles assembled at the
water–oil interfaces and were stable during the whole polymerization
procedure.

Embedding inorganic nanoparticles will endow polymeric matrices with
versatile functionality, such as optical, magnetic and electric properties.
Aqueous or oil multi-cores will enable the polymer microspheres to encapsulate an abundance of substances, such as drug, protein, enzyme and
other biomolecules. Hence, these kinds of hybrid multihollow polymer
microspheres are expected to have wide potential applications in both
materials science and biotechnology.

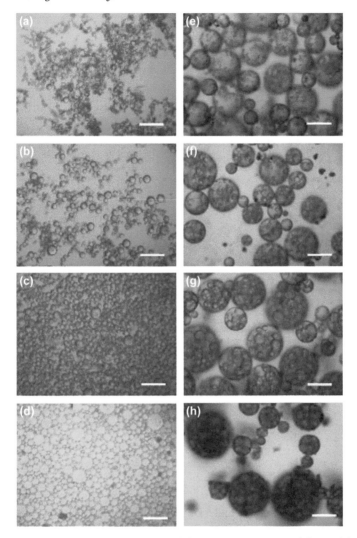

Figure 8.12 Optical microscope images of the primary W_1/O emulsions with W_1:O of 1:8 (a), 1:4 (b), 1:1 (c), 4:1 (d) and the corresponding double $W_1/O/W_2$ emulsions with $(W_1 + O)$:W_2 of 1:4, respectively. The scale bars are 100 μm. Taken from reference 94 with the permission of Elsevier.

8.3.2 Capsule Clusters

As we know, a double Pickering emulsion droplet contains many smaller primary emulsion droplets and there are two kinds of interfaces, including W/O and O/W interfaces. For single Pickering emulsions, hollow spheres can be conveniently fabricated by interfacial reaction. What kind of structure can we get if the interfacial reaction happens in these two kinds of interfaces within double Pickering emulsions?

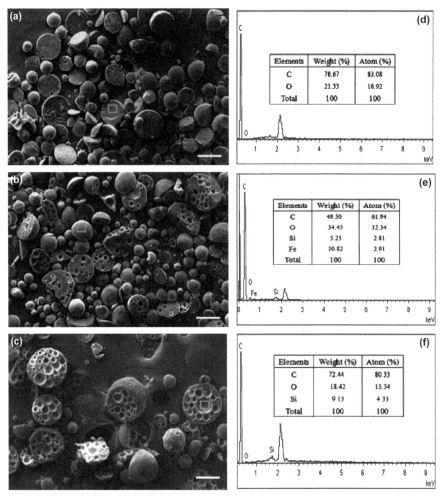

Figure 8.13   SEM images (a–c) and EDS spectra (d–f) of the whole and sectioned microspheres with W_1:O of 1:8 (a and d), W_1:O of 1:4 (b and e), and W_1:O of 1:1 (c and f). The corresponding double $W_1/O/W_2$ emulsions with $(W_1 + O)$:W_2 of 1:4. The scale bars are 100 μm.
Taken from reference 94 with the permission of Elsevier.

Based on O/W/O Pickering emulsions, our group[106] has presented a facile, controlled and large-scale way of fabricating (gram-sized quantities) of novel multi-compartment capsules with a capsule-in-capsule structure, defined as capsule clusters. More interestingly, different substances with controlled proportions can be encapsulated in the formed capsule simultaneously or separately. As shown in Figure 8.15, different organic core materials are emulsified into the water phase by mechanical stirring to form O/W Pickering emulsions. A mixture of single emulsions is obtained by proportionally blending two or more emulsions with different oil components. The hybrid

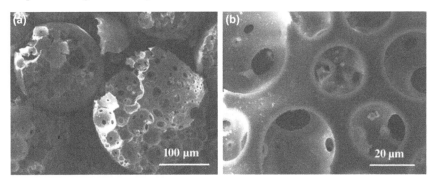

Figure 8.14 SEM images of the sectioned microspheres in low (a) and high (b) magnification with W_1:O of 4:1 and $(W_1 + O)$:W_2 of 1:4.
Taken from reference 94 with the permission of Elsevier.

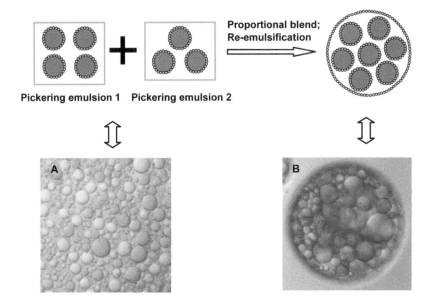

Figure 8.15 Schematic representation of the formation of capsule clusters. (A) Optical micrograph of two hybrid Pickering emulsions (O_1/W). (B) Optical micrograph of DBP-in-water-in-toluene $(O_1/W/O_2)$ Pickering emulsions containing two kinds of single droplets.
Taken from reference 106 with the permission of The Royal Society of Chemistry.

emulsion is further emulsified into the oil phase by hand shaking to produce the O/W/O double emulsion. Subsequently, polymerization is carried out at two oil–water interfaces to form outer and inner capsule shells and the capsule clusters with multiple components are achieved after accomplishment of the interfacial reaction.

Cluster dimension is controllable because the size of Pickering emulsion droplets can be adjusted by varying the volume ratio of the dispersed phase to the continuous phase. So, the structure of $O_1/W/O_2$ Pickering emulsions droplets can be controlled by the volume ratio of $O_1:W$ and $(O_1+W):O_2$. Furthermore, the morphology of the corresponding capsule clusters can be tuned by these two parameters. Figure 8.16A presents SEM images of the capsule clusters prepared by varying $(O_1+W):O_2$ while fixing $O_1:W$ at 1:3. The average sizes of the capsule clusters decrease from about 500 μm to 150 μm with decreasing $(O_1+W):O_2$ from 1:1 to 1:6. However, the average sizes of the inner capsules within the clusters are maintained at 35 μm because the inner capsule size is mainly determined by the $O_1:W$ volume ratio when the stabilizer usage of single emulsions is fixed. However, the amount of the inner capsules within the cluster decreases with decreasing $(O_1+W):O_2$.

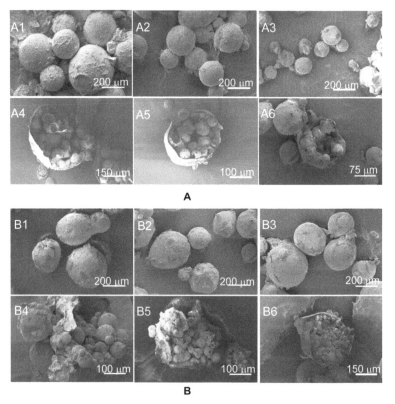

Figure 8.16 SEM images of capsule clusters: (A) The volume ratio of $O_1:W$ is fixed at 1:3 and the volume ratio of $(O_1+W):O_2$ in A1 and A4, A2 and A5, and A3 and A6 is 1:1; 1:3 and 1:6, respectively. (B) The volume ratio of $(O_1+W):O_2$ is fixed at 1:3 and the volume ratio of $O_1:W$ in B1 and B4, B2 and B5, and B3 and B6 is 1:1, 1:6 and 1:10, respectively.
Taken from reference 106 with the permission of The Royal Society of Chemistry.

Figure 8.16B presents SEM images of the capsule clusters prepared by varying the volume ratio of O_1:W while fixing $(O_1 + W)$:O_2 at 1:3. The average sizes of the capsule clusters are close to 200 μm because of the same $(O_1 + W)$:O_2 volume ratio. However, the average sizes of the inner capsules within the clusters decrease from about 70 μm to 10 μm with decreasing O_1:W from 1:1 to 1:10.

Such capsule clusters can not only encapsulate multiple reactive materials in a single capsule simultaneously, but also protect each of them from each other and from the outside environment. The sizes of the inner capsule and the whole cluster, the amount of the inner capsules, the thickness of the inner capsule and outer cluster shells, and the ratio of the different inner ingredients can be tuned precisely and independently by simply changing corresponding parameters. Moreover, the functional capsule clusters with stimuli-responsiveness are developed (Figure 8.17). The co-encapsulated different droplets can act as separate compartments for synergistic delivery of incompatible actives or chemicals, or as microreactor vessels for biochemical or chemical reactions. The approach presented here provides a new method for fabrication and applications of tunable multi-compartment capsules.

Subsequent research showed that capsule clusters with a capsule-in-capsule structure could be fabricated based on a novel capsule-in-water-in-oil Pickering emulsion.[107] To be specific, first, dibutyl phthalate (DBP)-loaded polyurea (PU) capsules were prepared via interfacial polymerization templated from oil-in-water Pickering emulsions stabilized by hydrophilic silica nanoparticles. Then, the aqueous dispersion of the DBP-loaded PU capsules was emulsified into toluene as a water-in-oil emulsion using hydrophobic silica nanoparticles as the emulsifier. Finally, the capsule clusters were obtained by polymerization of melamine and formaldehyde in the water phase. These capsule clusters consisted of poly(melamine formaldehyde) (PMF)/silica hybrid shells and DBP-loaded PU capsules as the cluster core.

Firstly, DBP-loaded PU capsules were successfully fabricated by adding DETA to the above O/W Pickering emulsions to initiate an interfacial

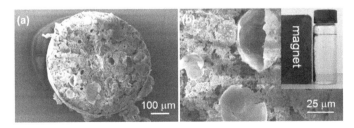

Figure 8.17 SEM images of magnetic capsule clusters filled with Fe_3O_4 nanoparticles: (a) low magnification and (b) high magnification. The insert in (b) is the capsule clusters in water under an external magnetic field.
Taken from reference 106 with the permission of The Royal Society of Chemistry.

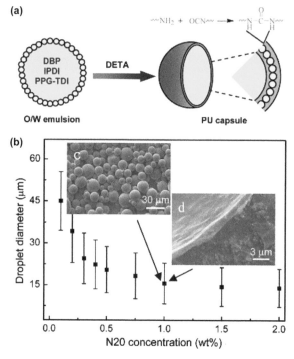

Figure 8.18 (a) Schematic illustration of the preparation of DBP-loaded PU capsules by interfacial polymerization of IPDI and DETA based on O/W Pickering emulsions. (b) Relationship between hydrophilic SiO_2 nanoparticle concentration and mean droplet diameter of Pickering emulsions. The insets in (b) are SEM images of (c) DBP-loaded PU capsules and (d) their shell prepared via the emulsion stabilized by 1.0 wt% hydrophilic SiO_2 nanoparticles in the water phase.
Taken from reference 107 with the permission of The Royal Society of Chemistry.

polymerization between IPDI in the oil phase and DETA in the water phase (Figure 8.18a). SEM was used to observe the morphology of the obtained capsules (Figure 8.18c). The PU capsules maintain a spherical shape and remain separate from each other. Meanwhile, the shell wall thickness is roughly uniform and about 1.2 μm (Figure 8.18d), which acts as an appropriate barrier against leakage and provides enough mechanical stiffness from rupture during post-processing.

Then, the capsule clusters were fabricated by suspension polymerization based on C/W/O Pickering emulsions. The schematic illustration of the preparation is presented in Figure 8.19. Various PU capsules containing different core materials were prepared by the above method. In this work, only two kinds of PU capsules, which were dyed with Sudan I and oil-soluble blue, were prepared. Figure 8.20a$_1$–c$_1$ show the optical micrographs of the C/W/O emulsions prepared by various PU capsule concentrations of 10, 20 and 40 wt% in the water phases, respectively. The droplets exhibited a

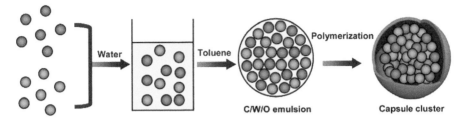

Figure 8.19 Schematic illustration of the preparation of capsule clusters based on capsule-in-water-in-oil Pickering emulsions.
Taken from reference 107 with the permission of The Royal Society of Chemistry.

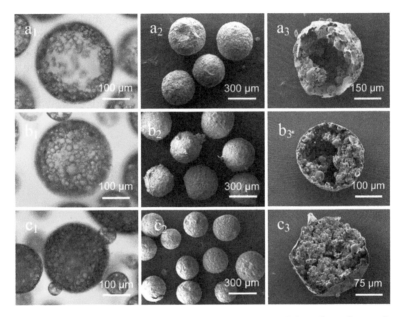

Figure 8.20 Optical microscopy images of the C/W/O emulsions (a_1–c_1); SEM images of the relative capsule clusters (a_2–c_2) and their cross-sections (c_1–c_3). The emulsions contained the PU capsule concentrations of 10 (a), 20 (b) and 40 (c) wt% in the water phases. The toluene to water volume ratio was 3:1 and the concentration of hydrophobic SiO_2 nanoparticles was 1.0 wt%.
Taken from reference 107 with the permission of The Royal Society of Chemistry.

spherical shape. The inner PU capsules tended to distribute at the oil–water interfaces. The SEM images of the corresponding capsule clusters fabricated based on the above C/W/O emulsions are shown in Figure 8.20a_2–c_2, respectively. All capsule clusters were revealed to be of a nearly spherical shape with a bit of wrinkling on the surfaces. It is worth noting that there was no mechanical stirring to help in the cluster formation process. However, the capsule clusters exhibited no adhesion to each other. This is because there are enough hydrophobic SiO_2 nanoparticles adsorbed at the outer oil–water interface acting as the barrier against coalescence and adhesion to the neighbouring droplets. This is one of the biggest advantages of the Pickering emulsion templates for capsule formation studies, which is difficult to achieve for surfactant-stabilized emulsion templates. Furthermore, the number of small capsules in a big capsule structure and the volume percentage of voids in capsule clusters can be calculated. For example, it turns out that there are about 1600 PU capsules in a capsule cluster, and the volume percentage of voids in capsule clusters is about 50%.

As the microcontainers of the reactive materials, the release properties of these capsule clusters are also studied. Herein, Sudan I dyed DBP-loaded PU capsules were encapsulated in the capsule clusters and the respective diffusion rates of Sudan I in THF, ethanol, hexane and water were investigated. Furthermore, the release properties of the capsule clusters which simultaneously encapsulated Sudan I and oil-soluble blue that were separate from each other were also studied. With the hydrophilic outer PMF/SiO_2 shell and inner PU/SiO_2 shell, the diffusion rate of encapsulated Sudan I in THF and ethanol is relatively high at the beginning of diffusion. The diffusion process is dramatically fast in ethanol, and the cumulative diffusion fraction achieved 96.3% after Sudan I diffusing only for 60 min, and in ethanol the fraction also achieved 84.1% (Figure 8.21a). The diffusion of Sudan I in hexane was very slow as the hydrophobic hexane cannot wet the capsule clusters (Figure 8.21b). Although the water can wet the capsule clusters, Sudan I cannot diffuse into the water because of the water insolubility of Sudan I. No diffussion could be observed (Figure 8.21b).

Furthermore, the diffusion of Sudan I from the capsule clusters containing 10, 20 and 40 wt% PU capsules was tested in the medium of ethanol. For the purposes of comparison, the diffusion of Sudan I in the original PU capsules was also studied. It is shown in Figure 8.21c that the diffusion of Sudan I from the original PU capsules is the fastest with 80.3% for 5 min, due to the lack of outer PMF shells. With increasing inner capsule content, the diffusion of Sudan I from the capsule clusters slowed down. As the higher PU capsule contents lead to a more compact inner capsule-in-capsule structure, this would retard the inner Sudan I diffusion to the outer ethanol.

Finally, Sudan I and oil-soluble blue were encapsulated in the PU capsules separately. Then the two kinds of capsules were proportionally blended, and the mixed capsules were encapsulated in the above capsule clusters to imitate the multidrug-loaded capsule clusters. Afterwards, the release properties of these multidrug-loaded capsule clusters were investigated by measuring the diffusion of the two dyes in ethanol. The UV absorbance at 480 nm is for Sudan I and 640 nm for oil-soluble blue. Figure 8.21d shows that both dyes have similar diffusion. It indicates that such novel capsule clusters are inherently permeable, exhibiting excellent multidrug controlled release properties in polar organic solvents.

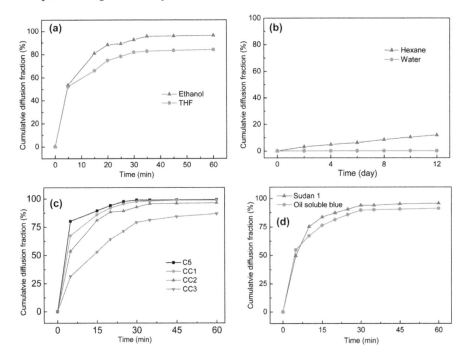

Figure 8.21 Diffusion profiles of Sudan I from the capsule clusters in (a) THF and ethanol, (b) hexane and water. Diffusion profiles of Sudan I in ethanol from (■) the capsules and the capsule clusters which contain (●) 10, (▲) 20 and (▼) 40 wt% of the PU capsules, respectively. Diffusion profiles of Sudan I and oil-soluble blue from the capsule clusters in ethanol. Sudan I and oil-soluble blue were simultaneously and separately encapsulated in the capsule clusters.
Taken from reference 107 with the permission of The Royal Society of Chemistry.

In summary, capsule clusters with capsule-in-capsule structure can be successfully prepared based on the stable O/W/O double Pickering emulsion templates or the novel capsule/W/O Pickering emulsion templates. The functional Pickering stabilizers can endow some specific properties (magnetic, optic or catalytic) to the capsule clusters. More importantly, different ingredients can be simultaneously encapsulated in the same cluster, exhibiting great potential applications in many fields.

8.3.3 Colloidosomes

Colloidosome refers to those microcapsules whose shells are composed of densely packed colloidal particles.[108–124] A hallmark of colloidosomes is their robust structure, as their building blocks are solid particles, which endow them with enhanced mechanical strength against coalescence. Another advantage associated with colloidosomes is that the permeability can be adjusted by changing the size of the particles. This is because the shells of colloidosomes are intrinsically porous rather than dense owing to the existence of interstitial voids between the packed particles. Hence, the size of the interstitial pores is theoretically proportional to the size of particles applied. Both of these two characteristics make them good candidates for encapsulation and controlled release systems.[125]

The traditional method for preparing colloidosomes is complex and time-consuming. In general, it involves several steps before achieving the final targets.[126-131] First, single water-in-oil (W/O) emulsions should be prepared; thereafter, the particles assembled at the interface form a colloidal shell structure; finally, the most difficult step is to transfer this colloidal shell structure to an aqueous phase without any damage either by repeated washing or centrifugation. The strategy will be slightly different if the single emulsion is of oil-in-water (O/W) type. In this case, the difficulty is to find an efficient way to remove the inner oil phase or replace the interior of the capsule with the desired release cargo.

Fortunately, colloidosomes templated from double Pickering emulsions require neither centrifugation nor washing steps. They can be easily obtained by the solvent extraction method and this allows for not only high yield but encapsulation efficiency as well.

Weitz's group[132] has demonstrated a simple way of preparing nanoparticle colloidosomes using double emulsion templates (W/O/W), which are generated by applying a microfluidic device and stabilized by silica nanoparticles. In their work, a volatile organic solvent such as toluene or a mixture of toluene and chloroform is used as the middle oil phase, subsequently evaporating the organic solvent, forming the nanoparticle colloidosomes (Figure 8.22A). This approach offers many advantages over conventional methods. First, a microcapillary device can precisely control the dimensions of the double emulsions, which is not achievable by other methods (Figure 8.22B and C). Second, the thickness of the resultant colloidosome shell can be controlled by tuning the dimension of the double emulsions and the content of silica nanoparticles in the middle oil phase (Figure 8.22D and E). Third, this method is capable of preparing multi-component colloidosomes, or composite microcapusules; moreover, the selective permeability can also be adjusted to some extent by varying the size of nanoparticles used.

Then they used a similar method to prepare non-spherical colloidosomes with multiple compartments.[133] W/O/W double emulsions were also prepared by a glass capillary microfluidic device. The outer O/W interface was stabilized by a partially hydrolysed PVA in the continuous phase and the inner W/O interface was stabilized by 15 nm hydrophobic SiO_2 nanoparticles; the nanoparticles dispersed in the oil phase adsorbed to the two W/O interfaces as evidenced by the formation of wrinkles in the O/W interface during evaporation of the solvent due to the buckling of the adsorbed nanoparticle layers (Figure 8.23). W/O/W double emulsions with a different number of internal aqueous drops per oil drop (n) were generated

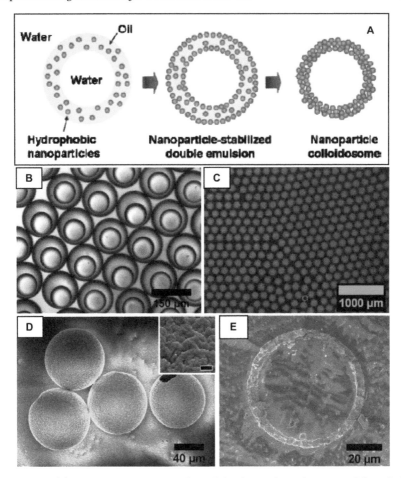

Figure 8.22 (A) Schematic representation of the formation of nanoparticle colloidosomes from nanoparticle-stabilized water-in-oil-in-water double emulsions. Hydrophobic nanoparticles are suspended in the oil phase. Evaporation of the oil phase leads to the formation of nanoparticle colloidosomes. (B) Optical and (C) fluorescence microscope images of double emulsions encapsulating 250 mg mL^{-1} FITC-dextran ($M_W = 70,000$). (D) SEM image of nanoparticle colloidosomes dried on a substrate. Inset is a high magnification image of colloidosome surface (scale bar = 600 nm). (E) Cross-section of a colloidosome obtained with freeze-fracture cryo-SEM.
Taken from reference 72 with the permission of Wiley-VCH.

by controlling the flow rates of three phases independently. Evaporation of solvent from double emulsions in Figure 8.24 results in the generation of colloidosomes with two internal compartments. Depending on the morphology of the double emulsions with $n = 2$, the colloidosomes have a different morphology. In the case of the spherical double emulsions with $n = 2$, the resultant colloidosomes are ellipsoids, as shown in the scanning electron

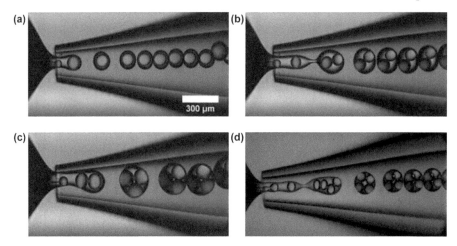

Figure 8.23   Generation of double emulsions with varying number of internal drops.
Taken from reference 133 with the permission of Wiley-VCH.

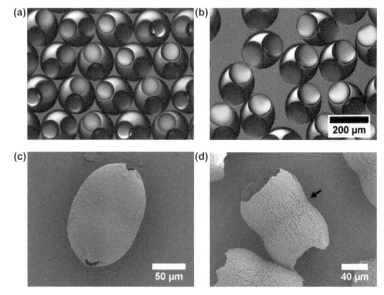

Figure 8.24   (a) and (b) Optical microscopy images of W/O/W double emulsions with
different volume ratios.
Taken from reference 133 with the permission of Wiley-VCH.

microscopy (SEM) image in Figure 8.24c. In the case of the ellipsoidal double
emulsions, the resultant colloidosomes resemble peanuts (Figure 8.24d).
Interestingly, these peanut-shaped colloidosomes have a saddle structure
(black arrow in Figure 8.24d), which would be unstable due to the negative
curvature. However, the structure is maintained because of the solid

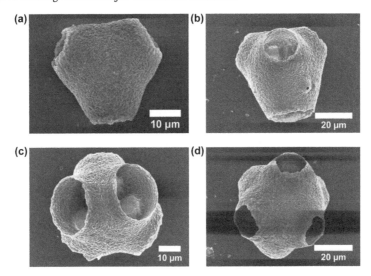

Figure 8.25 SEM images of non-spherical colloidosomes with $n = 3$ (a), 4 (b), 5 (c) and 6 (d) internal voids.
Taken from reference 133 with the permission of Wiley-VCH.

structure formed by the SiO_2 nanoparticles that become jammed during the evaporation of the oil phase. Evaporation of the oil (toluene) from double emulsions with $n > 2$ leads to non-spherical colloidosomes with interesting geometry as shown in Figure 8.25. By increasing n, colloidosomes with 3D structures can be prepared as illustrated in Figure 8.25b–d. The as-prepared non-spherical colloidosomes would have important applications in multi-drug delivery.

As discussed above, the size of the pores between closely packed nano-particles is determined by particle dimension. So from this point of view, changing the particle size can fulfil the purpose of varying the pore size. Apart from this method, there are other approaches to manipulating the pore size, and so adjusting the permeability. Based on double Pickering emulsion templates, Hehrens' group[134] has presented a novel type of aqueous core colloidosomes that combine the low capsule permeability with stimulus-response fast release under the desired environment. They have demonstrated three methods for controlling the capsule permeability, in-cluding: (a) colloidosome consolidation by partial dissolution and sintering, (b) deposition of additional polyelectrolytes and responsive particles via layer-by-layer adsorption onto the original colloidosomes, and (c) inclusion of the water-insoluble poly-(lactic-*co*-glycolic acid), PLGA, in the middle oil phase of the double emulsion at the time of the capsule preparation, re-sulting in pH-responsive PLGA–nanoparticle composite microcapsules. All these routes are capable of adjusting the capsule permeability, making it easy to realize the controlled release.

8.3.4   Hierarchical Porous Polymeric Microspheres

As shown in Figure 8.26, multihollow microspheres can be produced easily by solidification of the middle phase of double emulsions. In these cases, the middle phases are composed of polymerizable monomers without any solvent. One may wonder what would be obtained if phase separation happens during the polymerization process when non-polymerizable solvent is added in the middle oil phase? In the next section, we will present how to fabricate hierarchical porous polymeric microspheres as well as their application in adsorption and as catalytic carriers.

Porous materials are important candidates for adsorption and catalyst scaffold materials, and so they have gained considerable attention over the past decade.[135,136] A range of factors responsible for their outstanding performance including high porosity, low density, high specific surface area and permeability.[137–140] The approaches to preparing inorganic porous materials are well established while the synthesis of porous polymeric materials generally involves multiple and complex procedures, which seriously limit their upscaling and wide application. Recently, we developed a simple method for the synthesis of a novel kind of polymeric microsphere with a hierarchical porous structure based on double Pickering emulsion templates (aqueous Fe_3O_4 dispersion-in-oil-in-water, $W_F/O/W$), in which the components of the middle oil phase were carefully chosen and played a key factor in forming the resultant hierarchical porous structure.[140] In this case, the inner interface of W_F/O and outer interface of O/W of the double Pickering emulsions were emulsified by hydrophobic amorphous fumed silica powders (H30) and hydrophilic meso-pore silica nanoparticles, respectively. The middle oil phase consisted of styrene, divinylbenzene and non-polymerizable hexadecane.

Thanks to the stable double Pickering emulsion templates, without emulsion break-up, target hierarchical porous polymeric microspheres (HPPMs) can be facilely prepared by polymerization of the monomers in the middle phase, followed by removal of the solvent hexadecane (Figure 8.26).

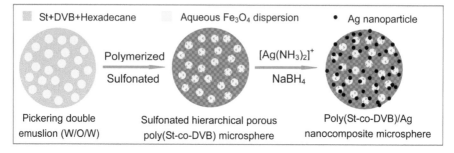

Figure 8.26   Schematic representation of the preparation of hierarchical porous polymeric microspheres.
Taken from reference 140 with the permission of The Royal Society of Chemistry.

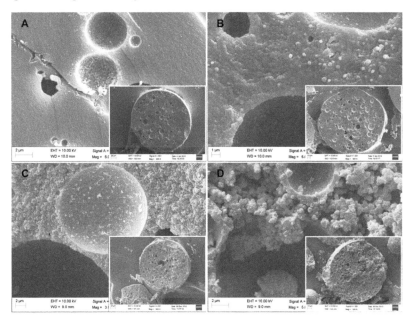

Figure 8.27 Effect of hexadecane content in the oil phase of double Pickering emulsions on the structure of the resulting microspheres: (A) $V_M:V_H = 8{:}1$; (B) $V_M:V_H = 4{:}1$; (C) $V_M : V_H = 2{:}1$; (D) $V_M : V_H = 1{:}1$. The insets represent the corresponding whole microspheres.
Taken from reference 140 with the permission of The Royal Society of Chemistry.

Unlike the system whose middle oil phase is composed of pure monomers, in this study, the skeleton structure of the resultant multihollow microsphere depends upon the oil composition. SEM images of the inner structure of the resultant samples produced under different experimental conditions are presented in Figure 8.27, which clearly shows that structural changes in the skeleton part are closely related to the use of hexadecane, which here plays the role of a porogen. In Figure 8.27A, a multihollow microsphere with solid skeleton is observed. As the volume ratio of monomer to hexadecane (V_M/V_H) decreases, the skeleton structure experiences a change from dense status to meshy structure (from Figure 8.27A to Figure 8.27D). This is reasonable because the meshy structure could be formed due to the phase separation of poly(St-*co*-DVB) from hexadecane.

To extend their application, these HPPMs were chemically modified with sulfuric acid, with the aim of endowing them with a sulfonic acid group, which can not only adjust their amphipathicity, but make further noble metal loading feasible as well. It is well known that direct application of noble metal nanoparticles in catalysis is often difficult as a result of their ultrasmall size and high tendency toward agglomeration because of van der Waals forces.[141] Herein, the hierarchical porous structure can not only offer many loading sites for the Ag particles, but also protect the Ag particles from

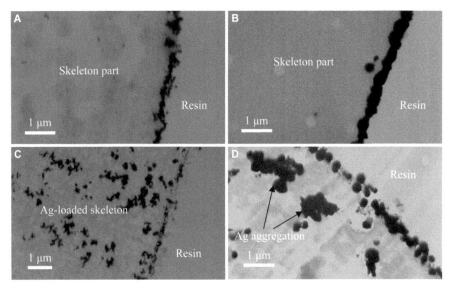

Figure 8.28   TEM images of poly(St-*co*-DVB)/Ag nanocomposite microspheres: (A) Ag-S-HPPM-1; (B) Ag-S-HPPM-2; (C) Ag-S-HPPM-3; (D) Ag-S-HPPM-4. Taken from reference 140 with the permission of The Royal Society of Chemistry.

aggregation, which may remarkably decrease the catalytic efficiency. Figure 8.28 apparently demonstrates the diverse Ag-loading statuses in different S-HPPMs.

The adsorption isotherms for rhodamine 6G (R6G), a cationic dye widely used in chemical industries but harmful to humans, on the sulfonated HPPMs (S-HPPMs) with various skeleton structures demonstrate an outstanding adsorptive ability (see Figure 8.29A and 8.29B). At the same time, the Ag-loaded S-HPPMs (Ag-S-HPPMs) exhibit excellent catalytic ability, as shown in Figure 8.29C and 8.29D. It can be concluded that the unique hierarchical porous structure, which provides the loading position for Ag nanoparticles and meanwhile effectively prevents the Ag nanoparticles from agglomeration, plays the most important role in determining catalytic ability as well as recyclability.

Binks' group have also contributed some outstanding works on fabrication of various polymeric materials templated from double Pickering emulsions stabilized solely by dichlorodimethylsilane (DCDMS)-modified amorphous silica nanoparticles of well-controlled surface hydrophobicities or organomodified Laponite clay nanoparticles.[142–144] Based on a water-in-styrene-in-water-in-styrene (W/O/W/O) triple Pickering emulsion stabilized by 3 wt% of 50% SiOH silica nanoparticles in the inner oil phase, and 2 wt% of the same silica type in the outer oil phase and oil globules, was stabilized by 2 wt% of 80% SiOH silica based in the second water phase, Dyab *et al.* have gained a new porous hierarchical structure after the polymerization of the oil phase.[145]

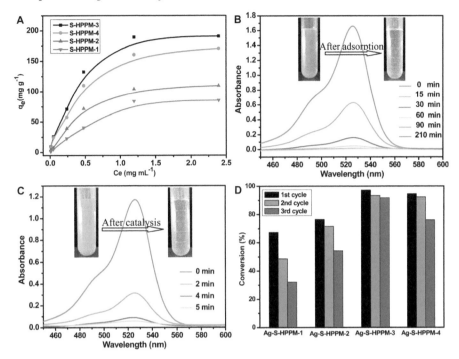

Figure 8.29 (A) Adsorption isotherms for R6G on a series of HPPMs at room temperature, in each experiment, HPPMs: 10 mg, R6G aqueous solution: 10 mL. (B) UV-visible spectra of R6G aqueous solution adsorbed by S-HPPM-3 as a function of reaction time, S-HPPM-3: 10 mg; R6G aqueous solution: 2×10^{-5} mol L^{-1}, 10 mL. The inset indicates photographs of R6G aqueous solution before and after adsorption. (C) UV-visible spectra of R6G aqueous solution reduced by KBH_4 combined with Ag-S-HPPM-3 as a function of reaction time, Ag-S-HPPM-3: 20 mg; R6G aqueous solution: 2×10^{-5} mol L^{-1}, 5 mL; KBH_4: 1×10^{-2} mol L^{-1}, 1 mL. The inset shows the photographs of R6G aqueous solution before and after catalysis. (D) Recyclable catalytic ability of various Ag-S-HPPMs for the reduction of R6G with KBH_4.
Taken from reference 140 with the permission of The Royal Society of Chemistry.

8.3.5 Janus Microspheres

Janus particles, whose two sides or surfaces are different in terms of chemical and/or physical properties, have gained significant attention in recent years due to their novel morphologies and unique natures, endowing them with diverse potential applications such as switchable display devices,[146] interface assembles,[147-150] optical sensors[151] and anisotropic building blocks for complex structures.[152-156]

A novel kind of Janus microparticle with dual anisotropy of porosity and magnetism based on Pickering-type double emulsions was obtained by our group.[157,158] The key advantage of this method is that it is simple, and is

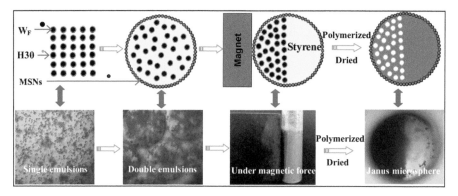

Figure 8.30 Schematic representation of the method developed to prepare Janus microspheres. The top row presents the cartoons while the bottom row indicates the corresponding photographs.
Taken from reference 157 with the permission of the American Chemical Society.

capable of being scaled up to produce a large quantity of the final product. Meanwhile, it allows one to precisely control the interior structure of the Janus microspheres. Figure 8.30 shows the process that has been used for the fabrication of the Janus microparticles with tunable interior structure. Firstly, a stable aqueous Fe_3O_4 dispersion-in-oil-in-water ($W_F/O/W$) double Pickering emulsion was generated by using amorphous fumed silica powders (H30) and our recently synthesized mesoporous silica nanoparticles (MSNs) as stabilizers for the primary inner W_F/O droplets and outer O/W droplets, respectively. Note that the aqueous Fe_3O_4 dispersion (W_F) offers single emulsion droplets with magnetism, making them movable under a magnetic field. After that, polymerization of the middle oil phase of the double emulsions was started under an external magnetic field. Finally, the Janus microspheres with multi-hollow structure possessing magnetite nanoparticles concentrated within one side of the sphere were obtained after completing the polymerization.

It is very interesting that we are able to precisely control the Janus balance of the products by changing the volume ratio of inner water phase to middle oil phase (W_F/O). Figure 8.31 shows the as-prepared anisotropic microspheres with different distributions of the multi-hollow inside the microspheres. It can be seen that the volume of the multi-hollow part increases gradually as the W_F/O volume ratio increases from 1:10 to 1:2. Meanwhile, the multi-hollow part becomes darker and darker due to the diffuse scattering caused by the increased voids in the microsphere. From the corresponding SEM images, the precise distribution of voids within the big microspheres under different W_F/O ratio conditions can be clearly seen.

More importantly, it is feasible to get anisotropic microspheres with desired Janus balance by setting the initial W_F/O. In other words, we can

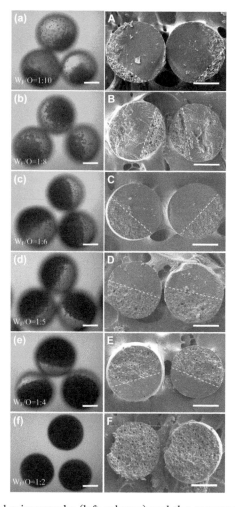

Figure 8.31 Optical micrographs (left column) and the corresponding SEM images (right column) of microspheres produced under various W_F/O volume ratios. All scale bars represent 150 μm.
Taken from reference 157 with the permission of the American Chemical Soicety.

get any Janus balance we like. To achieve this goal, we have to figure out the relationship between Janus balance and W_F/O volume ratio by mathematic calculation. Actually, the Janus balance depends on the relative volumes of the multi-hollow part and the solid part in each microsphere (see Figure 8.32). Therefore, we firstly calculate the relative volumes of the multi-hollow and solid parts ($V_{multi-hollow}/V_{solid}$) of the Janus microsphere prepared under different W_F/O volume ratios, and then we can indirectly obtain the relationship between Janus balance and W_F/O volume ratio.

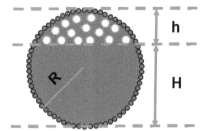

Figure 8.32 Schematic representation of a Janus microsphere.
Taken from reference 157 with the permission of the American Chemical Society.

$V_{h,e}/V_{H,e}$ can be directly obtained by measuring the height of the multi-hollow part and the solid part of the prepared Janus microspheres.

$$\text{Because } V_h = \frac{1}{3}\pi h^2(3R - h), \; V_H = \frac{1}{3}\pi H^2(3R - H)$$

$$\text{so } \frac{V_{h,e}}{V_{H,e}} = \frac{h^2(3R - h)}{H^2(3R - H)} \tag{8.1}$$

while $V_{h,t}/V_{H,t}$ can be deduced based on maximum packing fraction theory. For a high internal phase emulsion (HIPE), the maximum packing fraction of the monodispersed spheres is 0.74.[159] Accordingly, we assumed that all single emulsion droplets are of the same size and closely packed within one side of the double emulsion droplets, so we can then achieve the relationship between $V_{h,t}/V_{H,t}$ and W_F/O volume ratio.

Assuming $\dfrac{W_F}{O} = \dfrac{1}{\gamma}$, in one single double emulsion droplet, $W_F = A$ mL

$$\Rightarrow O = A\gamma \text{ mL}, \; A + A\gamma = V_T = \frac{4}{3}\pi R^3 = \frac{V_T}{1+\gamma} = \frac{\frac{4}{3}\pi R^3}{1+\gamma}$$

According to the maximum packing fraction theory (0.74):

$$V_{h,t} \times 0.74 = A = \frac{V_T}{1+\gamma}$$

$$\Rightarrow V_{h,t} = \frac{V_T}{0.74 \times (1+\gamma)}$$

and $V_{h,t} + V_{H,t} = V_T, \; \Rightarrow V_{H,t} = V_T - V_{h,t}$

$$\Rightarrow \frac{V_{h,t}}{V_{H,t}} = \frac{\dfrac{V_T}{0.74 \times (1+\gamma)}}{V_T - \dfrac{V_T}{0.74 \times (1+\gamma)}} = \frac{1}{0.74\gamma - 0.26}$$

$$\text{So, } \frac{V_{h,t}}{V_{H,t}} = \frac{1}{0.74\gamma - 0.26} \tag{8.2}$$

In which h and H express the hemispherical heights, where R is the radius of the Janus microsphere. V_h and V_H refer to the volume of the spherical segment with a height of h and H, respectively; $V_{h,e}$ and $V_{H,e}$ represent the experimental volume of the spherical segment with a height of h and H, respectively; $V_{h,t}$ and $V_{H,t}$ represent the theoretical volume of the spherical segment with a height of h and H, respectively; V_T and γ indicate the total volume of the sphere and the ratio value of O/W$_F$, respectively.

Figure 8.33 shows the relationship between $V_{h,e}/V_{H,e}$ and $V_{h,t}/V_{H,t}$ under various W$_F$/O volume ratios. Thus, by checking the relationship between $V_{h,e}/V_{H,e}$ and W$_F$/O volume ratio, we can easily and precisely fabricate the microspheres of any Janus balance we want by choosing the corresponding W$_F$/O volume ratio. This work is fantastic because it is impossible for other methods to precisely fabricate Janus particles with the desired Janus balance so easily.

Figure 8.34 illuminates the asymmetric distribution of elements within the Janus microsphere. The multi-hollow part is composed of C, O, Si and Fe while the solid side is composed of C, O and Si. It is no surprise to find that these multi-hollow Janus microspheres can be magnetically orientated due to the special internal structure and magnetic anisotropy. The magnetic orientation process is clearly demonstrated in Figure 8.35.

Owing to the unique structure and magnetic anisotropy, these Janus microspheres may find potential applications in the areas of biomedical systems, asymmetric catalysis, and one-side drug delivery. Besides, its

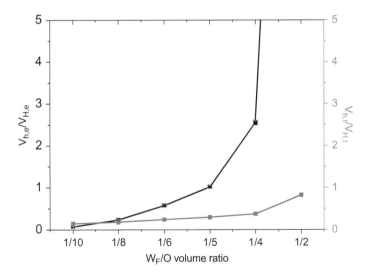

Figure 8.33 The relationship between $V_{h,e}/V_{H,e}$ and $V_{h,t}/V_{H,t}$ under different conditions of W$_F$/O volume ratios. The black line demonstrates the relationship between $V_{h,e}/V_{H,e}$ and W$_F$/O volume ratio, while the blue line shows the relationship between $V_{h,t}/V_{H,t}$ and W$_F$/O volume ratio. Taken from reference 157 with the permission of the American Chemical Society.

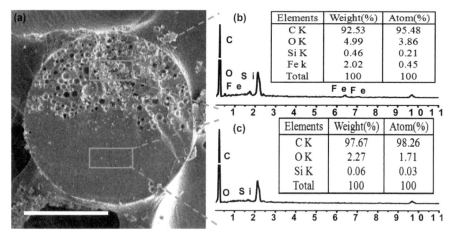

Elements	Weight(%)	Atom(%)
C K	92.53	95.48
O K	4.99	3.86
Si K	0.46	0.21
Fe k	2.02	0.45
Total	100	100

Elements	Weight(%)	Atom(%)
C K	97.67	98.26
O K	2.27	1.71
Si K	0.06	0.03
Total	100	100

Figure 8.34   (a) SEM image of the inner structure of the Janus microsphere (Run 10, Table S1); elemental analysis of the sample at multi-hollow parts (b) and solid sites (c) illuminates the asymmetric distribution of the Fe element. The scale bar represents 150 µm.
Taken from reference 157 with the permission of the American Chemical Society.

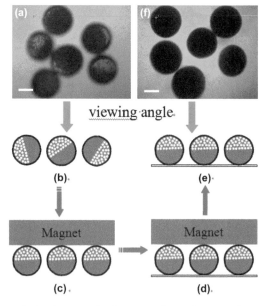

Figure 8.35   (a) Optical micrograph of Janus microspheres before orientation (Sample 13, Table S1, Supporting Information); (b) random pattern; (c) Janus microspheres being orientated under a magnet; (d) transporting onto a transparent adhesive tape; (e) orientated Janus microspheres being attached to a transparent adhesive tape; (f) optical micrograph of the orientated Janus microspheres. The scale bar represents 150 µm.
Taken from reference 157 with the permission of the American Chemical Society.

magnetic orientation property makes it a good candidate for synthesizing Janus membranes.

8.3.6 Novel Oil-in-Water-in-Air Double Pickering Emulsions

Colloidal particles can be irreversibly adsorbed at fluid interfaces, such as oil–water and air–water interfaces.[160–164] The particle adsorption leads to stabilization of dispersed systems of two immiscible fluids, and Pickering emulsions or foams can be prepared. When the liquid is water, a water-in-air (w/a) material, called dry water, is produced by aerating water in the presence of extremely hydrophobic particles.[165–169] The dry water is a free-flowing powder that can contain significant quantities of water as micron-sized drops. One may imagine that when particle-stabilized emulsions with water as the continuous phase, *i.e.* oil-in-water (O/W) emulsions, are aerated with similar hydrophobic particles, the surfaces between the emulsion drops and air can be encased in the hydrophobic particles and oil-in-water-in-air (O/W/A) materials may be prepared (Figure 8.36a). Binks *et al.*[170] have stabilized O/W/A materials using colloidal particles alone and found that control of the coalescence of oil droplets in O/W emulsions on aeration is crucial in the ability to produce powdered emulsions successfully. To be specific, by analogy with the stabilization mechanism of dry water, globules of water containing oil drops within from O/W emulsions can be coated by hydrophobic particles when aerated, producing O/W/A materials. The O/W/A materials containing micrometre-sized drops may be called 'powdered emulsions', as they consist of a high concentration of water globules containing micrometre-sized oil droplets and display free-flowing behaviour (see Figure 8.36b).

The optical microscopy image of W/A material dispersed in n-dodecane (Figure 8.37a) shows water drops of varying size in the micron range, indicating that the original dry material is of the type W/A, *i.e.* dry water. An O/W emulsion was prepared by mixing 5 mL of n-dodecane and 5 mL of an aqueous dispersion of partially hydrophobic silica nanoparticles (possessing 37% SiOH) at a concentration of 1 wt% relative to water ($w_{37\% \text{ SiOH}}$). After adding 15 mL of water to the emulsion, the diluted emulsion (total volume = 25 mL) was aerated in the presence of 3 g of very hydrophobic silica particles (possessing 20% SiOH) using a blender, in the same way as for the preparation of the aforementioned dry water (20,000 rpm for 15 s). The aerated material took up 70 vol% of the diluted emulsion and showed a similar appearance to dry water. However it was agglomerated and less powdery compared to dry water in terms of flowability. Microscopy observations revealed that the material contains water drops, although their inner volume appears structured, but individual oil droplets cannot be ascertained (Figure 8.37b).

The effect of mixing speed during aeration on the preparation of O/W/A materials was investigated. When the mixing speed was lowered to 6000 rpm, the prepared materials exhibit free-flowing behaviour as shown in

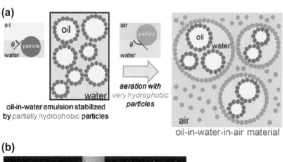

Figure 8.36 Stabilization of oil-in-water-in-air materials. (a) Preparation of oil-in-water-in-air materials. An oil-in-water emulsion stabilized by partially hydrophobic particles (blue) can be aerated by mixing with very hydrophobic particles (red). Oil-in-water-in-air materials consist of oil drops dispersed in water globules themselves dispersed in air, with very hydrophobic particles adsorbing at the air–water surfaces. Excess particles that are not adsorbed retain their powdery nature. (b) Photograph of oil-in-water-in-air material prepared by aerating 10 mL of a n-dodecane-in-water emulsion stabilized by 1 wt% of partially hydrophobic (37% SiOH) silica nanoparticles diluted with 15 mL of pure water in the presence of 3 g of very hydrophobic silica nanoparticles (20% SiOH) at 6000 rpm, showing free-flowing behaviour.
Taken from reference 170 with the permission of Wiley-VCH.

Figure 8.36b and the flowability is better than the ones prepared at 20,000 rpm. The material consists of agglomerated granules with a size ranging from tens to hundreds of microns (Figure 8.37c). In Figure 8.37d, distinct oil droplets are now visible in the water globules and their size is comparable to those of the oil droplets in the original O/W emulsion (inset in Figure 8.37d). It was found that a decrease in the mixing speed leads to a reduction in the shear-induced coalescence of oil droplets but is sufficient to break up bulk water into globules. Figure 8.37e shows that an O/W/A material prepared at $w_{100\% \text{ SiOH}} = 5.3$ wt% contains more oil droplets than that at $w_{100\% \text{ SiOH}} = 0$ wt% (Figure 8.37d). It is presumed that O/W/A powdered emulsions prepared without coalescence of inner oil droplets behave like dry water in terms of the flowability of the powder. Optical microscopy images of the O/W/A material prepared after dispersing in n-dodecane clearly show that water globules of an O/W/A material prepared at $w_{37\% \text{ SiOH}} = 2$ wt% and $w_{100\% \text{ SiOH}} = 5.3$ wt% contain an abundant amount

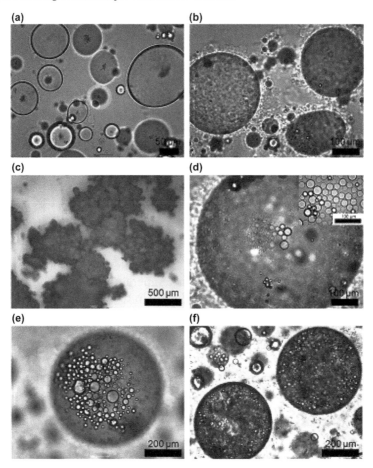

Figure 8.37 Optical microscopy images of particle-stabilized dispersed systems. (a, b) Materials prepared by aerating water (a) and n-dodecane-in-water emulsion stabilized by 1 wt% of silica nanoparticles (37% SiOH) (b) at 20,000 rpm. Both materials were subsequently dispersed in n-dodecane. (c, d) O/W/A materials prepared by aerating O/W emulsions stabilized by 1 wt% of silica nanoparticles (37% SiOH) at 6000 rpm: O/W/A material as prepared (c) and dispersed in n-dodecane (d). The inset is an image of the precursor O/W emulsion. (e, f) O/W/A materials prepared by aerating O/W emulsions at 6000 rpm for different concentrations of particulate emulsifier (37% SiOH silica): 1 wt% (e) and 2 wt% (f). Before aeration, O/W emulsions were diluted with an aqueous dispersion of 5.3 wt% silica nanoparticles (100% SiOH).
Taken from reference 170 with the permission of Wiley-VCH.

of relatively small oil droplets that are densely packed inside (Figure 8.37f). The O/W/A materials prepared at $w_{37\%\ \text{SiOH}} = 2$ wt% show slightly lower response angle than the materials at $w_{37\%\ \text{SiOH}} = 1$ wt%, indicating improved flowability. For all the O/W/A materials, we have not observed any separation of oil or water up to 6 months after preparation if stored in a sealed vessel, implying that significant coalescence of oil droplets and water globules does not occur.

In summary, O/W/A powdered emulsions could be successfully fabricated by using nano-sized silica particles of different wettability: partially hydrophobic particles were used to prepare oil-in-water emulsions and water-repellent particles were used to aerate them and stabilize water-in-air globules. In preparing the powdered O/W/A materials it is crucial to decrease the extent of coalescence of oil droplets in O/W emulsions. Generally, by careful selection of nanoparticles of the required wettability, oil–water and air–water interfaces can be stabilized sequentially, yielding a powdered emulsion.

8.3.7   Novel Double Pickering Emulsions of Room Temperature Ionic Liquids

Ionic liquids are a unique collection of liquid materials composed solely of ions. Under ambient conditions, room temperature ionic liquids (RTILs) stay as liquids whereas conventional salts are in the crystalline state. This is because, in RTILs, the Coulombic attractions of ion pairs are damped by the large ion size and the lattice packing is frustrated by the sterical mismatch of irregular shaped ions. With a combination of many unique properties such as negligible volatility, non-flammability, thermal and chemical stability, and high ionic conductivity, as well as potential broad applications, increasing amounts of attention have been paid to ionic liquids, especially RTILs.[143,171–175] The properties of RTILs can be varied from hydrophilic (*i.e.* water-miscible but immiscible with low polarity solvents (we use the term 'oil' here)) to hydrophobic (*i.e.* water-immiscible but oil-miscible).

Binks[176] considered combinations of only three generic solvent types which are mutually immiscible (water w, oil o and ionic liquid i), giving rise to the range of possible simple and multiple emulsion types listed in Table 8.1. Table 8.1 also summarizes the emulsion types prepared as part of this study using nanoparticles as stabilizers. They used fumed silica nanoparticles with a mean primary particle diameter of ca. 20 nm and which were coated to different extents with dimethyldichlorosilane to stabilize emulsions. The ionic liquids used were 1-butyl-3-methylimidazolium bis(trifluoromethylsulfon-yl)-imide (BmimNT), 1-butyl-3-methylimidazolium tetrafluoro-borate (BmimBF), 1-ethyl-3-methylimidazolium tetrafluoro-borate (EmimBF, Fluka, > 97%), 1,2-dimethyl-3-propyl-imidazolium bis(trifluoromethylsulfonyl)imide (DpimNT, Fluka, > 97%), 1-ethyl-3-methylylimidazolium bis(trifluoromethylsulfonyl)-imide (EmimNT, Fluka, > 97%) and 1-ethyl-3-methylimidazolium chloride (EmimCl, Fluka, > 97%). EmimCl was solid at room temperature and was used as a 10 wt% water mixture that was liquid at room temperature. The BF and Cl salts were miscible with water and immiscible with the oils used here whereas the NT salts were immiscible with both water and the oils.

Figure 8.38 shows a representative example of a three-component multiple emulsion (double emulsion). Both two- and three-component multiple

Table 8.1 Summary of the possible simple and multiple emulsion types involving ionic liquids, including those that have been successfully prepared. The entries o, i and w refer to oil, ionic liquid and water, respectively. The heading 'outer' refers to the continuous phase of the emulsion, 'inner' refers to the innermost dispersed droplet phase and 'middle' refers to the phase separating the inner and outer phases within a multiple emulsion. Taken from reference 176 with the permission of The Royal Society of Chemistry.

Type/number	Inner	Middle	Outer	Made?
Simple				
1	o	—	i	yes
2	i	—	o	yes
2	w	—	i	yes
4	i	—	w	yes
Two component multiple				
5	o	i	o	yes
6	i	o	i	yes
7	w	i	w	yes
8	i	w	i	no[a]
Three component multiple				
9	o	i	w	yes
10	o	w	i	no[a]
11	i	o	w	yes
12	w	o	i	yes
13	i	w	o	yes
14	w	i	o	no[a]

[a]Not tried.

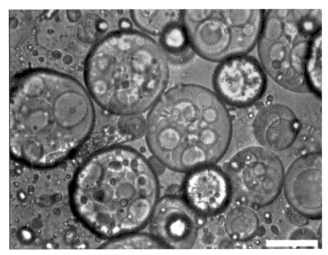

Figure 8.38 Optical micrograph of a Miglyol 810N-in-DpimNT-in-water three-component multiple emulsion with 4 vol% inner phase, 16 vol% middle phase and 80 vol% outer phase. The inner interfaces are stabilized with 76% SiOH silica initially dispersed in the ionic liquid. The outer interfaces are stabilized with the same silica, which was initially dispersed in water. The emulsion showed no observable change after 45 days. The scale bar is 50 μm.
Taken from reference 176 with the permission of The Royal Society of Chemistry.

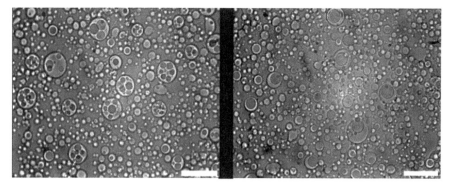

Figure 8.39 Optical micrographs of an EmimBF-in-Miglyol 810N-in-DpimNT multiple emulsion with 4 vol% inner liquid, 16 vol% middle phase and
80 vol% outer phase. The inner interfaces are stabilized with 47% SiOH
silica and the outer interfaces with 76% SiOH silica. The left-hand
image was recorded 5 min after preparation, the right-hand image after
2 h. The scale bars are 50 μm.
Taken from reference 176 with the permission of The Royal Society of
Chemistry.

emulsions generally show excellent stability with respect to coalescence
and Ostwald ripening and most of the different types can again be
prepared with several different oils and ionic liquids. Altering the volume
fractions of the liquid components and the concentrations of the stabilizing
silica particles enables a degree of control of both the inner and outer
droplet sizes. For both simple and multiple emulsions, increasing the
concentration of the silica nanoparticles generally leads to a decrease in
emulsion drop size.

Figure 8.39 shows a micrograph of a further emulsion type not listed in
Table 8.1. The multiple emulsion contains two different ionic liquids that
are mutually miscible but which are separated by oil within the multiple
emulsion. This particular emulsion showed the intriguing behaviour that
the inner drops totally coalesced together within 2 hours or so to yield a
stable multiple emulsion in which no oil globule contained more than a
single drop of the second ionic liquid.

Introducing the room temperature ionic liquid to the Pickering emulsion
system greatly increases possible applications in different fields, such as
material synthesis, extraction, chemical reaction, and encapsulation.

8.4   Conclusions and Perspective

As discussed above, multiple Pickering emulsions can offer a feasible way to
fabricate multi-hollow nanocomposite microspheres, capsule clusters, colloidosomes, Janus microspheres, hierarchical porous polymeric microspheres, dry water and ionic liquid encapsulation due to their multiple
phases, multiple interfaces and multiple parameters.

The research into multiple Pickering emulsions is still in its initial stages. Its huge potential in the fabrication of functional structural materials is attracting more and more scientists. It is so interesting that slight changes in any factor could result in a different structure, and consequently lead to excellent applications.

Acknowledgements

This work was supported by the National Basic Research Program of China (973 Program, 2012CB821500), the National Natural Science Foundation of China (21274046), and the Natural Science Foundation of Guangdong Province (S20120011057).

References

1. W. Seifriz, *J. Phys. Chem.*, 1925, **29**, 738.
2. I. C. Kwon, Y. H. Bae and S. W. Kim, *Nature*, 1991, **354**, 291.
3. C. Laugel, A. Baillet, M. P. Y. Piemi, J. P. Marty and D. Ferrier, *Int. J. Pharm.*, 1998, **160**, 109.
4. C. Laugel, P. Rafidison, G. Potard, L. Aguadisch and A. Baillet, *J. Control. Release*, 2000, **63**, 7.
5. B. Raghuraman, N. Tirmizi and J. Wiencek, *Environ. Sci. Technol.*, 1994, **28**, 1090.
6. A. Edris and B. Bergnstahl, *Nahrung/Food*, 2001, **45**, 133.
7. V. M. Balcao, C. I. Costa, C. M. Matos, C. G. Moutinho, M. Amorim, M. E. Pintado, A. P. Gomes, M. M. Vila and J. A. Teixeira, *Food Hydrocolloids*, 2013, **32**, 425.
8. A. Benichou, A. Aserin and N. Garti, *Polym. Adv. Technol.*, 2002, **13**, 1019.
9. W. Ramsden, *Proc. R. Soc. London*, 1903, **72**, 156.
10. S. U. Pickering, *J. Am. Chem. Soc.*, 1907, **91**, 2001.
11. S. Fujii, Y. L. Cai, Jonathan V. M. Weaver and S. P. Armes, *J. Am. Chem. Soc.*, 2005, **127**, 7304.
12. H. X. Liu, C. Y. Wang, Q. X. Gao, J. X. Chen, B. Y. Ren, X. X. Liu and Z. Tong, *Int. J. Pharm.*, 2009, **376**, 92.
13. C. Y. Wang, C. J. Zhang, Y. Li, Y. H. Chen and Z. Tong, *React. Funct. Polym.*, 2009, **69**, 750.
14. H. X. Liu, C. Y. Wang, Q. X. Gao, J. X. Chen, X. X. Liu and Z. Tong, *Mater. Lett.*, 2009, **63**, 884.
15. H. X. Liu, C. Y. Wang, Q. X. Gao, X. X. Liu and Z. Tong, *Acta Biomater.*, 2010, **6**, 275.
16. Z. J. Wei, C. Y. Wang, S. W. Zou, H. Liu and Z. Tong, *Polymer*, 2012, **53**, 1229.
17. Y. H. Chen, C. Y. Wang, J. X. Chen, X. X. Liu and Z. Tong, *J. Polym. Sci., Part A: Polym. Chem.*, 2009, **47**, 1354.

18. Q. X. Gao, C. Y. Wang, H. X. Liu, C. H. Wang, X. X. Liu and Z. Tong, *Polymer*, 2009, **50**, 2587.
19. R. Aveyard, B. P. Binks and J. H. Clint, *Adv. Colloid Interface Sci.*, 2003, **100**, 503.
20. B. P. Binks and J. H. Clint, *Langmuir*, 2002, **18**, 1270.
21. B. P. Binks and S. O. Lumsdon, *Langmuir*, 2001, **17**, 4540.
22. E. Vignati, R. Piazza and T. P. Lockhart, *Langmuir*, 2003, **19**, 6650.
23. S. Fujii, E. S. Read, B. P. Binks and S. P. Armes, *Adv. Mater.*, 2005, **17**, 1014.
24. N. P. Ashby and B. P. Binks, *Phys. Chem. Chem. Phys.*, 2000, **2**, 5640.
25. B. P. Binks and T. S. Horozov, *Angew. Chem., Int. Ed.*, 2005, **44**, 3722.
26. S. Cauvin, P. J. Colver and S. A. F. Bon, *Macromolecules*, 2005, **38**, 7887.
27. Z. Du, M. P. Bilbao-Montoya, B. P. Binks, E. Dickinson, R. Ettelaie and B. S. Murray, *Langmuir*, 2003, **19**, 3106.
28. S. Abend, N. Bonnke, U. Gutschner and G. Lagaly, *Colloid Polym. Sci.*, 1998, **276**, 730.
29. S. Melle, M. Lask and G. G. Fuller, *Langmuir*, 2005, **21**, 2158.
30. T. Chen, P. J. Colver and S. A. F. Bon, *Adv. Mater.*, 2007, **19**, 2286.
31. S. A. F. Bon and P. J. Colver, *Langmuir*, 2007, **23**, 8316.
32. D. J. Voorn, W. Ming and A. M. van Herk, *Macromolecules*, 2006, **39**, 2137.
33. D. Suzuki, S. Tsuji and H. Kawaguchi, *J. Am. Chem. Soc.*, 2007, **129**, 8088.
34. T. Ngai, S. H. Behrens and H. Auweter, *Chem. Commun.*, 2005, **3**, 331.
35. S. Fujii, S. P. Armes, B. P. Binks and R. Murakami, *Langmuir*, 2006, **22**, 6818.
36. Q. Lan, C. Liu, F. Yang, S. Y. Liu, J. Xu and D. J. Sun, *J. Colloid Interface Sci.*, 2007, **310**, 260.
37. B. Liu, W. Wei, X. Z. Qu and Z. Z. Yang, *Angew. Chem., Int. Ed.*, 2008, **47**, 3973.
38. B. P. Binks and C. P. Whitby, *Colloids Surf., A*, 2005, **253**, 105.
39. T. Ngai, H. Auweter and S. H. Behrens, *Macromolecules*, 2006, **39**, 8171.
40. H. Strohm and P. Löbmann, *J. Mater. Chem.*, 2004, **14**, 2667.
41. I. Akartuna, A. R. Studart, E. Tervoort, U. T. Gonzenbach and L. J. Gauckler, *Langmuir*, 2008, **24**, 7161.
42. S. Sacanna, W. K. Kegel and A. P. Philipse, *Phys. Rev. Lett.*, 2007, **98**, 158301.
43. B. P. Binks and J. A. Rodrigues, *Angew. Chem., Int. Ed.*, 2005, **44**, 441.
44. J. A. Hanson, C. B. Chang, S. M. Graves, Z. B. Li, T. G. Mason and T. J. Deming, *Nature*, 2008, **455**, 85.
45. L. Z. Hong, G. Q. Sun, J. G. Cai and T. Ngai, *Langmuir*, 2012, **28**, 2332.
46. Y. Nonomura, N. Kobayashi and N. Nakagawa, *Langmuir*, 2001, **27**, 4557.
47. A. S. Miguel, J. Scrimgeour, J. E. Curtis and S. H. Behrens, *Soft Matter*, 2010, **6**, 3163.
48. Y. Q. He, F. Wu, X. Y. Sun, R. Q. Li, Y. Q. Guo, C. B. Li, L. Zhang, F. B. Xing, W. Wang and J. P. Gao, *ACS Appl. Mater. Inter.*, 2013, **5**, 4843.

49. P. V. Rijin, H. H. Wang and A. Böker, *Soft Matter*, 2011, 7, 5274.
50. A. K. F. Dyab, *Macromol. Chem. Phys.*, 2011, 7, 5274.
51. M. Matos, A. Timgren, M. Sjöö, P. Dejmek and M. Rayner, *Colloids Surf., A*, 2013, **423**, 147.
52. A. T. Poortinga, *Colloids Surf., A*, 2013, **419**, 15.
53. S. Yang, H. Liu and Z. C. Zhang, *Langmuir*, 2008, **24**, 10395.
54. A. S. Utada, E. Lorenceau, D. R. Link, P. D. Kaplan, H. A. Stone and D. A. Weitz, *Science*, 2005, **308**, 537.
55. T. Nisisako, *Chem. Eng. Technol.*, 2008, **31**, 1091.
56. A. R. Studart, H. C. Shum and D. A. Weitz, *J. Phys. Chem. B*, 2009, **113**, 3914.
57. P. W. Ren, X. J. Ju, R. Xie and L. Y. Chu, *J. Colloid Interface Sci.*, 2010, **343**, 392.
58. M. C. Draper, X. Z. Niu, S. Cho, D. I. Jarnes and J. B. Edel, *Anal. Chem.*, 2012, **84**, 5801.
59. J. D. Wan, L. Shi, B. Benson, M. J. Bruzek, J. E. Anthony, P. J. Sinko, R. K. Prudhomme and H. A. Sone, *Langmuir*, 2012, **28**, 13143.
60. A. Abbaspourrad, W. J. Duncanson, N. Lebedeva, S. H. Kim, A. P. Zhushma, S. S. Datta, P. A. Dayton, S. S. Sheiko, M. Rubinstein and D. A. Weitz, *Langmuir*, 2013, **29**, 12352.
61. W. Wang, M. J. Zhang, R. Xie, X. J. Ju, C. Yang, C. L. Mou, D. A. Weitz and L. Y. Chu, *Angew. Chem., Int. Ed.*, 2013, **52**, 8084.
62. A. Abbaspourrad, N. J. Carroll, S. H. Kim and D. A. Weitz, *Adv. Mater.*, 2013, **25**, 3215.
63. L. Mazutis, J. Gilbert, W. L. Ung, D. A. Weitz, A. D. Griffiths and J. A. Heyman, *Nat. Protoc.*, 2013, **8**, 870.
64. L. L. A. Adams, T. E. Kodger, S. H. Kim, H. C. Shum, T. Franke and D. A. Weitz, *Soft Matter*, 2012, **8**, 10719.
65. W. J. Duncanson, T. Lin, A. R. Abate, S. Seiffert, R. K. Shah and D. A. Weitz, *Lab Chip*, 2012, **12**, 2135.
66. J. I. Park, Z. H. Nie, A. Kumachev, A. I. Abdelrahman, B. P. Binks, H. A. Stone and E. Kumacheva, *Angew. Chem., Int. Ed.*, 2009, **48**, 5300.
67. Z. H. Nie, J. I. Park, W. Li, S. A. F. Bon and E. Kumacheva, *J. Am. Chem. Soc.*, 2008, **130**, 16508.
68. W. Li, H. H. Pham, Z. H. Nie, B. MacDonald, A. Guenther and E. Kumacheva, *J. Am. Chem. Soc.*, 2008, **130**, 9935.
69. S. Dubinsky, H. Zhang, Z. H. Nie, I. Gourevich, D. Voicu, M. Deetz and E. Kumacheva, *Macromolecules*, 2008, **41**, 3555.
70. W. Li, Z. H. Nie, H. Zhang, C. Paquet, P. Garstecki and E. Kumacheva, *Langmuir*, 2007, **23**, 8010.
71. Z. H. Nie, S. Q. Xu, M. Seo, P. C. Lewis and E. Kumacheva, *J. Am. Chem. Soc.*, 2005, **127**, 8058.
72. D. Lee and D. A. Weitz, *Adv. Mater.*, 2008, **20**, 3498.
73. C. Miesch, I. Kosif, E. Lee, J. K. Kim, T. P. Russell, R. C. Hayward and T. Emrick, *Angew. Chem., Int. Ed.*, 2012, **51**, 145.

74. Y. Y. Yang, H. H. Chia and T. S. Chung, *J. Controlled Release*, 2000, **69**, 81.
75. C. H. Villa, L. B. Lawson, Y. M. Li and K. D. Papadopoulos, *Langmuir*, 2003, **19**, 244.
76. J. Cheng, J. F. Chen, M. Zhao, Q. Luo, L. X. Wen and K. D. Papadopoulos, *J. Colloid Interface Sci.*, 2007, **305**, 175.
77. M. M. Dragosavac, R. G. Holdich, G. T. Vladisavljević and M. N. Sovilj, *J. Membr. Sci.*, 2012, **392-393**, 122.
78. M. Kanouni, H. L. Rosano and N. Naouli, *Adv. Colloid Interface Sci.*, 2002, **99**, 229.
79. L. X. Wen and K. D. Papadopoulos, *J. Colloid Interface Sci.*, 2001, **235**, 398.
80. Q. Wang, G. Tan, L. B. Lawson, V. T. John and K. D. Papadopoulos, *Langmuir*, 2010, **26**, 3225.
81. A. S. Hussein, N. Abdullah and A. Fakru'l-razi, *Adv. Polym. Technol.*, 2013, **32**, E486.
82. D. Terwagne, T. Gilet, N. Vandewalle and S. Dorbolo, *Langmuir*, 2010, **26**, 11680.
83. A. Schuch, P. Deiters, J. Henne, K. Köhler and H. P. Schuchmann, *J. Colloid Interface Sci.*, 2013, **402**, 157.
84. D. Chognot, M. Léonard, J. L. Six and E. Dellacherie, *Colloids Surf., B*, 2006, **51**, 86.
85. W. Yafei, Z. Tao and H. Gang, *Langmuir*, 2006, **22**, 67.
86. B. Wu and H. Q. Gong, *Microfluid. Nanofluid.*, 2013, **14**, 637.
87. M. F. Zambaux, F. Bonneaux, R. Gref, P. Maincent, E. Dellacherie, M. J. Alonso, P. Labrude and C. Vigneron, *J. Controlled Release*, 1998, **50**, 31.
88. M. Pradhan and D. Rousseau, *J. Colloid Interface Sci.*, 2012, **386**, 398.
89. S. H. Wu, Y. Hung and C. Y. Mou, *Chem. Mater.*, 2013, **25**, 352.
90. J. Zhang, X. Ge, M. Wang, J. Yang, Q. Wu, M. Wu, N. Liu and Z. Jin, *Chem. Commun.*, 2010, **46**, 4318.
91. H. Maeda, M. Okada, S. Fujii, Y. Nakamura and T. Furuzono, *Langmuir*, 2010, **26**, 13727.
92. Q. Gao, C. Wang, H. Liu, Y. Chen and Z. Tong, *Polym. Chem*, 2010, **1**, 75.
93. S. W. Zou, C. Y. Wang, Q. X. Gao and Z. Tong, *J. Dispersion Sci. Technol.*, 2013, **34**, 173.
94. S. W. Zou, H. Liu, Y. Yang, Z. J. Wei and C. Y. Wang, *React. Funct. Polym.*, 2013, **73**, 1231.
95. Z. J. Wei, C. Y. Wang, H. Liu, S. W. Zou and Z. Tong, *J. Appl. Polym. Sci.*, 2012, **125**, 358.
96. H. Liu, C. Y. Wang, S. W. Zou, Z. J. Wei and Z. Tong, *Langmuir*, 2012, **28**, 11017.
97. Y. Yang, Z. J. Wei, C. Y. Wang and Z. Tong, *ACS Appl. Mater. Inter.*, 2013, **5**, 2495.
98. Y. Hu, Y. Yang, Y. Ning, C. Y. Wang and Z. Tong, *Colloids Surf., B*, 2013, **112**, 96.

99. H. Liu, C. Y. Wang, S. W. Zou, Z. J. Wei and Z. Tong, *Polym. Bull.*, 2012, **69**, 765.

100. S. W. Zou, Y. Yang, H. Liu and C. Y. Wang, *Colloids Surf., A*, 2013, **436**, 1.

101. Z. J. Wei, C. Y. Wang, S. W. Zou, H. Liu and Z. Tong, *Colloids Surf., A*, 2011, **39**, 116.

102. Z. J. Wei, C. Y. Wang, H. Liu, S. W. Zou and Z. Tong, *Colloids Surf., B*, 2012, **91**, 97.

103. Z. J. Wei, Y. Yang, R. Yang and C. Y. Wang, *Green Chem.*, 2012, **14**, 3230.

104. H. Liu, Q. X. Gao, H. X. Liu and C. Y. Wang, *J. Macromol. Sci., Phys.*, 2014, **53**, 52.

105. Y. Yang, Z. J. Wei, C. Y. Wang and Z. Tong, *Chem. Commun.*, 2013, **49**, 7144.

106. Y. Yang, C. Y. Wang and Z. Tong, *RSC Advances*, 2013, **3**, 4514.

107. Y. Yang, Y. Ning, C. Y. Wang and Z. Tong, *Polym. Chem.*, 2013, **4**, 5407.

108. M. L. Pang, A. J. Cairns, Y. L. Liu, Y. Belmabkhout, H. C. Zeng and M. Eddaoudi, *J. Am. Chem. Soc.*, 2013, **135**, 10234.

109. J. W. Kim, A. Fernandez-Nieves, N. Dan, A. S. Utada, M. Marquez and D. A. Weitz, *Nano Lett.*, 2007, **7**, 2876.

110. G. P. Liu, S. Y. Liu, X. Q. Dong, F. Yang and D. J. Sun, *J. Colloid Interface Sci.*, 2010, **345**, 302.

111. A. S. Miguel, J. Scrimgeour, J. E. Curtis and S. H. Behrens, *Soft Matter*, 2010, **6**, 3163.

112. S. Simovic and C. A. Prestidge, *Langmuir*, 2008, **24**, 7132.

113. Q. C. Yuan, O. J. Cayre, S. Fujii, S. P. Armes, R. A. Williams and S. Biggs, *Langmuir*, 2010, **26**, 18408.

114. H. L. Wang, X. M. Zhu, L. Tsarkova, A. Pich and M. Moller, *ACS NANO*, 2011, **5**, 3937.

115. L. A. Fielding and S. P. Armes, *J. Mater. Chem.*, 2012, **22**, 11235.

116. J. O. You, M. Rafat and D. T. Auguste, *Langmuir*, 2011, **27**, 11282.

117. K. L. Thompson, E. C. Giakoumatos, S. Ata, G. B. Webber, S. P. Armes and E. J. Wanless, *Langmuir*, 2012, **28**, 16501.

118. W. C. Mak, J. Bai, X. Y. Chang and D. Trau, *Langmuir*, 2009, **25**, 769.

119. Y. Pan, J. H. Gao, B. Zhang, X. X. Zhang and B. Xu, *Langmuir*, 2010, **26**, 4184.

120. K. L. Thompson, S. P. Armes, J. R. Howse, S. Ebbens, I. Ahmad, J. H. Zaidi, D. W. York and J. A. Burdis, *Macromolecules*, 2010, **43**, 10466.

121. O. J. Cayre, P. F. Noble and V. N. Paunov, *J. Mater. Chem.*, 2004, **14**, 3351.

122. S. Jiang and S. Granick, *Langmuir*, 2008, **24**, 2438.

123. H. N. Yow and A. F. Routh, *Soft Matter*, 2006, **2**, 940.

124. P. F. Noble, O. J. Cayre, R. G. Alargova, O. D. Velev and V. N. Paunov, *J. Am. Chem. Soc.*, 2004, **126**, 8092.

125. R. Langer, *Nature*, 1998, **392**, 5.

126. A. Zhuo, Z. F. Li, G. Z. Zhang and T. Ngai, *Colloids Surf., A*, 2011, **384**, 592.

127. C. Y. Wang, H. X. Liu, Q. X. Gao, X. X. Liu and Z. Tong, *ChemPhysChem*, 2007, **8**, 1157.
128. P. Dommersnes, Z. Rozynek, A. Mikkelsen, R. Castberg, K. Kjerstad, K. Hersvik and J. O. Fossum, *Nat. Commun.*, 2013, **4**, 2066.
129. S. Easwaramoorthi, P. Kim, J. M. Lim, S. Song, H. Suh, J. L. Sessler and D. Kim, *J. Mater. Chem.*, 2010, **20**, 9684.
130. P. H. Keen, N. K. H. Slater and A. F. Routh, *Langmuir*, 2012, **28**, 1169.
131. R. K. Kumar, M. Li, S. N. Olof, A. J. Patil and S. Mann, *Small*, 2013, **9**, 357.
132. A. Boker, J. He, T. Emrick and T. P. Russell, *Soft Matter*, 2007, **3**, 1231.
133. D. Lee and D. A. Weitz, *Small*, 2009, **5**, 1932.
134. A. S. Miguel and S. H. Behrens, *Soft Matter*, 2011, 7, 1948.
135. A. J. Karkamkar, S. Kim, S. D. Mahanti and T. J. Pinnavaia, *Adv. Funct. Mater.*, 2004, **14**, 507.
136. B. Liu and H. C. Zeng, *J. Am. Chem. Soc.*, 2004, **126**, 16744.
137. F. Caruso, R. A. Caruso and H. Möhwald, *Science*, 1998, **282**, 1111.
138. Y. Yin, R. M. Rioux, C. K. Erdonmez, S. Hughes, G. A. Somorjai and P. Alivisatos, *Science*, 2004, **304**, 711.
139. X. W. Lou, L. A. Archer and Z. Yang, *Adv. Mater.*, 2008, **20**, 3987.
140. Y. Ning, Y. Yang, C. Y. Wang, T. Ngai and Z. Tong, *Chem. Commun.*, 2013, **49**, 8761.
141. C. W. Chen, T. Serizawa and M. Akashi, *Chem. Mater.*, 2002, **14**, 2232.
142. B. P. Binks, A. K. F. Dyab, P. D. I. Fletcher and H. Barthel (Wacker-Chemie GmbH), US 7722891 B2, 2010.
143. B. P. Binks, A. K. F. Dyab and P. D. I. Fletcher, *Phys. Chem. Chem. Phys.*, 2007, **9**, 6391.
144. A. K. F. Dyab, PhD thesis, The University of Hull, 2004.
145. A. K. F. Dyab, *Macromol. Chem. Phys.*, 2012, **213**, 1815.
146. T. Nisisako, T. Torii, T. Takahashi and Y. Takizawa, *Adv. Mater.*, 2006, **18**, 1152.
147. F. Wurm and A. F. M. Kilbinger, *Angew. Chem., Int. Ed.*, 2009, **48**, 8412.
148. N. Glaser, D. J. Adams, A. Böker and G. Krausch, *Langmuir*, 2006, **22**, 5227.
149. T. M. Ruhland, A. H. Gröschel, A. Walther and A. H. E. Müller, *Langmuir*, 2011, **27**, 9807.
150. A. Walther, K. Matussek and A. H. E. Müller, *ACS Nano*, 2008, **2**, 1167.
151. M. D. McConnell, M. J. Kraeutler, S. Yang and R. J. Composto, *Nano Letters*, 2010, **10**, 603.
152. S. C. Glotzer and M. J. Solomon, *Nat. Mater.*, 2007, **6**, 557.
153. A. Walther, M. Drechsler and A. H. E. Müller, *Soft Matter*, 2009, **5**, 385.
154. A. Walther, M. Drechsler, S. Rosenfeldt, L. Harnau, M. Ballauff, V. Abetz and A. H. E. Müller, *J. Am. Chem. Soc.*, 2009, **131**, 4720.
155. K. P. Yuet, D. K. Hwang, R. Haghgooie and P. S. Doyle, *Langmuir*, 2009, **26**, 4281.
156. S. Jiang, Q. Chen, M. Tripathy, E. Luijten, K. S. Schweizer and S. Granick, *Adv. Mater.*, 2010, **22**, 1060.

157. Y. Ning, C. Y. Wang, T. Ngai and Z. Tong, *Langmuir*, 2013, **29**, 5138.

158. Y. Ning, C. Y. Wang, T. Ngai, Y. Yang and Z. Tong, *RSC Advances*, 2012, **2**, 5510.

159. N. R. Cameron and D. C. Sherrington, *Adv. Polym. Sci.*, 1996, **126**, 163.

160. B. P. Binks and R. Murakami, *Nat. Mater.*, 2006, **5**, 865.

161. F. Leal-Calderon and V. Schmitt, *Curr. Opin. Colloid Interface Sci.*, 2008, **13**, 217.

162. B. P. Binks and A. Rocher, *Phys. Chem. Chem. Phys.*, 2010, **12**, 9169.

163. A. C. Martinez, E. Rio, G. Delon, A. Saint-Jalmes, D. Langevin and B. P. Binks, *Soft Matter*, 2008, **4**, 1531.

164. B. P. Binks, B. Duncumb and R. Murakami, *Langmuir*, 2007, **23**, 9143.

165. K. Saleh, L. Forny, P. Guigon and I. Pezron, *Chem. Eng. Res. Design*, 2011, **89**, 537.

166. A. D. Dinsmore, M. F. Hsu, M. G. Nikolaides, M. Marquez, A. R. Bausch and D. A. Weitz, *Science*, 2002, **298**, 1006.

167. B. P. Binks, A. J. Johnson and J. A. Rodrigues, *Soft Matter*, 2010, **6**, 126.

168. R. Murakami and A. Bismark, *Adv. Funct. Mater.*, 2010, **20**, 732.

169. B. O. Carter, W. Wang, D. J. Adams and A. I. Cooper, *Langmuir*, 2010, **26**, 3186.

170. R. Murakami, H. Moriyama, M. Yamamoto, B. P. Binks and A. Rocher, *Adv. Mater.*, 2012, **24**, 767.

171. R. Hayes, G. G. Warr and R. Atkin, *Phys. Chem. Chem. Phys.*, 2010, **12**, 1709.

172. H. X. Gao, J. C. Li, B. X. Han, W. N. Chen, J. L. Zhang, R. Zhang and D. D. Yan, *Phys. Chem. Chem. Phys.*, 2004, **6**, 2914.

173. Z. Qiu and J. Texter, *Curr. Opin. Colloid Interface Sci.*, 2008, **13**, 252.

174. T. L. Greaves and C. J. Drummond, *Chem. Soc. Rev.*, 2008, **37**, 1709.

175. H. Ma and L. L. Dai, *Langmuir*, 2011, **27**, 508.

176. B. P. Binks, A. K. F. Dyab and P. D. I. Fletcher, *Chem. Commun.*, 2003, **54**, 2540–2541.

CHAPTER 9

Particle-Stabilized Emulsions as Templates for Hollow Spheres and Microcapsules

SIMON BIGGS*[a] AND OLIVIER CAYRE[b]

[a] Faculty of Engineering, Architecture & Information Technology, The University of Queensland, Brisbane QLD 4072; [b] Institute of Particle Science and Engineering, The University of Leeds, Leeds LS2 9JT, UK
*Email: simon.biggs@uq.edu.au

9.1 Introduction

Colloidosomes are an interesting class of hollow core-shell microcapsules typically produced from Pickering emulsion templates. The term colloidosome was first introduced by Dinsmore *et al.*[1] in their seminal work in this area and arises from the similarities to liposome or polymerosome capsules. In this case, the capsule wall consists of close packed colloidal particles that have been permanently locked using methods reviewed in this chapter.

The interest in colloidosomes arises from a range of unique features. For example, the shell thickness can be manipulated by choosing different sized colloids. Similarly, the porosity of the shell can be adjusted by changing the colloidal particle size through an associated change in the size of the interstitial pores between the particles in a close packed film. Another key feature is the inherent flexibility in these systems offered by the extremely large range of particle types that can be used to form the capsule shell across metallic, inorganic, organic and natural colloidal materials. Additionally, the enhanced stability of Pickering emulsions (as compared to surfactant- or

RSC Soft Matter No. 3
Particle-Stabilized Emulsions and Colloids: Formation and Applications
Edited by To Ngai and Stefan A. F. Bon
Published by the Royal Society of Chemistry, www.rsc.org

polymer-stabilized emulsions) lends itself well to their use as precursors for microcapsules through 'tough' additional processing steps.

In the following sections, we review briefly the current state of knowledge in this area including approaches to the manufacture of these capsules, routes to manipulate the porosity and permeability of the shell, and examples of systems that have been successfully encapsulated.

9.2 Early Research

The field of microcapsules produced from particle-stabilized emulsion templates has grown dramatically following the seminal work of Weitz and coworkers, in which they coined the term colloidosome for these types of capsules.[1] However, there is a significant amount of work that both preceded and inspired this heavily cited article.

For example, from 1996 to 1999, Velev and coworkers published a series of articles that demonstrated the possibility of assembling small particles at the surface of emulsion droplets.[2–4] This procedure was carried out in multiple steps where the surface of the emulsion droplets was initially coated with an appropriate material to enable an electrostatic attraction between the emulsion droplet and the particle to drive the particle coating. This initial work demonstrated the possibility of using a combination of colloidal particles and emulsion droplets to create complex microstructures where the packing of the particles on the surface of the liquid droplets can be controlled.

Later, Tsapis *et al.* demonstrated the possibility of synthesizing thin-walled macroscale hollow structures, where the shell was entirely composed of 170 nm latex particles by spray drying suspensions.[5] The synthesis of these structures was motivated by enabling a more efficient controlled delivery of drugs in the form of nanoparticles *via* inhalation, without the potential for sustained release in the lungs (Figure 9.1).

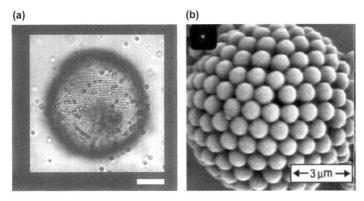

(a)   **(b)**

←3 μm→

Figure 9.1 (a) Optical micrograph of 2D-hexagonal packing of adsorbed latex particles on an octanol droplet. Taken from reference 2. Scale bar: 10 μm. (b) 'Colloidosome' microcapsules created from water-in-oil emulsions stabilized by latex particles. Taken from reference 28. Scale bar: 3 μm.

In the late 1990s, two strongly developing areas of interest contributed to the progress made in designing colloidal-shell hollow spheres from emulsion droplet templates.

Firstly, the resurgence of interest in the stabilization of emulsion droplets with colloidal particles led to several studies on understanding this phenomenon and this is extensively reviewed elsewhere.[6–8] For a comprehensive review of the understanding of these systems at that time, interested readers are directed to reference 6. An improved understanding of the behaviour of colloidal particles at liquid–liquid interfaces subsequently facilitated the development of new methods for the synthesis of microcapsules with a shell of colloidal particles.

Secondly, at the same time several groups were working on the use of particle templates, subsequently dissolved, for creating soft, functional shells to be used as capsule membranes. From the initial articles by Decher[9] and Möhwald and coworkers,[10] this work rapidly evolved to integrate colloidal particles within the designed shell. This was undertaken as a way of both expanding the range of material incorporated within the multilayer shells and introducing additional functionality to these systems.

The field evolved to look for additional functionality to these systems. This chapter will review the processes used to synthesize microspheres and microcapsules with the potential for adding functionality.

9.3   Hollow Spheres/Microcapsules

Designing hollow spheres with a shell containing colloidal particles has been achieved in several different ways,[11] some of which are mentioned in the previous section.[10]

However, a more limited number of methods have been developed for using particle-stabilized emulsions (*i.e.* in the absence of another emulsifier) as templates for the preparation of hollow microspheres and microcapsules. In reviewing these different methods in this chapter, we propose to classify them on the basis of the procedure used to 'freeze' the structure, yielding a hollow core with a colloidal shell permanently fixed on the surface of the microcapsules. Three main categories are devised.

Firstly, we describe methods where a polymerization or a sol–gel reaction occurring either in the droplet core or at the interface induces the creation of a membrane entrapping the colloidal monolayer at the interface. Secondly, the adsorption/precipitation of polymers already present in the initial system are reviewed. Lastly, we examine methods of modifying the adsorbed colloidal monolayer itself to create a permanent shell.

9.3.1   Polymerization/Sol–Gel Reactions

Polymerization and sol–gel reactions in heterogeneous systems have long been used for the preparation of solid-core inorganic and latex particles. Over the last 10–15 years, they have been adapted for solid-stabilized emulsion droplets to prepare particles with an embedded particle mono-layer/film on their surface. For example, several articles have demonstrated the possibility of using polymerization of monomer formulations to obtain a solid polymer core surrounded by a film of the stabilizing particles (or ar-moured latex particles). The core of the complex particles obtained with this method can be chosen to be a crosslinked polymer mesh[12] or a swellable microgel,[13] for instance.

However, in this chapter we will concentrate on reviewing the literature where a hollow core is achieved, which we define as a liquid core surrounded by a polymeric or inorganic film entrapping the stabilizing particle monolayer.

9.3.1.1 Reactions Occurring Within Emulsion Droplet Templates

Here we review methods where specific systems are engineered to enable a chemical reaction to occur within the solid-stabilized emulsion droplets leading to the subsequent formation of a shell at the interface. The chemical reaction here is either a polymerization or a sol–gel precipitation. In both cases, this enables the creation of a complex shell composed of the created material and the particle monolayer initially adsorbed at the interface.

For example, Gao *et al.* developed a polymerization method to generate hollow microspheres from inverse (water core) emulsions stabilized by SiO_2 nanoparticles in hexane.[14] These authors polymerized N-isopropyl acrylamide (NIPAM) monomer dissolved in the emulsion internal aqueous phase above the lower critical solution temperature of its polymer ($\sim 32\ ^\circ C$). As the polymer formed, it precipitated within the droplet core and adsorbed at the interface to form a film trapping the SiO_2 nanoparticles. This work is analogous to a study from Duan *et al.*[13b] who prepared microgel particles decorated with SiO_2 nanoparticles.

9.3.1.2 Interfacial Reactions

This method involves the growth of a film that entraps the colloidal monolayer stabilizing the emulsion droplets to enable the formation of a permanent shell. This can be achieved in two ways. In one case, a poly-merization can be initiated from the surface of the stabilizing particles.[15] This requires the particle surface to be initially modified with a moiety that can participate in the polymerization reaction, generally a chain transfer agent for controlled living radical polymerization. A different example in-volves the creation of a film through an interfacial reaction between prod-ucts dispersed in the oil phase and the water phase.[16–18]

Both of these methods allow for locally growing the film at the interface between oil and water and have been shown to successfully entrap the stabilizing particles.

An example of the first method was demonstrated by Chen *et al.*[15] These authors modified the surface of SiO_2 nanoparticles with an atom transfer radical polymerization initiator and used them to stabilize a paraffin oil-in-water emulsion, where the particles were initially dispersed in the oil phase. By subsequently adding the monomer 2-hydroxyethyl methacrylate (HEMA) to the aqueous phase and polymerizing the system at 35 °C using copper(I) chloride/bipyridine as catalyst (initiated from the particle surface), they were able to create a polymer film embedding the stabilizing particles. The ability of the growing polymer to entangle and potentially crosslink within the particle monolayer was key in this case to obtain fully formed capsules with a permanent binary colloidal/polymeric shell.

Croll *et al.* used an interfacial reaction between isocyanates dissolved in the oil phase and amines added in the continuous aqueous phase to form a film of polyurea. The film formation was activated by the addition of the amines once latex or microgel particles had adsorbed onto the oil–water interface. This process resulted in the particles remaining embedded in the formed polyurea capsule shells. The authors subsequently compared the release of xylene from the capsule cores for samples synthesized with microgel particle stabilizers of different crosslinking degrees.[16]

Recently, two research groups have also developed interfacial sol–gel methods for creating an inorganic film of silica on the surface of Pickering emulsions. This film was shown to efficiently capture the stabilizing particles on the surface of the created microcapsule shell.[17,18]

9.3.2   Existing Polymer Precipitation by Solvent Extraction

The ability to precipitate a polymer by extraction of a solvent from within the dispersed phase was used to create different types of microstructures from solid-stabilized emulsion templates. Several examples have shown that this method can lead to 'solid' cores with an entrapped colloidal monolayer.[19,20]

Again, we concentrate here on reports where hollow cores surrounded by a colloidal shell are created. In this case, upon extraction of the solvent from the dispersed phase, a polymer film is precipitated at the droplet interface, where it entraps the stabilizing colloidal monolayer. This was demonstrated with both organic[21–23] and aqueous dispersed phases for emulsions stabilized by a broad range of particles, including silica,[21,22] latex,[23] gold,[21] clay[21] and microgel particles.[24,25]

For example, Berger *et al.* prepared emulsions of chloroform in water stabilized by microgel particles made from the temperature-responsive polymer poly(N-vinylcaprolactam). The dispersed phase also contained a water-insoluble polymer, which precipitated upon extraction of the chloroform from the emulsion droplets. The polymer formed a film on the surface of the droplets and entrapped the microgel particles adsorbed on the surface of the emulsion droplets.[24]

We have used a similar method to trap a range of colloidal particles on the surface of microcapsules. We prepared solid-stabilized emulsion droplets of a mixture of a water-insoluble polymer and two oils (a good solvent and a bad solvent for the polymer). By extraction of the good solvent, the polymer

(a) (b) (c)

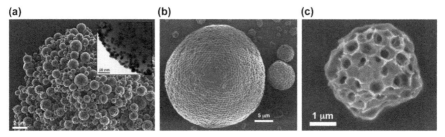

Figure 9.2 (a) Scanning electron micrograph of microcapsules obtained from a solvent extraction method leading to the precipitation of a polymer shell on the surface of oil-in-water droplets. As the film precipitates, it can trap particles initially adsorbed at the interface. Here gold nanoparticles are initially driven to the emulsion droplet surface and remain embedded in the polymer film, as demonstrated in the transmission electron micrograph (inset).[21] (b) Microcapsules obtained from the method described in reference 21, while using microgel particles as the emulsifier. Microcapsules have a shell of microgel particles initially adsorbed at the oil–water interface and subsequently captured on the surface of a forming PMMA film as a co-solvent is extracted from the droplets.[25] (c) Scanning electron micrograph of a typical structure of the microcapsules from (b) after swelling of the microgel particles at the appropriate pH. A PMMA film 'skeleton' with indented pores where the microgels were attached is clearly observable.[22]

was precipitated at the interface and the particle monolayer was retained on the surface of the created microcapsules.[21]

On this basis we also performed the same experiments using poly(2-vinylpyridine) microgel particles as the Pickering emulsifiers. We demonstrated that these particles could also be retained on the surface of the formed polymer film, forming a tightly packed monolayer (Figure 9.2b). Furthermore, by switching the pH to values where the microgel particles were swollen, we were able to force the microgels to desorb from the film, leaving a 'skeleton-like' structure with well-defined pores (Figure 9.2c).[25]

This procedure allows preparation of binary membranes where the thickness and porosity of the polymer film can be adjusted by simply varying the initial polymer concentration in the dispersed phase and the solvent extraction rate, respectively.[19]

9.3.3 Locking the Particle Monolayer at the Oil–Water Interface

One of the initial ways of securing the particle monolayer at the oil–water interface was to perform a physical or chemical modification of the particle surface (or in some cases core). This technique evolved from a simple fusing of adjacent particles *via* heating above the glass transition temperature (T_g) of the polymer particles to the use of chemical crosslinkers binding adjacent particles to one another.

9.3.3.1 Particle Monolayer Sintering

The concept of sintering the particle monolayer on the surface of the emulsion droplets was introduced by Weitz and coworkers.[1] They assembled 900 nm polystyrene particles as a monolayer on an emulsion drop surface and raised the temperature of the system above the T_g of the polymer particles. This resulted in sintering of adjacent particles and formation of bridges within the monolayer at the oil–water interface, thus rendering it permanently bound.

Using the same principle, Laib and Routh prepared latex particles of poly(styrene-co-butylacrylate) with the aim of reducing the T_g of the particles used to stabilize the initial emulsion droplets. Using these particles, they were able to permanently sinter the particle monolayer at the interface at 45 °C.[26] A similar result was achieved when using particles with a polymer shell of pDMAEMA as the Pickering emulsifier.

The ability of some oils to swell polymeric particles has also been used to enable interpenetration of adjacent particles within the stabilizing monolayer to achieve the same permanent locking. For example, the use of tricaprylin oil at temperatures around 75 °C was found to swell polystyrene latex particles. By preparing water-in-tricaprylin emulsions at this temperature, colloidosome capsules were obtained with a particle shell fused to different degrees, depending on the time left for the monolayer to sinter at this temperature. An optical micrograph of a typical microcapsule obtained by this process is shown in Figure 9.3c.[27]

Examples of this technique have been published, where the effect of sintering times on the structure of the permanent particle shell is studied.[27,28] These studies have all shown the formation of microcapsules with a colloidal shell of apparent smaller porosity as the particle monolayer is sintered to a

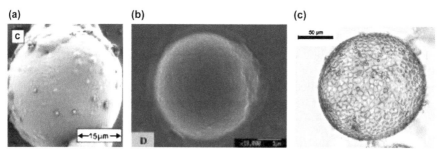

Figure 9.3 (a) Microcapsule obtained from a sunflower oil-in-water emulsion stabilized by 1.1 μm DVB-crosslinked polystyrene particles sintered at 105 °C for 2 h.[28] (b) Microcapsule obtained from a water-in-dodecane emulsion where poly(styrene-co-butylacrylate) particles were assembled at the interface by the addition of a non-ionic surfactant. The particle monolayer was sintered above the T_g of the latex particles at 60 °C for 15 min.[26] (c) Aqueous gel core microcapsule obtained from a water-in-tricaprylin emulsion at 75 °C as a result of swelling and fusing of the particle monolayer.[27]

higher degree when heating for longer periods. In addition, it has been demonstrated that above a critical time of sintering, the particle shell appears entirely non-porous in the limit of resolution of electron microscopes used to characterize these systems.

In addition, a recent study from Salari *et al.* demonstrated the successful use of a polymer stabilizer for the emulsion droplets to drastically decrease inter-droplet fusing while heating the system above the T_g of the particles used as stabilizers.[29] This article provides a potential solution to an often overlooked issue by enabling the process of sintering the particle monolayer at the interface of the droplets for concentrated systems.

This is a key aspect of colloidosome microcapsule research because their potential impact as encapsulation/delivery systems depends greatly on the possibility of manufacturing them at industry-relevant scales.

9.3.3.2   Particle Crosslinking

This process was developed as an alternative to sintering at high temperature and allows one to work with large emulsion droplet concentrations (*i.e.* 50% volume fraction of dispersed phase and potentially above) if the crosslinker is initially dissolved in the dispersed phase. This avoids the inter-droplet crosslinking/fusing highlighted above in the case of particle monolayer sintering above the polymer particle T_g at large dispersed phase volume fraction.

Several research groups have used this method to lock the particle monolayer on the microcapsule shells to study encapsulation and release of model species,[30,31] pH-triggered morphological changes[32,33] or the interaction of approaching colloidosome surfaces.[34] With this method, a range of particles has been used from sterically-stabilized latex,[30,31,35,36] microgel particles,[33] silica[37] or metallic nanoparticles and epoxy resin microrods.[38]

For example, we have used polystyrene latex particles synthesized by emulsion polymerization in the presence of a diblock copolymer stabilizer, where the water-soluble block is pH-responsive. These particles show significant variation in hydrodynamic diameter due to the response of the polymer shell to variations in pH. We created oil-in-water emulsions stabilized by these particles and crosslinked the polymer stabilizer on adjacent particles in the adsorbed monolayer.[35,36] We replaced the oil core of the prepared capsules with water using a co-solvent miscible with both phases and obtained flattened 'deflated' microcapsules. We subsequently showed that it was possible to re-swell these capsules by equilibration of a created chemical potential difference between the bulk and the inner core of the microcapsules.[31]

9.3.3.3   Addition of One (Multiple) Layer(s) of Adsorbing Polymer

Another method for rendering the stabilizing particle monolayer is to adsorb one or more additional layer(s) of polymer/polyelectrolyte on the surface of the Pickering emulsion droplets.[30,39,40]

This has two main advantages. Firstly it ensures that adjacent particles are bridged on the surface of the emulsion droplets. Secondly it provides a way of controlling further the permeability of the microcapsule shells created. Generally this is used to decrease the shell porosity in the case where the structures are designed to encapsulate molecules of low molecular weight.

However, one major drawback of this technique is the necessity to perform multiple additional washing steps to remove both the excess stabilizing particles (if originally dispersed in the continuous phase) and polymer/polyelectrolyte in solution necessary to avoid inter-droplet bridging.

For example, Li and Stöver have used emulsions stabilized by silica nanoparticles as templates for performing the adsorption of multiple layers of oppositely charged polyelectrolytes.[40]

Thompson *et al.* initiated the polymerization of a pyrrole monomer within the continuous aqueous phase of microcapsules with a crosslinked particle monolayer. This procedure resulted in the coating of the microcapsules with polypyrrole, which improved the retention of a fluorescein dye initially dissolved in the sunflower oil core of the emulsion templates.[30]

9.4 Adding Functionality/Structural Strength and Applications

9.4.1 Methods for Consolidating the Microcapsule Core

During early development of these microcapsules, some research groups noted the potential issues posed by the sometimes harsh processing conditions used to enable the removal of the oil phase (whether dispersed or continuous phase) from the system and obtain water-filled microcapsules. Two main methods were then used to increase the mechanical strength of the capsule cores.

i. The introduction of a gelling polysaccharide to the aqueous dispersed phase of the Pickering emulsion templates.[27,38] In this case, the emulsions were prepared above the melting point of the polysaccharide and subsequently cooled down to enable gelling of the droplets. The gelation of the microcapsule cores served two purposes in this case. Firstly it provided a higher mechanical strength to the capsules and allowed for a higher rate of survival to the processing conditions used to obtain water-in-water microcapsules (*i.e.* when removing the continuous oil phase). Secondly, it was claimed that a gelled aqueous core would allow for additional control over release of encapsulated species.

Subsequently, other research groups introduced a gelling agent within the microcapsule aqueous cores as a way of increasing their mechanical strength.[41,42] In one case, this procedure enabled the prepared robust microcapsules – with magnetic nanoparticles as the colloidal shell – to be washed and separated with a magnet without damaging their structure.[41] Alternatively, it also allowed for a robust aqueous-based template to be prepared and transferred to appropriate solvents to conduct a layer-by-layer polymer adsorption to increase the retention of water-soluble encapsulated species.[42]

ii. Wang *et al.*[43] combined the use of $CaCO_3$ nanoparticles as the stabilizers of a water-in-sunflower oil emulsion with sodium alginate dissolved in the aqueous dispersed phase. This allowed for gelation of (some of) the aqueous core by complexation between the alginate and the calcium divalent ions. The authors demonstrated that not all nanoparticles at the interface were consumed by gelation of the aqueous core and that the structure remained with a colloidal shell. A short study on release of a dye initially dissolved in the aqueous dispersed phase showed increased dye retention in these colloidosomes as compared to retention in standard alginate gel spheres.

9.4.2   Adding Functionality

One of the main advantages of microcapsules made from Pickering emulsion templates is the variety of the particles (and polymers) that can be used to build the surface shells. Several examples have demonstrated the versatility of these systems by incorporating functional particles/polymers within the microcapsule shells.

9.4.2.1   *Magnetic Microcapsules*

For example, the use of magnetic nanoparticles as the Pickering emulsifiers for the emulsion template has allowed the preparation of microcapsules that can be manipulated with a magnet.[41] This serves multiple purposes.

- The microcapsules can easily be isolated within the suspension in a specific area of a vessel. This enables easier treatment of the prepared emulsions, where excess particles are washed away and, potentially, where the dispersed phase is replaced.
- When considering an application for these microcapsules, they could be manipulated towards a target of interest (*e.g.* a cell wall, a surface) using a magnet to direct the microcapsules. This could be used to release encapsulated active ingredients at a specific location where they are needed. This is a clear advantage to avoid the common large losses of actives as a result of non-targeted release with current commercialized technologies (Figure 9.4).

9.4.2.2   *Responsive Polymers/Particles*

Several methods have incorporated the properties of environment-responsive polymers into the shell of the microcapsules derived from Pickering emulsions. In most cases, these polymers are initially integrated in the

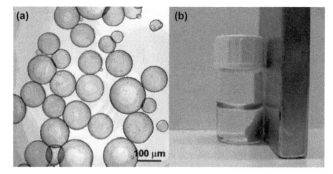

Figure 9.4 (a) Aqueous dispersion of magnetic colloidosome microcapsules prepared with 8 nm Fe_3O_4 particles trapped on the surface of the capsules by
the gelled aqueous core. (b) Demonstration of the magnetic properties of
the microcapsules. This was used as a simple way of isolating the
microcapsules while the oil continuous phase was replaced by water.[41]

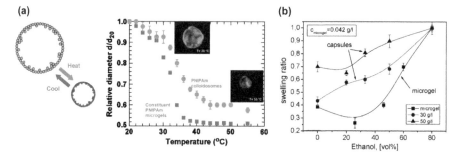

Figure 9.5 Examples of the possibility of controlling the volume contained within
microcapsules when using microgel particles as the colloidal shell.
(a) Microcapsules prepared from a Pickering emulsion initially stabilized
by PNIPAm microgel particles. The particles are crosslinked, allowing
swelling and contraction of the colloidosome microcapsules to be controlled by varying the temperature.[33] (b) Swelling ratios of microcapsules
with a shell of poly(N-vinylcaprolactam) microgels suspended in various
water/ethanol mixtures.[24]

particle emulsifier design in the form of a microgel or a solid core-polymer
shell particle. Review articles on the design of these particles and their
ability to selectively stabilize interfaces are available (Figure 9.5).[44]

9.5   Release Studies

9.5.1   Small Encapsulated Molecules ($<$5 kDa)

In their original article, Dinsmore *et al.*[1] postulated that colloidosomes
could be used to encapsulate a wide range of different materials including
drug actives, nutrients, colloids and living cells. Of most widespread interest

for new commercial encapsulation technologies are approaches for low molecular weight molecules that are highly volatile (*e.g.* perfumes), poorly soluble or extremely sensitive to their environment. A number of papers have been published over the last decade or so examining the efficacy of colloidosome capsules for low molecular weight molecule applications. The use of colloidal particles to fabricate the shell of a colloidosome provides these capsules with an inherent rigidity that can be adjusted somewhat through the choice of the colloidal particle material properties. Such rigid shells may, for some applications, offer advantages over the softer liposome or polymersome technology. As we have noted above, the inherent porosity of the colloidosome is controlled by interstitial pore size and this can be altered either by varying the size of the colloidal particles used to make the shell or through some post-formation annealing process (see Section 9.3.3). In general, standard colloidosomes have a high permeability even to polymer molecules as a consequence of the large number of relatively large pores. The porosity can be reduced somewhat by annealing the shell when using polymer latex particles as the building block, but early work suggested that an impermeable shell was never achieved through this route and pinhole defects were always present, providing a leakage route for small molecules.

The rigid nature of the colloidosome shell requires a breakage mechanism to allow release of the capsule contents. Initial approaches suggested the use of mechanical forces. However, more sophisticated methods that may provide routes for a triggered release have also been suggested. For example, Kim *et al.*[45] explored the use of a microgel scaffold as a template onto which a colloidosome can be constructed. They showed how the swelling/collapse response of the microgel to changing conditions could be used to tune the porosity of the shell. Using the uptake and release of fluorescein, they demonstrated how the permeability of the shell could be adjusted by altering the degree of swelling for the microgel template. No conditions were found, however, even when using a low T_g latex and very strong annealing, that completely prevented release of the small molecule fluorophore. Lee and Weitz[46] utilized a microcapillary manufacturing technique to produce a double emulsion stabilized with silica colloids. Evaporation of the oil led to the formation of relatively thick-walled colloidosome capsules with a permeability that is controlled by the colloidal particle size and the shell thickness. Once again, tests with a low molecular weight fluorescent probe, calcein, showed that the colloidosome shell produced did not provide an effective barrier (Figure 9.6).

In a subsequent investigation, Yow and Routh also explored the release of fluorescein from colloidosome capsules produced using colloidal poly(styrene-*co*-butyl acrylate) latex particles as the building blocks for the shell.[47] This latex has a relatively low T_g and sintering of the shell at low temperatures can be easily controlled to vary the permeability. Using a range of annealing times these authors were able to reduce the release rates of fluorescein from within the capsules as a function of increased sintering. Again, however, even after extended sintering the capsules were never

(a)   (b)

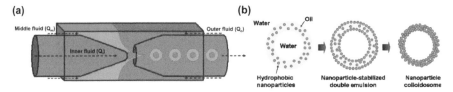

Figure 9.6   Schematic illustration of the microcapillary manufacturing approach taken by Lee and Weitz for the preparation of colloidosomes from water-in-oil-in-water (W/O/W) double emulsions.[46]

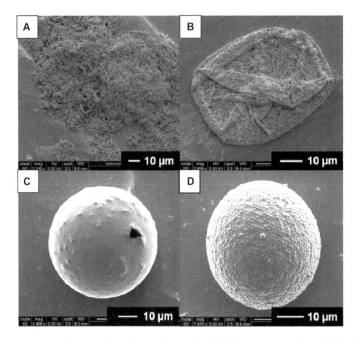

Figure 9.7   SEM micrographs of a series of colloidosomes produced by chemical crosslinking and/or annealing. Sample A is not crosslinked or annealed, resulting in the absence of capsules after drying. Sample B was cross-linked but not annealed, leading to collapsed capsules being observed on drying. Sample C was annealed but not crosslinked. Sample D was crosslinked and annealed. Here, it was shown that the crosslinking prior to annealing results in a more robust character to the latex particle monolayer.[30]

impermeable to this small molecule. A similar effect was also seen by Thompson *et al.*[30] when using sterically stabilized polystyrene latex particles to produce a colloidosome sample. In this case, the authors used chemical crosslinking between the steric stabilizers on the latex particles to produce a permanent colloidosome shell. Once again, annealing was used to alter the shell permeability. As before, leakage of fluorescein was seen even after extended periods of annealing that produced smooth shell capsules (Figure 9.7).

Thompson *et al.*[30] modelled the surface of the colloidosome and demonstrated that for small spheres coating a larger droplet defects and pores will always be present. To overcome this, the authors coated these annealed colloidosomes with another polymer layer of polypyrrole. Whilst this did hinder the dye release, it did not lead to an impermeable shell. In the same article, and using their modelling results, they also explored the effect of particle size on release rates; smaller particles are predicted to produce shells with fewer defects and smaller interstitial holes. The benefits of using the smaller particles in terms of dye retention were minimal and it was postulated that the better shell properties are cancelled out by a reduction in the shell thickness and hence a higher inherent permeability.

Following on from this earlier work, Zhao *et al.*[48] have recently shown how aggregated colloidal Ludox particles can be used to form the shell of a colloidosome. The key advantage of using a fractal colloidal aggregate is that the inherent tortuosity within the aggregate is then incorporated into the colloidosome shell. Such colloidosomes are shown to have a reduced permeability as a result of the thicker shells caused by using the aggregated Ludox, and the higher tortuosity through the shell. However, once again these colloidosomes remain permeable to small molecules and do not provide an impermeable barrier.

To date, many attempts have been made to encapsulate small molecules within colloidosomes. It has been repeatedly demonstrated that such capsules can be loaded with these molecules. Unfortunately, however, a reliable route to produce capsules with an impermeable shell has not been realized. This currently limits the value of this technology for many 'real-world' applications.

9.5.2 Large Molecules (>5 kDa)

Given that the production of an impermeable shell has proven difficult to date, it is perhaps more useful to utilize the variable pore size feature of colloidosomes to explore their use as capsules for larger molecular weight materials. For example, in their paper exploring colloidosomes produced from double-walled emulsions,[46] Lee and Weitz demonstrated the effective barrier properties of these capsules to high molecular weight fluorescently labelled dextran ($M_w \sim 2 \times 10^6$ Da), which has a radius of gyration that is approximately 2–3 times the pore size in the capsule shell.

In our research, we have explored the use of sterically stabilized latex particles as the fundamental building block for colloidosome production. By careful choice of polymer stabilizer for the latex particles, a chemical crosslinking reaction can be used to render the particle shell permanent. Our choice of stabilizer was a responsive PDMAEMA polymer, which can reversibly swell/contract as a function of pH. This allowed us to build colloidosomes where the pore sizes in the shell can be adjusted as a function of the pH.[31,35,36] Using such a system, we have shown how the release of a fluorescently labelled dextran ($M_w \sim 70$ kDa) can be triggered using pH to

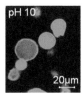

Figure 9.8   Confocal microscopy images of colloidosome microcapsules prepared by chemical crosslinking of a sterically stabilized latex. Capsules were filled with fluorescently labelled 70 kDa dextran. After cleaning they were left for 5 hours at either pH 10 or pH 3. The influence of shell porosity is clearly evident and also confirmed through fluorescence intensity measurements.[31]

swell the stabilizer which, in turn, expands the pore dimensions.[31] In this case, a larger fluorescence intensity decrease ($>50\times$) when varying pH than what is recorded from the fluorescein-tagged polymer alone ($\sim 4\times$) under the same conditions evidenced release from the capsule cores (Figure 9.8).

In related research, Berger *et al.*[24] have demonstrated the encapsulation and release of dextran ($M_w \sim 70$ kDa) from a colloidosome produced using microgel particles as the key component of the shell. The microgels were based on poly(N-vinylcaprolactam) and are sensitive to both temperature and co-solvents. These microgels exhibit a volume phase transition at 32 °C, driving a large associated swelling upon cooling. Measurements of the release of the fluorescently labelled dextran from these colloidosomes showed how the release rate could be altered with temperature, with high rates recorded at the lower temperatures where the microgel particles are strongly swollen.

9.5.3   Nanoparticles

A further opportunity to exploit the size exclusion nature of the colloidosome capsule shell exists with the encapsulation of nanoparticles. Möhwald and coworkers have demonstrated the encapsulation of 4 nm CdTe particles within a colloidosome.[41] Given the size of these particles, 5–8 nm Fe_3O_4 nanoparticles were used to produce the shell of the colloidosome, giving interstitial pores small enough to prevent rapid leakage. The colloidosomes were reinforced by using an agarose core to lock the Fe_3O_4 particles onto the surface and provide mechanical strength to the overall structure (see Section 9.4.1). By varying the Fe_3O_4 nanoparticle size, these authors were able to adjust the pore size and hence the degree of leakage for the CdTe particles.

9.6   Outlook

It has been pointed out by several studies that microcapsules from Pickering emulsions are not a solution to the encapsulation of small molecular species. However, it has been demonstrated that additional steps in their manufacture, such as the coating of polymer/polyelectrolyte layers on the surface of the capsules or the creation of thick inorganic shells, increases the retention time of actives within the microcapsules. This generally offsets the benefits of using Pickering emulsions as robust capsule templates that can be made by existing manufacturing routes at high volume fractions on industry-relevant scales. Instead, simple routes for the creation of colloidosome microcapsules are more likely to find applications in areas where the encapsulation and controlled release of large molecular species or small particulates is needed owing to the size exclusion properties of their colloidal shells.

The variety of available Pickering stabilizer materials such as, for example, latex particles made from responsive polymers, will continue to provide inspiration for researchers to design smart microcapsules from Pickering emulsions and to tailor them for specific applications.

References

1. A. D. Dinsmore, M. F. Hsu, M. G. Nikolaides, M. Marquez, A. R. Bausch and D. A. Weitz, Colloidosomes: selectively permeable capsules composed of colloidal particles, *Science*, 2002, **298**, 1006.
2. O. D. Velev, K. Furusawa and K. Nagayama, Assembly of latex particles by using emulsion droplets as templates. 1. Microstructured hollow spheres, *Langmuir*, 1996, **12**, 2374.
3. O. D. Velev, K. Furusawa and K. Nagayama, Assembly of latex particles by using emulsion droplets as templates. 2. Ball-like and composite aggregates, *Langmuir*, 1996, **12**, 2385.
4. O. D. Velev and K. Nagayama, Assembly of latex particles by using emulsion droplets. 3. Reverse (water ion oil) system, *Langmuir*, 1997, **13**, 1856.
5. N. Tsapis, D. Bennett, B. Jackson, D. A. Weitz and D. A. Edwards, Trojan particles: Large porous carriers of nanoparticles for drug delivery, *Proc. Natl. Acad. Sci. U.S.A.*, 2002, **99**, 12001.
6. B. P. Binks, Particles as surfactants – similarities and differences, *Curr. Opin. Colloid Interface Sci.*, 2002, 7, 21.
7. F. Leal-Calderon and V. Schmitt, Solid-stabilized emulsions, *Curr. Opin. Colloid Interface Sci.*, 2008, **13**, 217.
8. R. J. G. Lopetinsky, J. H. Masliyah and Z. Xu, Solids-stabilized emulsions: A review, in B. P. Binks, P. Bernard and S. Tommy (eds), *Colloidal Particles at Liquid Interfaces*, 1st edn, Cambridge University Press, Cambridge, 2006, pp. 186–224.
9. G. Decher, Fuzzy nanoassemblies: Toward layered polymeric multicomposites, *Science*, 1997, **277**, 1232.
10. E. Donath, G. B. Sukhorukov, F. Caruso, S. A. Davis and H. Möhwald, Novel hollow polymer shells by colloid-templated assembly of polyelectrolytes, *Angew. Chem., Int. Ed.*, 1998, **37**, 2202.

11. (a) L. M. Croll and H. D. H. Stöver, Composite tectocapsules *via* the self-assembly of functionalized poly(divinylbenzene) microspheres, *Pure Appl. Chem.*, 2004, **76**, 1365; (b) L. M. Croll, H. D. H. Stöver and A. P. Hitchcock, Composite tectocapsules containing porous polymer microspheres as release gates, *Macromolecules*, 2005, **38**, 2903.

12. (a) S. Cauvin, P. J. Colver and S. A. F. Bon, Pickering stabilized miniemulsion polymerization: Preparation of clay armored latexes, *Macromolecules*, 2005, **38**, 7887; (b) S. A. F. Bon and P. J. Colver, Pickering miniemulsion polymerization using laponite clay as a stabilizer, *Langmuir*, 2007, **23**, 8316.

13. (a) S. A. F. Bon, S. Cauvin and P. J. Colver, Colloidosomes as micron-sized polymerisation vessels to create supracolloidal interpenetrating polymer network reinforced capsules, *Soft Matter*, 2007, **3**, 194; (b) L. Duan, M. Chen, S. Zhou and L. Wu, Synthesis and characterization of poly(N-isopropylacrylamide)/silica composite microspheres *via* inverse Pickering suspension polymerization, *Langmuir*, 2009, **25**, 3467.

14. Q. Gao, C. Wang, H. Liu, C. Wang, X. Liu and Z. Tong, Suspension polymerization based on inverse Pickering emulsion droplets for thermo-sensitive hybrid microcapsules with tunable supracolloidal structures, *Polymer*, 2009, **50**, 2587.

15. Y. Chen, C. Wang, J. Chen, X. Liu and Z. Tong, Growth of lightly crosslinked PHEMA brushes and capsule formation using Pickering emulsion interface-initiated ATRP, *J. Polym. Sci., Part A: Polym. Chem.*, 2009, **47**, 1354.

16. (a) L. M. Croll and H. D. H. Stöver, Composite tectocapsules *via* the self-assembly of functionalized poly(divinylbenzene) microspheres, *Pure Appl. Chem.*, 2004, **76**, 1365; (b) L. M. Croll, H. D. H. Stöver and A. P. Hitchcock, Composite tectocapsules containing porous polymer microspheres as release gates, *Macromolecules*, 2005, **38**, 2903.

17. H. Wang, X. Zhu, L. Tsarkova, A. Pich and M. Möller, All-silica colloidosomes with a particle-bilayer shell, *ACS Nano*, 2011, **5**, 3937.

18. J. van Wijk, J. W. O. Salari, N. Zaquen, J. Meuldijka and B. Klumperman, Poly(methyl methacrylate)–silica microcapsules synthesized by templating Pickering emulsion droplets, *J. Mater. Chem. B*, 2013, **1**, 2394.

19. S. Fujii, M. Okada, H. Sawa, T. Furuzono and Y. Nakamura, Hydroxyapatite nanoparticles as particulate emulsifier: Fabrication of hydroxyapatite-coated biodegradable microspheres, *Langmuir*, 2009, **25**, 9759.

20. S.-H. Kim, C.-J. Heo, S. Y. Lee, G.-R. Yi and S.-M. Yang, Polymeric particles with structural complexity from stable immobilized emulsions, *Chem. Mater.*, 2007, **19**, 4751.

21. O. J. Cayre and S. Biggs, Hollow microspheres with binary porous membranes from solid-stabilised emulsion templates, *J. Mater. Chem.*, 2009, **19**, 2724.

22. O. J. Cayre and S. Biggs, Hollow microspheres with binary colloidal and polymeric membrane: Effect of polymer and particle concentrations, *Adv. Powder Technol.*, 2010, **21**, 19.

23. Z. Ao, Z. Yang, J. Wang, G. Zhang and T. Ngai, Emulsion-templated liquid core polymer shell microcapsule formation, *Langmuir*, 2009, **25**, 2572.
24. S. Berger, H. Zhang and A. Pich, Microgel-based stimuli-responsive capsules, *Adv. Funct. Mater.*, 2009, **19**, 554.
25. O. J. Cayre, S. Fujii and S. Biggs, unpublished data.
26. S. Laib and A. F. Routh, Fabrication of colloidosomes at low temperature for the encapsulation of thermally sensitive compounds, *J. Colloid Interface Sci.*, 2008, **317**, 121.
27. O. J. Cayre, P. F. Noble and V. N. Paunov, Fabrication of novel colloidosome microcapsules with gelled aqueous cores, *J. Mater. Chem.*, 2004, **14**, 3351.
28. M. F. Hsu, M. G. Nikolaides, A. D. Dinsmore, A. R. Bausch, V. D. Gordon, X. Chen, J. W. Hutchinson and D. A. Weitz, Self-assembled shells composed of colloidal particles: Fabrication and characterization, *Langmuir*, 2005, **21**, 2963.
29. J. W. O. Salari, J. van Heck and B. Klumperman, Steric stabilization of Pickering emulsions for the efficient synthesis of polymeric microcapsules, *Langmuir*, 2010, **26**, 14929.
30. K. L. Thompson, S. P. Armes, J. R. Howse, S. Ebbens, I. Ahmad, J. H. Zaidi, D. W. York and J. A. Burdis, Covalently crosslinked colloidosomes, *Macromolecules*, 2010, **43**, 10466.
31. O. J. Cayre, J. Hitchcock, M. S. Manga, S. Fincham, A. Simoes and S. Biggs, pH-responsive colloidosomes and their use for controlling release, *Soft Matter*, 2012, **8**, 4717.
32. J.-O. You, M. Rafat and D. T. Auguste, Cross-linked, heterogeneous colloidosomes exhibit pH-induced morphogenesis, *Langmuir*, 2011, **27**, 11282.
33. R. K. Shah, J.-W. Kim and D. A. Weitz, Monodisperse stimuli-responsive colloidosomes by self-assembly of microgels in droplets, *Langmuir*, 2010, **26**, 1561.
34. K. L. Thompson, E. C. Giakoumatos, S. Ata, G. B. Webber, S. P. Armes and E. J. Wanless, Direct observation of giant Pickering emulsion and colloidosome droplet interaction and stability, *Langmuir*, 2012, **28**, 16501.
35. S. Biggs, R. A. Williams, O. Cayre and Q. Yuan, Microcapsules and methods, WO/2009/037482, PCT/GB2008/003197.
36. Q. C. Yuan, O. J. Cayre, S. Fujii, S. P. Armes, R. A. Williams and S. Biggs, Responsive core-shell latex particles as colloidosome microcapsule membranes, *Langmuir*, 2010, **26**, 18408.
37. L. A. Fielding and S. P. Armes, Preparation of Pickering emulsions and colloidosomes using either a glycerol-functionalised silica sol or core–shell polymer/silica nanocomposite particles, *J. Mater. Chem.*, 2012, **22**, 11235.
38. P. F. Noble, O. J. Cayre, R. G. Alargova, O. D. Velev and V. N. Paunov, Fabrication of 'hairy' colloidosomes with shells of polymeric microrods, *J. Am. Chem. Soc.*, 2004, **126**, 8092.

39. I. Akartuna, E. Tervoort, A. R. Studart and L. J. Gauckler, General route for the assembly of functional inorganic capsules, *Langmuir*, 2009, **25**, 12419.
40. J. Li and H. D. H. Stöver, Pickering emulsion templated layer-by-layer assembly for making microcapsules, *Langmuir*, 2010, **26**, 15554.
41. H. Duan, D. Wang, N. S. Sobal, M. Giersig, D. G. Kurth and H. Möhwald, Magnetic colloidosomes derived from nanoparticle interfacial self-assembly, *Nano Lett.*, 2005, **5**, 949.
42. W. C. Mak, J. Bai, X. Y. Chang and D. Trau, Matrix-assisted colloidosome reverse-phase layer-by-layer encapsulating biomolecules in hydrogel microcapsules with extremely high efficiency and retention stability, *Langmuir*, 2009, **25**, 769.
43. C. Wang, H. Liu, Q. Gao, X. Liu and Z. Tong, Facile fabrication of hybrid colloidosomes with alginate gel cores and shells of porous $CaCO_3$ microparticles, *Chem. Phys. Chem.*, 2007, **8**, 1157.
44. (a) O. J. Cayre, N. Chagneux and S. Biggs, Stimulus responsive core-shell nanoparticles: synthesis and applications of polymer based aqueous systems, *Soft Matter*, 2011, 7, 2211; (b) A. Boker, J. He, T. Emrick and T. P. Russell, Self-assembly of nanoparticles at interfaces, *Soft Matter*, 2007, **3**, 1231; (c) M. Motornov, Y. Roiter, I. Tokarev and S. Minko, Stimuli-responsive nanoparticles, nanogels and capsules for integrated multifunctional intelligent systems, *Prog. Polym. Sci.*, 2010, **35**, 174.
45. J.-W. Kim, A. Fernandez-Nieves, N. Dan, A. S. Utada, M. Marquez and D. A. Weitz, Colloidal assembly route for responsive colloidosomes with tunable permeability, *Nano Lett.*, 2007, 7, 2876.
46. D. Lee and D. A. Weitz, Double emulsion-templated nanoparticle colloidosomes with selective permeability, *Adv. Mater.*, 2008, **20**, 3498.
47. H. N. Yow and A. F. Routh, Release profiles of encapsulated actives from colloidosomes sintered for various durations, *Langmuir*, 2009, **25**, 159.
48. Y. Zhao, N. Dan, Y. Pan, N. Nitin and R. V. Tikekar, Enhancing the barrier properties of colloidosomes using silica nanoparticle aggregates, *J. Food Eng.*, 2013, **118**, 421.

CHAPTER 10

Particle-Stabilized Food Emulsions

R. PICHOT, L. DUFFUS, I. ZAFEIRI, F. SPYROPOULOS* AND
I. T. NORTON

Centre for Formulation Engineering, Department of Chemical Engineering,
University of Birmingham, Edgbaston, Birmingham B15 2TT, UK
*Email: f.spyropoulos@bham.ac.uk

10.1   Introduction

A large number of food products are emulsions or emulsion-based. Common examples of natural or manufactured food emulsions are mayonnaise, butter, margarine, ice cream, salad dressing or milk. Even though these products are all different in taste, texture, appearance, *etc.*, they are formed of small droplets dispersed within a continuous phase. Emulsions can be classified into two types: a dispersion of water droplets in oil, water-in-oil emulsions, *e.g.* butter or margarine, or conversely, a dispersion of oil droplets in water, oil-in-water emulsions, *e.g.* milk or mayonnaise. Droplets, made of either water or oil, are usually referred to as the dispersed phase, and the medium in which droplets are dispersed is referred to as the continuous phase. A wide range of droplet sizes is found in food emulsions, from 0.1 to 100 μm.

When oil and water are mixed, the two phases separate very quickly to form superposed layers because this is the most thermodynamically stable configuration. For this reason, emulsions are thermodynamically unstable systems. However, they can remain kinetically stable for months or even years when one or more constituents, called stabilizers or emulsifiers, are added to the oil and water phases.[1] Most of the emulsifiers used in the food

RSC Soft Matter No. 3
Particle-Stabilized Emulsions and Colloids: Formation and Applications
Edited by To Ngai and Stefan A. F. Bon
© The Royal Society of Chemistry 2015
Published by the Royal Society of Chemistry, www.rsc.org

industry are either water- or oil-soluble compounds; they can be low molecular weight surfactants, *e.g.* polysorbate or lecithin, polymeric surfactants, *e.g.* polyglycerol polyricinoleate, or surface-active biopolymers, *e.g.* proteins. These emulsifiers are amphiphilic compounds, *i.e.* they possess hydrophilic and hydrophobic moieties. Because they have the ability to interact with both water and oil due to their amphiphilic character, these emulsifiers have a natural tendency to be located at the oil–water interface. Their presence at the interface increases the interactions between oil and water molecules, which tends to reduce the tension across the interface and favours droplet break-up mechanisms during emulsification.[2] The type of emulsions formed depends upon the hydrophilic-lipophilic balance (HLB) of the emulsifiers. Low HLB surfactants usually dissolve in oil and favour the formation of water-in-oil emulsions; conversely, high HLB surfactants dissolve in water and form oil-in-water emulsions.

Besides these soluble emulsifiers, solid particles have proved to be excellent emulsion-stabilizing agents. Unlike surfactants or proteins, they are not soluble in either water or oil and possess very different interfacial properties. Despite some similarities between particles and classic emulsifiers, stabilization mechanisms differ in the fact that particles form a solid layer around the dispersed phase droplets, which plays the role of a steric barrier that prevents droplet coalescence.[3] In spite of the fact that solid particles are often present in food formulations, they rarely contribute to oil–water interfacial stabilization. Only a few food products, such as butter or margarine, are examples of solid particle stabilized emulsions, where fat crystal particles behave as emulsifiers.

The food industry is facing new challenges and problems, such as the growing number of obese people in Western countries, and companies are now expected to somewhat provide solutions. Solid-stabilized emulsions, also called Pickering emulsions in tribute to the person who published the first extended study on the subject,[4] have the potential for new food applications and have received increasing levels of interest over the last three decades. This chapter gives an insight into the current knowledge of Pickering emulsions for food applications. The various solid particles that are either used in the food industry or are food grade with the potency for industrial applications are first described. The focus lies on the origin of the various particle types as well as their production methods and interfacial properties as Pickering particles. Current knowledge and understanding regarding solid particle stabilized food emulsions are then discussed, emphasizing the stabilization mechanisms associated with different solid particles and the potential applications of such emulsions for the food industry.

10.2   Food-Grade Solid Particles

Renewed interest in Pickering emulsions since the 1980s mainly arose with the emergence of nanotechnologies. The development of solid particles with a broader range of properties (size, hydrophobicity, shape, *etc.*), has allowed researchers to gain a better understanding of the mechanisms involved in the stabilization of Pickering emulsions. Designing Pickering particles for topical applications has become a challenge over the last decade. This section reviews some of the knowledge regarding the main particles with potential food applications.

10.2.1 Inorganic Particles

Since the discovery of solid particle stabilized emulsions by Pickering[4] in 1907, many particles such as silica,[5] titanium dioxide,[6] clays,[7] latex,[8] copper-based particles,[4] *etc.* have been used. Of these, silica particles remain the most researched, developed and possibly the most understood type of inorganic particles to be used for Pickering stabilization applications. Moreover, some silica particles are considered as food grade and have been utilized in food formulations as thickeners.

The basic molecular formula for amorphous silica is SiO_2 and particle sizes for commercially available silica particles lie between 5 and a few hundreds of nanometres.[9] The bulk structure of silica particles is derived from simple random packing of $[SiO_4]^{4-}$ molecules, however, the extensive research on silica particles concludes that the physical and chemical nature of the silica surface is complex. The main groups present at the surface are siloxane groups ($\equiv$Si-O-Si$\equiv$) and hydroxyl groups in the form of silanol groups ($\equiv$Si-OH). Of the silanol groups, there are three different forms: isolated silanols ($\equiv$Si-OH), vicinal silanols ($\equiv$Si-O-H-OH-Si$\equiv$) and geminal silanols ($=$Si-O$_2$-H$_2$).[10]

Various methods are used to produce silica particles. Silica hydrosols, *i.e.* dispersions of silica particles in an aqueous medium, are mainly produced from aqueous sodium silicate via ion exchange through neutralization of the solution by an acid (usually sulphuric acid). Primarily, the solution is thermally treated to produce small silica nuclei, which is then followed by the addition of dilute sodium silicate solution and dilute acid to maintain a basic pH. Under these conditions, silicic acid forms and is deposited upon the silica nuclei.[11] Another method developed by Stöber *et al.*[12] consists of the polymerization of orthosilicic acid induced by hydrolysis and condensation reactions. This method involves rapid nucleation followed by particle growth, but this particle growth reaction has not been clearly explained yet as a few different models were suggested.[13,14] Irrespective of the growth model, the characteristics of the colloidal silica particles produced via the Stöber process are mainly dependent upon the reaction conditions. For instance, particle diameter can be controlled by a number of factors relating to the process parameters, such as the reaction time, the pH of the solution or the reaction temperature;[15] for example, small porous particles are formed at acidic pH and when the hydrolysis reaction rate is significantly higher than the condensation rate. The main advantage of this method is that it allows the production of tailored colloidal silica particles (size, spherical shape and monodispersity).

Synthesized colloidal silica particles are generally hydrophilic due to the presence of silanol groups at the particle surface,[16] but its level of hydrophilicity can be modified by chemical or thermal treatment. Under heating, water molecules adsorbed at the surface are removed, favouring the formation of hydrophobic siloxane groups. By controlling the temperature and the reaction time, the hydrophilicity can be adjusted accordingly.[11,17] Chemical treatment involves the substitution of silanol groups by hydrophobic organofunctional groups by either condensation or hydrolysis reactions.[18] Typically, the reactants used are chloro-, amino- and alkoxysilanes. Modification of particle hydrophilicity plays an important role in the wettability of these particles. The positioning of a silica particle at an oil–water interface is governed by their level of hydrophilicity; purely hydrophilic or hydrophobic particles remain within the water or oil phase, respectively.[19] Nonetheless, despite the fact that silica particles can sit at the oil–water interface, recent studies have shown that the interfacial tension between oil and water was not modified by the presence of either hydrophilic[20] or hydrophobic[21] silica.

10.2.2 Protein Particles

Proteins are biological linear polymers that fulfil crucial functions in essentially all biological processes and can be derived from plant, vegetable or animal sources. Amongst the most common animal-sourced proteins relative to the food industry are those derived from milk, meat, fish, poultry and eggs. Proteins are composed of various functional groups including alcohols, thiols, thioethers, carboxylic acids, carboxamides, as well as a variety of basic groups, each one contributing accordingly to protein function. The type, number and sequence of amino acids contained in the polypeptide chain forming the protein molecules determine a number of protein properties, such as molecular weight, conformation, electrical charge, hydrophobicity, physical interactions and chemical reactivity. Proteins in water, similarly to any other systems, seek to achieve a minimum overall free energy, which leads them to adopt a folding structure where hydrophobic residues cluster and the polar charged chains are exposed to the aqueous surroundings.[22] Caseinates, for example, form aggregates or micelles of $\sim$ 100–200 nm in water at very low concentrations (Figure 10.1A). Protein micelles, despite their solid particle appearance and the fact that they are insoluble in water, are not considered as solid particles because they unfold at a liquid–liquid interface (typically oil and water); protein aggregates dissociate and rearrange to form a stabilizing adsorbed layer.[23]

A few attempts have been made to modify protein properties in order to form solid protein particles. Solid protein particles can be formed from zein,[27] the major storage protein of corn ($\sim$ 45–50% of the protein in corn), via an anti-solvent precipitation method.[28] Typically, zein powder is first dissolved in highly concentrated ethanol solution (80 v/v% in water). This solution is then diluted in water at an ethanol concentration of 20 v/v% to

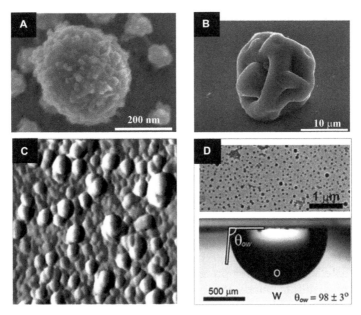

Figure 10.1 Micrographs of various protein particles. (A) SEM image of an individual casein micelle (reprinted from Dalgleish *et al.*[24] with permission from Elsevier). (B) SEM image of a soy protein particle trapped on PDMS from the air–water interface (reprinted from Paunov *et al.*[25] with permission from Elsevier. (C) AFM image (1 μm×1 μm) of heat-treated lactoferrin particles (reprinted from David-Birman *et al.*[26] with permission from Elsevier. (D) top, TEM of zein colloidal particles, and bottom, contact angle of soy bean oil droplets attached to zein films immersed in an aqueous sub-phase (pH 4.0) (reproduced from de Folter *et al.*[27]).

favour solvent evaporation. Subsequent particle characterization revealed small spherical particles, $\sim$ 70–80 nm, with an isoelectric point around 6.5. At high or low pH, particles were strongly charged, negatively or positively, respectively, but no change in size was observed. Particles were wetted by both water and soybean oil, exhibiting a contact angle of $\sim 90°$ (Figure 10.1D), which confers to the particles the ability to adsorb at the interface.

Globular proteins, such as β-lactoglobulin, are denaturated above a certain temperature. Proteins initially unfold, exposing their hydrophobic parts. Due to the hydrophobic effect and physiochemical interactions, proteins curl up so that hydrophobic elements of the protein are buried deep inside the structure and hydrophilic elements are located on the outside. Using this specific property, Santipanichwong *et al.*[29] showed protein particles could be formed from β-lactoglobulin by heating to 80° C for 15 min followed by cooling to room temperature. Large aggregates were formed close to the isoelectric point ($\sim$ 4.7), but at high or low pH particles were within the nano-range ($\sim$ 100 nm) due to strong electrostatic repulsion. In this study, which aimed at forming protein/polysaccharide complexes, no indication regarding interfacial properties of β-lactoglobulin particles was provided. This method, consisting of protein thermal denaturation and pH adjustment, was used by other researchers[26,30] to form particles from lactoferrin, a globular glycoprotein widely found in various secretory fluids, such as milk, saliva or tears. Particles were very small (∼50 nm), but with a relatively broad size distribution as can be seen in Figure 10.1C. No insight regarding the interfacial properties of the lactoferrin nanoparticles was provided by these authors but evidence can be found in the literature that these particles are wetted by both water and oil as they are able to strongly adsorb at the oil–water interface.[31] Spray-dried soy protein particles are another type of protein particulates with interesting interfacial properties[25] (Figure 10.1B). Nonetheless, there is no description in the literature regarding the production of these particles.

Another type of protein that may be characterized as solid particles is hydrophobin, although whether hydrophobins are solid particles is still being debated. Unlike other proteins, hydrophobins possess an extended network of disulfide bonds that stabilizes the structure and confers the protein a unique configuration, where one part of the surface consists almost entirely of hydrophobic chains, the other part being almost entirely hydrophilic.[32] Investigations regarding their interfacial properties revealed that hydrophobins act in a very similar way to Janus particles.[33] Using tensiometry and surface shear rheology, Cox *et al.* showed that hydrophobins are very surface active proteins, reducing the surface tension to approximately 30 mN m^{-1} and exhibiting a strong elastic behaviour at the air–water surface (the elasticity being much higher than any other type of proteins). Compared to other protein particles, hydrophobins do not require any specific heat or chemical treatment to be produced as they self-assemble as particles in aqueous media.

10.2.3   Polysaccharide Particles

Polysaccharides are essentially polymeric carbohydrates and are a class of naturally occurring polymers, most abundant in agricultural feedstock and crustacean shells. Polysaccharides such as cellulose and chitin are found in nature as structural building blocks and others such as starch provide fuel for biological cells by storing solar energy in the form of sugars.

Starch is most abundant in plants, specifically in its granular form within plant cells. Within the starch granules is contained a macromolecular complex of two polymeric components; linearly structured amylose (with primary α-1,4 linkages) and branched chain amylopectin (containing approximately 5% α-1,6 linked branch points). The proportions of each component differ depending upon which plant species they are derived from. Similarly, cellulose is prevalent in plant systems; it is the main structural polymer in wood and is constituted of chains of many glucose molecules. It supplies more than 50% of the carbon within vegetation and is the structural component of the cell walls within plants such as ramie (Boehmeria Nivea) and cotton (Gossypium).[34] Typical sizes of native starch and cellulose particles lie within the range of several microns down to diameters in the nano range, with some of the smallest native starches being rice starch ($<$ 5 μm) and quinoa starch (500 nm to 3 μm).[35,36] Similarly, microcrystalline cellulose has individual particle diameters ranging from several hundred nanometres to several microns. Zoppe *et al.*[37] however found that by carrying out strong acid hydrolysis, rod-shaped cellulose nanocrystals could be produced, measuring approximately 3–15 nm in width and 50–250 nm in length. Figure 10.2 shows a few examples of size, shape and other physical features of starch and cellulose that depend upon the plant cells they are isolated from.

These polysaccharides are theoretically not water soluble. Whistler[42] found that this insolubility in water arises from the preference of these polysaccharide molecules for partial crystallization. Nonetheless, depending upon their size and the amylose/amylopectin ratios, the starch particles can be dispersed into an aqueous solution when the water has been heated above a certain temperature, as are cellulose particles. Polysaccharides are used as thickeners and gelling agents in industry because they are able to

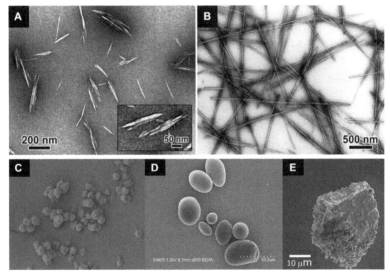

Figure 10.2 Examples of cellulose and starch particles. (A and B) TEM micrographs of cellulose nanocrystals following sulfuric acid hydrolysis for cotton and tunicate cellulose, respectively (reprinted with permission from Elazzouzi-Hafraoui *et al.*,[38] © 2007 American Chemical Society; (C–E) SEM micrographs of various starch granules, (C) OSA-modified quinoa starch (reprinted from Timgren *et al.*[39] with permission from Elsevier; (D) native potato starch (reprinted from Li *et al.*[40] with permission from Elsevier; and (E) OSA-modified tapioca starch (reprinted from Yusoff and Murray[41] with permission from Elsevier.

form gel networks throughout the bulk to inhibit movement of both dispersed and continuous phases.[43,44] Besides their gelling properties, some polysaccharide structures can be tailored to adsorb at oil–water interfaces and to reduce the interfacial tension, by addition of hydrophobic chemical groups to the main hydrophilic backbone of the polysaccharide molecule. This is the case with hydrophobically modified starches and celluloses.[43] Modification of starch hydrophobicity is achieved by either physical or chemical methods, or combinations of these two.[45,46] The physical modification of starch can involve the application of heat, shear stress or moisture to alter the properties of native starch. Using dry heat, the surface character of the naturally hydrophilic starch particles can be changed and made partially hydrophobic, which confers to the particles a greater ability to adsorb to an oil–water interface. Chemical modification can be performed via crosslinking, conversion or substitution reactions and most commonly, starch and cellulose modification occurs via chemical substitution reaction. Traditionally, substitution of the hydroxyl groups on starch particles with different alkenyl succinyl anhydrides produces modified starch with amphiphilic properties.[47] In the case of cellulose, modified cellulose particles are produced via the substitution of hydroxyl groups within cellulose molecules with ether groups.

Native starch is most commonly treated with octenyl succinic anhydride (OSA), via an esterification reaction, as this hydrophobic group increases the starch particle's affinity for oil. The main advantage of using OSA may reside in the fact that it has been approved for food use in the USA since 1972. Restrictions on concentration levels are in place, however, and the maximum level of OSA treatment allowed is 3% and the amount of free OSA present in food formulations should not exceed 0.3%.[48] Rayner *et al.*[49] studied the effect of the type of hydrophobic modification on quinoa starch granules; particles had received either dry heating or OSA treatment. It was deduced that not only were the dry heat treated particles larger (2.22 µm) than the OSA modified starch (1.74 µm) and native starch (1.65 µm) but also that they possessed a significantly lower stabilizing ability than the OSA treated starch. They concluded that OSA modification was substantially more efficient than dry heat treatment in providing hydrophobic character to quinoa starch particles.

The addition of hydrophobic and hydrophilic groups along with the natural hydrophilicity of the native starch allows for the production of modified starch particles with a degree of amphiphilicity and as such the particles can be used as an effective emulsifier. The hydrophobicity of the modified starch and cellulose particles is dependent on the degree of substitution. This is defined as the average number of hydroxyl groups substituted per glucose unit and increasing this average leads to increased particle hydrophobicity. In the instance of cellulose, cellulose fibres have been hydrophobically modified in order to increase particle amphiphilicity and hence allow adsorption to an oil–water interface. Habibi and Dufresne[50] studied the water contact angle of starches and cellulose nanocrystals obtained via acid hydrolysis from waxy maize starch granules and ramie fibres, respectively. A proportion of these nanocrystals were then modified by grafting poly-caprolactone (PCL) to them via isocyanate-mediated reaction. The dynamic particle contact angles with water of the grafted and non-grafted nanocrystals were measured and compared. Data collected showed that the non-grafted starch and cellulose contact angles were approximately 30° and 40°, rising to 75° and 80°, correspondingly. It was therefore concluded that by grafting PCL chains onto the nanocrystals, the modified starch and cellulose substrates exhibited increased hydrophobicity.

10.2.4   Protein/Polysaccharide Complexes

As discussed previously, both proteins and polysaccharides can generate solid particles exhibiting interesting interfacial properties. The combination of these two constituents to form particulates has received a lot of interest over the last three decades, and the mechanisms underlying the production of such particles are nowadays well understood. Given the wide range of proteins and polysaccharides available, there are almost an infinite number of possible arrangements, and many have been investigated. This section aims to describe the different methods of producing protein/polysaccharide particles and their main characteristics. An extensive review of these particles could indeed be the subject of a full chapter/review.

Protein/polysaccharide particles (examples are given in Figure 10.3) originate from the significant attraction between the two different molecules when they both carry net opposite electrical charges. These interactions encouraging attraction between the two molecules can be electrostatic, steric or hydrophobic in nature, as well as hydrogen bonding and van der Waals interactions.[52] The electrostatic interactions are said to be the main driving force for the formation of the protein/polysaccharide complex and the hydrophobic and hydrogen bonding contributions provide a secondary stability to the complex aggregates.[53,54] The strength of these interactions dictates whether coacervates or precipitates (complexes) form. The type, composition and strength of the complexes, as well as the complexation process required, are dependent upon several factors.[55] These are process parameters such as pH, ionic strength and temperature, and material properties such as protein/polysaccharide type and molecular ratio.

The charge carried by protein molecules is dependent upon the pH of the medium within which they are dispersed. Protein molecules possess a positive overall zeta potential at pH lower than the isoelectric point and a negative zeta potential at pH above the isoelectric point. The polysaccharide molecules become anionic at a pH above their pKa and hence depending on the pH of the system, protein/polysaccharide complexes will form via electrostatic forces. However, at or very close to the isoelectric point of the protein solution, the surface charge on the protein molecules is close to zero and so much weaker complexes form. Additionally, if the net charges on both types of molecules are similar, repulsion rather than complexation

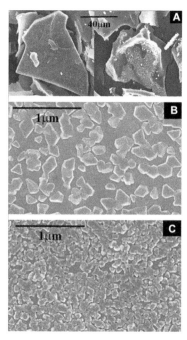

Figure 10.3 SEM images of protein/polysaccharide complexes. (A) α-lactalbumin/ chitosan complex (left) and β-lactoglobulin/chitosan complex (right); (B) α-lactalbumin/chitosan complex; and (C) β-lactoglobulin/chitosan complex (reprinted from Lee and Hong[51] with permission from Elsevier).

will occur. To this end, pH can be used to control complexation of the molecules. Hydrogen bonding between the two classes of molecules arises from the $-CO_2H$ and $-NH_3$ groups on the protein molecule and the $-CO_2H$ group on the polysaccharide and once again the strength of these interactions depend upon the pH and other process parameters of the system. Aside from the ionic zones on the two molecules, other non-polar regions on each molecule can associate with one another in the form of hydrophobic interactions. These hydrogen bonding and hydrophobic interactions are temperature dependent. At high temperatures, protein denaturation and polysaccharide conformational changes cause the exposure of more reactive sites, which favours more complexation interactions between the functional groups on both molecules.[56] Increasing the temperature of the system promotes hydrophobic interactions and covalent bonding while at low temperatures, hydrogen bonding is encouraged.

Biopolymer ratio and concentration influence complexation. Naturally, maximum complexation is achieved at a specific ratio due to the charge balance within the system. However, when one biopolymer is in excess in comparison to the other, charges that have not been neutralized are present within the solvent and this allows for the production of soluble complexes.

Conversely, when this excess of one biopolymer compared to the other occurs at high biopolymer concentrations, more counterions are released into the solution that block the charges on the protein or polysaccharide molecules, which, in turn, increase the solubility of the complexes but inhibit complex formation.[57]

The complexation process generally occurs in the presence of a solvent and as the complexes form, the system can separate into two phases, one containing the products of complexation and the other, mostly solvent and partially biopolymer(s) or a one-phase system, can exist where the complexes are soluble in the solvent. Due to the effect of the process parameters on the resultant complexes discussed previously, there are two main ways in which to prepare the complexes (Figure 10.4).

Method one produces complexes by mixing both molecule types at pH conditions far from the optimal pH at which maximum complexation is expected to occur. The pH of the system is then progressively altered to a pH below the isoelectric point of the protein. This gradual change allows for greater control over the complexation process and hence the complexes produced. The second method involves separate preparation of the biopolymers in solvent dispersions. The individual systems are then pH adjusted to their optimal pH of interaction and then combined and mixed (Method 2, Figure 10.4). This process is faster and complexes are formed quickly, however it does generally result in larger complexes than the first method.

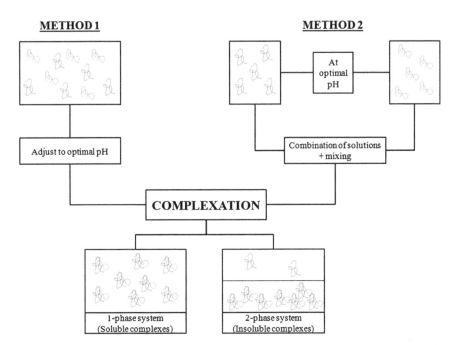

Figure 10.4 Simple schematic diagram representing the two main methods for producing protein–polysaccharide complexes.

Protein–polysaccharide complexes have a wide range of industrial applications and they have been used in food and beverage, pharmaceutical, medical and cosmetic industries, however their main uses lie within the food industry. They have been used to stabilize acidified milk-based drinks in the beverage industry as Tuninier *et al.*[58] found that the charged zones on pectin molecules (polysaccharide), at a specific pH, adsorbed onto the surface of casein micelles, leaving the uncharged areas on the pectin molecule to extend into the milk solution. This interaction aided stabilization of the solution via steric repulsion. Weinbreck *et al.*[59] demonstrated that these complexes could also be successfully used for encapsulation purposes. They formed lemon oil droplets surrounded by a shell comprised of complexes of whey protein and gum arabic. These droplets were introduced to cheese products and the shell-complex was able to stabilize the lemon oil droplets for in excess of 1 month. Yeo *et al.*[60] similarly constructed gelatin and gum arabic complexes to encapsulate flavours for use in frozen baked food. They were able to effectively encapsulate flavours using the complexes and control their release via salt addition and thermal treatment.

10.2.5 Lipid Particles

Lipids are a large group of naturally occurring organic compounds, consisting mainly of fats, fatty acid salts, waxes, glycerides, phospholipids, sterols and glycolipids. They are used by organisms for energy storage purposes and as a structural component of cell membranes. Triglycerides (TAGs), also called triacylglycerols, are the principal components of fats and oils. Other than their biological functions, they are widely used in the food industry as the major components in cream, margarine, and confectionery fats. A TAG consists of three fatty acid residues esterified to a glycerol backbone (Figure 10.5). The second most abundant type of lipids are phospholipids (*i.e.* lecithin), which occur in animal and plant cell membranes, and are molecularly different to TAGs in that they additionally contain phosphoric acid and a low molecular weight alcohol.

One of the most common types of lipid solid particles are lipid crystals. Some lipids such as TAGs undergo liquid–solid phase transition, resulting in the formation of fat crystals. Fat crystallization is a very complex phenomenon that has been investigated for a long time.[61–66] This section aims to

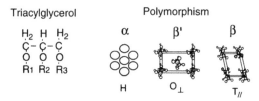

Figure 10.5 TAG molecular formula (left) and the main polymorphic forms of TAG (right) (reproduced from Sato and Ueno[61] with permission from Elsevier).

describe the main properties of fat crystals, as well as their interfacial properties. More detailed information can be found in reviews published by many researchers. TAG crystals are present within three main polymorphic forms (Figure 10.5), α (hexagonal subcell), β' (orthorhombic-perpendicular subcell) and β (triclinic-parallel subcell). β', due to the crystal morphology, as well as its ability to form crystal networks, is the most common polymorph in margarine and shortening, while β type crystals are most desirable in chocolate confectionery due to their melting behaviour and surface appearance. The formation and stability of these polymorphs are largely influenced by the TAG fatty acid composition. Polymorphic crystallization is mainly determined by the rate of nucleation, which depends on many factors such as cooling rate, supersaturation or shear rate. For example, under cooling, Sato and Kuroda[67] showed that the α tripalmitin form appears first, followed by the β' form and then the most stable form β (Figure 10.6a). Melting of tripalmitin crystals occurs as follows: the less stable α form melts first, followed by the nucleation and growth of more stable β' forms, and the mass transfer in the liquid formed by melting of the less stable form. Similarly, increasing the shear rate results in changing the polymorphic form of cocoa butter crystals (Figure 10.6b).[66,68]

Fat crystal properties were investigated in detail by Johansson and Bergensthål in the 1990s, in particular the interactions between crystals and liquid emulsifiers and their effect on the emulsion stability.[69-73] These authors also paid special attention to the wettability of fat crystals by oil and water and the various parameters that can affect it.[74-76] In an attempt to characterize the effect of emulsifiers on the fat crystal wettability, Johansson and Bergensthål first showed that the different polymorphs were wetted differently; the

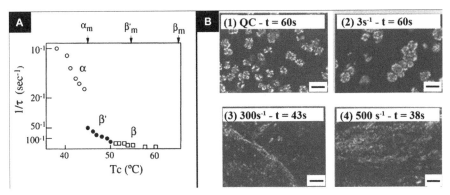

Figure 10.6 (A) Inverse of induction time (τ) for crystallization of tripalmitin. Induction is defined as the time until the occurrence of the first appearing crystals is optically detectable after cooling. α_m, β_m' and β_m are melting points of the polymorphs (reprinted from Sato[66] with permission from Elsevier). (B) Optical micrographs of cocoa butter crystal morphology as a function of shear rate (the bar represents 50 μm) (reprinted from Sonwai and Mackley[68] with kind permission from Springer Science and Business Media).

β polymorph consists of tight TAG molecules that create non-polar surfaces (only wetted by oil) while α and β′ form crystals are slightly more polar due to looser packing (wetted by both liquids). Crystal wettability can be tailored by adding oil-soluble emulsifiers; in particular, increasing the hydrophilic–lipophilic balance (HLB) of the surfactants increases the fat crystal affinity for water. This could be attributed to the decrease in interfacial tension that was found to be the resistive force for crystal adsorption to the interface; in the case of palm stearin, the authors demonstrated that fat crystals could only adhere strongly at the interface when the interfacial tension is low enough to decrease the energy to displace a particle from the interface to the oil phase.

Solid lipid nanoparticles (SLNs) can also be formed by the so-called 'emulsion route'; lipids such as TAGs are dispersed as droplets in an aqueous phase above their melting temperature using an emulsification device[77] (high shear mixer, homogenizer, sonicator, *etc.*) and rapidly cooled below their crystallization temperature to form solid particles dispersed in water. Research regarding the production of SLNs has been carried out for the last 50 years, mainly because of the biodegradable properties of lipids and their potential applications in many fields (food, pharmaceutical, cosmetics, *etc.*), particularly as drug carriers.[63,77–80] In spite of relatively simple process principles, various methods have been developed to produce SLNs. Two of them are given as examples in Figure 10.7.

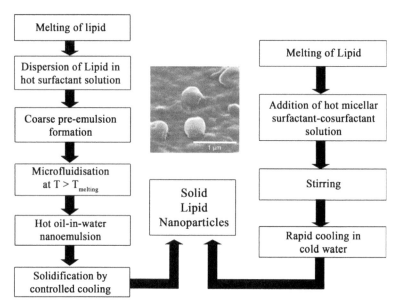

Figure 10.7   Flow chart representing the steps of two methods for producing solid lipid nanoparticles: (left) melt homogenization process and (right) melt micro-emulsification process (adapted from Weiss *et al.*[81] with kind permission from Springer Science and Business Media). Middle: REM picture of solid lipid nanoparticles made from Compritol stabilized with Poloxamer 188 (reprinted from Müller *et al.*[77] with permission from Elsevier).

With respect to the production of SLNs, one of the main challenges is to control both heating and cooling rates in order to control polymorphic transitions. Moreover, many lipids crystallize, in the bulk, at a temperature lower than their melting temperature (a phenomenon known as super-cooling) and this is even more pronounced when lipids are dispersed in water.[78] Even though their presence may not be required to stabilize liquid lipid droplets,[80] water-soluble surfactants are often used, which also affects the lipid crystallization temperature and the polymorphic transitions.[81] Awad *et al.*[82] showed that SLN stability could be improved by retarding the transition of emulsified tripalmitin from the α to β polymorphic form by increasing the cooling rate. Spherical particles, fully covered by surfactant, formed within the α crystal form, were less prone to aggregate than irregular crystals not uniformly covered by surfactant produced within the β form. Despite clear evidence that SLNs are able to adsorb at the oil–water inter-face,[80] the interfacial properties of these nanoparticles have not being investigated so far.

10.3   Particle-Stabilized Emulsions for Use in Food

10.3.1   Principles of Emulsion Stabilization with Solid Particles

Solid particles of various sizes (from few nanometres to micrometres) and shapes are often present in food products. By accumulating at liquid–liquid interfaces or remaining within the bulk phase, this particulate material often contributes to the long shelf stability of the product. A large amount of food systems are emulsion-based, offering a high liquid–liquid interfacial area for the solid particles to adsorb at. This is the idea behind the concept of Pickering stabilization; emulsion droplets coated by dispersed solid particles that form a steric barrier to prevent droplet coalescence.

Despite the relatively simple concept of Pickering stabilization, mechanisms involved in such stabilization are numerous and not trivial, including particle size and shape, wettability, and particle packing density at the interface. Since the pioneering studies of Ramsden[83] and Pickering[4] in the early 20th century, a lot of work has come out using Pickering particles, in particular for the last two decades, and the mechanisms of stabilization are now well understood. The increasing level of interest in Pickering emulsions is mainly due to the progress in material science and nanoparticle technology, and to their potential in the development of new emulsion-based products. Model systems have been investigated using several types of particles (clay, silica, metals, *etc.*) as well as applied systems for industrial applications, such as food or cosmetics. Scientific literature regarding particle-stabilized emulsions is abundant, including extensive reviews.[84–87]

The efficiency of particles to act as stabilizers resides in their ability to form steric barriers around the emulsion droplets. Considering forces acting on a spherical particle attached at the oil–water interface, Tambe and

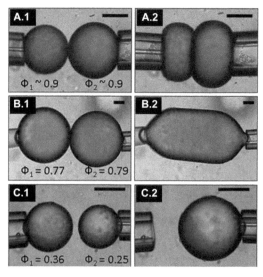

Figure 10.8 Coalescence behaviour as a function of the droplet surface coverage. (a) Total stability, (b) arrested coalescence and (c) total coalescence of Pickering droplets (reproduced from Pawar *et al.*[89]).

Sharma[88] demonstrated that coalescence is only possible when particles are either laterally displaced along the interface or detached from it, making the uncovered droplet surface large enough for the droplets to merge. As a consequence, droplets insufficiently covered tend to coalesce. This was confirmed by many studies, including the recent work by Pawar *et al.*[89] who investigated the coalescence of Pickering droplets through micromanipulation (Figure 10.8). Full coalescence was observed at droplet surface coverage below ~ 0.3, while no coalescence appears at high surface coverage (~ 0.9). At intermediate surface coverage, arrested (or partial) coalescence occurred. Nonetheless, evidence can be found in the literature showing stable emulsions at low surface coverage.[21,90] In particular, Vignati *et al.*[21] prepared stable emulsions with only 15% droplet coverage. Stability was assumed to be due to the lateral motion of the particles at the interface.

Particle wettability plays a major role in Pickering stabilization because it defines the particle position at the interface. The location of a single particle at the oil–water interface is best characterized by the contact angle θ. An idealized case of a spherical particle attached at an oil–water interface is represented in Figure 10.9a. The type of emulsion formed is related to the particle wettability; particles preferably wetted by water ($\theta < 90°$) tend to stabilize oil-in-water emulsions, while particles preferably wetted by oil ($\theta > 90°$) tend to stabilize water-in-oil emulsions (Figure 10.9b). From a thermodynamic point of view, the energy required to remove a single particle for the interface can be quantified by:

$$E = \pi r^2 \gamma_{ow}(1 \pm \cos \theta)^2 \tag{10.1}$$

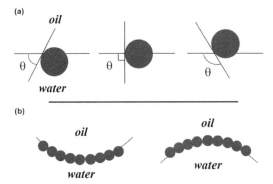

Figure 10.9 (a) Contact angle of a small particle at a planar oil–water interface; (b) particle position at a curved interface; O/W emulsion for $\theta < 90°$ (left), W/O emulsion for $\theta > 90°$ (right) (adapted from Binks[3] with permission from Elsevier).

where r is the radius of the particle, γ_{ow} the interfacial tension between oil and water. The sign inside the bracket is negative if the particle is removed into the water phase and positive if removed into the oil phase. For micro- or nanoparticles, this energy is extremely high and particles are considered irreversibly adsorbed $(E \gg 10 \text{ kT})$. Nonetheless, particle size also plays an important role in Pickering stabilization; in practice, particle average size must be an order of magnitude smaller than the droplet size to achieve emulsion stability.

10.3.2 Water-in-Oil Emulsions

Many food products such as butter, margarine or ice cream are formed of water droplets dispersed in oil and stabilized fully or partially by fat crystals. The type of stabilization depends on the ability of these crystals to adsorb at the oil–water interface. The term Pickering (or interfacial) stabilization is used for surface-active crystals that form a steric barrier around the water droplets preventing coalescence and film drainage.[91] This type of stabilization involves either the adsorption of discrete solid particles at the interface or the sintering of fat crystals at the interface. Crystals unable to reside at the interface remain in the continuous phase to form a network due to van der Waals interactions.[92] This network prevents water droplet collision, reducing the risk of coalescence, hence the term network stabilization. Both types of stabilization can also occur simultaneously under certain specific circumstances.[93]

10.3.2.1 Interfacial Stabilization

Interfacial, or Pickering, stabilization of water-in-oil emulsions has long been investigated using inorganic particles such as surface-modified silica.[5]

These particles are quite easy to tailor in terms of size and surface hydrophobicity, which makes them very advantageous for model studies. On the other hand, they are not suitable for the food industry and can only be used as model particles to get a better understanding of stabilization mechanisms. One of the challenges for the food industry has been to develop food-grade particles that can replace commonly used surfactants. However, despite the growing interest in Pickering emulsions and the need for food companies to develop new commercially attractive products, very few articles related to the use of discrete solid particles for the stabilization of water-in-oil emulsions can be found in the literature.

Lucassen-Reynders and Van Den Tempel[94] first published a work in which they studied the stabilization of water-in-paraffin oil with glycerol tristearate crystals. Emulsions exhibit long-term stability, provided that a small amount of surfactant is used to promote crystal dispersion within the oil phase. Interestingly, the authors showed that the incorporation of surfactant did not modify the surface activity of tristearin crystals. Since this pioneering work in the 1960s, only a few studies regarding the use of food-grade hydrophobic particles have been published and all followed the work by Johansson and Bergenståhl in understanding the formation and interfacial properties of fat crystals.[72–74] Water-in-oil emulsions formulated by Garti *et al.*[95] were unstable due to flocculation of tristearin particles; as also noticed by Lucassen-Reynders and Van Den Tempel, the addition of surfactant (polyglycerol polyricinoleate, PGPR) enhanced particle dispersion and emulsion stability. It appears clearly from these papers that the difficulty of using pre-formed fat crystals lies in (i) the wettability of these particles by the oil phase, resulting in crystal flocculation, and (ii) the weak particle interfacial activity, which does not allow efficient interfacial tension reduction and adsorption at the interface.

Nonetheless, fat crystals proved to be very efficient emulsifiers when formed directly at the water–oil interface.[96] Mechanisms of direct solidification at the interface are well understood. Oil-soluble surfactants that exhibit liquid–solid state transition such as glycerol monostearate (GMS) or glycerol monooleate (GMO) adsorb and self-assemble at the interface during emulsification. A temperature drop applied to the formed emulsion results in the crystallization of the surfactant molecules at the interface due to phase transition. The small crystals formed at the interface sinter to create a smooth solid crystalline shell around water droplets. The presence of a solid shell reduces dramatically the occurrence of droplet coalescence, thus providing emulsion long-term stability. Nonetheless, the presence of monoglycerides in the emulsion formulation increases oil phase viscosity, which may lead to water droplet flocculation and coalescence under either quiescent[73] or shear conditions.[97] Emulsion stability is optimal at a critical monoglyceride concentration, which is specific to a water/oil/monoglyceride system.

The ability of such surfactants to form a shell around water droplets depends on various parameters. The crystallization temperature is critical because it determines the processing conditions. For example, GMS has a transition temperature of ∼ 48 °C, which requires emulsions to be heated up during emulsification but allows rapid cooling that enhances shell formation. On the contrary, GMO transition temperature is ∼7 °C, which allows emulsification at room temperature but requires more effort to reach the crystallization temperature. This also affects the final emulsion-based food product as this would have to be stored at low temperature to keep its structure. Evidence can also be found regarding the different structures formed by GMO and GMS at the interface; GMS forms a smooth crystalline shell,[92] while GMO forms a liquid crystal layer.[95] This difference in interfacial structure is probably due to the intrinsic properties of these surfactants and the way they self-assemble at the interface. Saturated fatty acid chains of GMS allow these molecules to align along each other at the interface to form a packed layer, which favours molecule interactions and interfacial crystallization. GMO, on the other hand, is formed of an unsaturated fatty chain that does not allow close interfacial packing, leading to the formation of a liquid crystalline layer around the water droplets.[97]

Adsorption of monoglycerides at the interface, resulting in the formation of the crystal shell, is the consequence of their amphiphilic character; the hydrophilic polar moiety resides in the aqueous phase while the hydrophobic fatty acid chain remains in oil. Triglycerides also undergo solid–liquid phase transitions,[98] but despite their amphiphilic character, they present a low surface activity and are unable, when used as sole emulsifier, to form a solid shell around droplets.[96] This is due to the prominent size of the hydrophobic moiety as well as the low polarity of the head group. Nonetheless, evidence in the literature demonstrates that addition of liquid surfactant can promote interfacial crystallization of triglyceride. Gosh and Rousseau[99] showed that hydrogenated canola oil crystallizes at the interface in the presence of GMO. The authors suggested that GMO first adsorbed at the interface. Then triglyceride molecules interact with GMO due to molecular compatibility between the oleic acid of GMO and the stearic acid of triglycerides, enhancing triglyceride crystal nucleation at the interface (Figure 10.10). These results are in agreement with the conclusion made by Frasch-Melnik *et al.*,[100] who observed the formation of a solid crystalline shell of glycerol tripalmitin on the water droplet surface in the presence of monoglycerides, confering long-term stability to the emulsions (Figure 10.11a).

Interfacial crystallization of triglycerides to stabilize water droplets is a very promising technique but this field is at an early stage and open to many discussions. For example, Gosh and Rousseau observed that in the presence of polymeric surfactants such as PGPR, hydrogenated canola oil did not crystallize at the water–oil interface due to lack of interactions between these two components, while Norton and Fryer[101] presented results showing the formation of a TAG crystalline shell around water droplets of long-term stable cocoa butter emulsions (Figure 10.11b). The difference in the interfacial behaviour of TAG has not been explained so far.

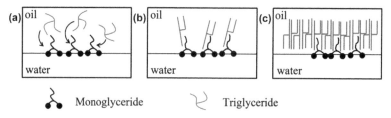

Figure 10.10 Formation of the triglyceride solid shell at an oil–water interface: (a) during emulsification, monoglycerides adsorb at the interface and interact with triglycerides; (b) due to a temperature drop, triglycerides undergo crystallization at the interface; (c) triglyceride crystals grow at the interface forming a solid barrier around the water droplets. Adapted from Rousseau[96] with permission from Elsevier.

Figure 10.11 Various types of Pickering stabilization. (A) Cryo-SEM micrograph of a 60% water-in-40% oil emulsion stabilized with both monoglyceride and triglyceride (reprinted from Frasch-Melnik *et al.*[100] with permission from Elsevier. (B) Cryo-SEM micrograph of a cocoa butter emulsion containing 30% water and 1% PGPR (reprinted from Norton and Fryer[101] with permission from Elsevier. (C) Polarized light microscopy micrograph of emulsions prepared with GMO and hydrogenated canola oil, and shear-cooled within the confined gap of the rheometer stage (reprinted with permission from Ghosh and Rousseau,[99] © 2011 American Chemical Society).

Besides adsorption of pre-formed crystals at the interface and interfacial crystallization of liquid surfactants, a third type of Pickering stabilization was reported by Ghosh and Rousseau.[99] Within the confined gap of a rheometer, sheared TAG crystals aggregated to form oval-shaped crystalline masses. In the presence of GMO, the molecular affinity between TAG and monoglyceride resulted in the encapsulation of water droplets with the crystalline masses labelled as 'cocoons' by the authors (Figure 10.11c).

10.3.2.2 Network Stabilization

When not adsorbed at the oil–water interface, surface-inactive solid particles can also contribute to the emulsion stability by either forming a steric barrier between emulsion droplets or forming a solid network that reduces or prevents droplet motion within the bulk. This is specifically the case in the food industry where triglycerides are often present in product formulation.[102,103] A lot of work has been undertaken to understand the role of particles, in particular fat crystals, on network stabilization of water-in-oil emulsions, the main contribution to the field being accredited to Rousseau and coworkers.[92,93,103–106] In a series of publications, Rousseau and Hodge investigated the effect of crystals formed either pre- or post-emulsification on the emulsion stability.[105,106] Emulsion stability was enhanced by the presence of both types of crystals. Nonetheless, crystals formed post-emulsification within rapid cooling were more efficient as they formed a stronger network that immobilized the dispersed phase, reducing dramatically droplet collision, and consequently coalescence. The formation of a fat crystal network in oil during post-crystallization was studied in more detail by Johansson and Bergenståhl,[72] who used a sediment volumetric technique to show the importance of cooling conditions and polymorphic forms on the sintering of fat crystals. The formation of solid bridges occurs between fat crystals of the same polymorphic form at a cooling rate that favours the nucleation of these specific crystals. For example, rapid cooling promotes the nucleation of β′ tristearin crystals that sinter to form solid bridges. The authors also observed that the sintering process could be affected by some emulsifiers. Monoolein and lecithin were shown to have opposite effects on the formation of crystal network; crystal sintering was promoted by monoolein, while lecithin weakened the network strength. Solid bridges were also observed by Frasch-Melnik et al.,[100] between crystals at the interface and in the bulk phase. It can be seen in Figure 10.11a that crystals sintered around water droplets ('2') are connected to those in the bulk phase ('1'), which enhances emulsion stability.

In a recent study, Ghosh et al.[92] compared the efficiency provided by Pickering and network stabilizations. Pickering stabilization was guaranteed by interfacial crystallization of GMS, while a mixture of GMO and hydrogenated canola oil provided network stabilization. Droplet size was similar for both types of emulsion but, based on the fact that the amount of free water was more important for network-stabilized emulsions, the authors concluded that Pickering stabilization was a more effective way of stabilizing water-in-oil emulsions. Nonetheless, this seems to be the only study comparing both stabilization mechanisms and this was done for a very specific system. Moreover, all mechanisms involved in network stabilization remain unclear and the optimization of fat network stabilization has received little (if any) interest so far. For example, no study has taken into account the conclusions made by Johansson and Bergenståhl regarding the effect of cooling rate and crystal polymorphic forms on the fat matrix properties. Thus, a substantial amount of work remains to be done before conclusions regarding the efficacy of Pickering and network stabilization can be drawn.

10.3.3 Oil-in-Water Emulsions

Many food products are oil-in-water emulsion-based, such as mayonnaise, salad dressing or milk. Unlike butter or margarine, water-in-oil emulsions stabilized by fat crystals, the main ingredients responsible for their stability are liquid-state surfactants, *e.g.* lecithins, proteins or hydrocolloids, which do not form Pickering structures. Solid-stabilized emulsions offer potential for developing new food products due to their different microstructures. In order to understand Pickering stabilization of oil droplets, many inorganic particles have been used, such as clay,[7] latex,[8] titanium dioxide[6] or silica,[5] most of them not being edible. Hydrophilic fumed silica can be produced as edible particles; its effect on the stability of 'food grade' oil-in-water emulsions was investigated by Pichot *et al.*[107,108] Despite their low surface activity, hydrophilic silica particles were able to provide long-term emulsion stability when prepared close to the isoelectric point. The absence of surface charge allows the particles to be closely packed at the oil–water interface. The authors also showed that addition of surfactant, either oil or water soluble, offers a better control of emulsion microstructure. A mixture of particles and surfactant as emulsifiers provides a wider range of droplet sizes and a better control over creaming. Interestingly, emulsion microstructure was modified from Pickering-like to surfactant stabilized-like by increasing the water-soluble surfactant concentration.

Despite the possibility of producing edible emulsions using inorganic particles, it seems unlikely that food companies will invest in the development of products containing 1% or more of silica particles. The main challenge for the food industry has been, over the last few years, to develop micro- or nano-size solid particles, insoluble in water, with sufficient surface activity to adsorb at the oil–water interface. Most research has focused on polysaccharide particles or complexes; only a few recent studies have demonstrated the ability of other species such as protein, spores, *etc.* to act as food particulate emulsifiers.

10.3.3.1 Oil-in-Water Emulsions Stabilized by Polysaccharide Particles

Starch is a well-accepted food ingredient and has many applications; it can be used as a thickener, gelling agent or water retention agent. The use of starch as particles in emulsion formulation has received growing interest in the last few years due to its potential for being used in food and the possibility of controlling surface activity, *i.e.* their level of hydrophobicity. Starch granules are cheap and can be produced from many different botanical sources.[49] Micrographs in Figure 10.3c–e show the size and shape of various types of starch particulates.

Native starch is mostly hydrophilic and is not suitable as a Pickering stabilizer. Nonetheless, Li *et al.*[40] demonstrated the efficacy of some native starches as Pickering emulsifiers. Amongst a variety of micron-size starch granules, only rice starch provided long-term stability; emulsions prepared with waxy maize, wheat and potato starch showed coalescence. Particle size was the determining parameter, the smallest forming the most stable emulsions; the large size (a few millimetres) of droplets formed with the largest particles favours rapid creaming, resulting in coalescence. The morphology and surface chemistry (particles had different levels of hydrophilicity) of the native particles did not influence emulsion stability. All the other studies available in the literature refer to modified starch. Dejmek and coworkers,[39,49] Yusoff and Murray[41] and Marku *et al.*[109] showed that emulsion stability could be achieved using a wide range of modified starch particles.

Emulsion physical properties and stability are affected by the method and the degree of modification of the starch granule hydrophobicity.[49] Chemical modification of quinoa starch particles was proved to be a more efficient method than heat treatment. When chemically modified, quinoa starch with ~ 3% of OSA (octenyl succinic anhydride) produced the most stable emulsions and the smallest emulsion droplets. The authors argued that, because all granule sizes were very similar, the level of stability was due to the surface activity of the modified particles. The interfacial properties of various OSA-modified starches were studied by Yusoff and Murray,[41] who showed that starch particulates do not swell when adsorbed at the interface, which demonstrated their Pickering character. The authors also demonstrated that besides the fact that starch particles strongly adsorb at the oil–water interface, they may also contribute to lowering the interfacial tension.

Droplets of emulsions prepared with starch granules are large, typically 100 µm to a few mm, due to the size of the particles. Efforts have been made to reduce the diameter of starch particles.[41,110] Tan *et al.*[110] prepared starch nanospheres of 200–300 nm by a nanoprecipitation technique, used to produce stable oil-in-water emulsions with droplet size ~ 10–30 µm. Interestingly, the authors also showed that phase inversion could be induced by adjusting the oil/water ratio or by modifying the pH, proving that starch particles could be suitable for stabilizing both oil-in-water and water-in-oil emulsions (Figure 10.12). However no evidence (*e.g.* contact angle) was shown to explain the phase inversion.

Another type of polysaccharide used as food-grade particles is cellulose, a polymer that can be found in many plants or algae and considered the most abundant organic compound on earth. Its availability, as well as its

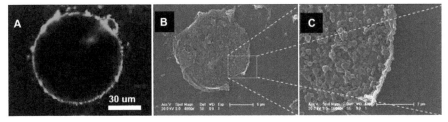

Figure 10.12   (A) Confocal laser scanning microscopy image of undecanol-in-water emulsion droplets prepared with fluorescent starch nanospheres. (B) SEM image of a single droplet of a toluene-in-water emulsion; small rectangle magnified in (C) showing starch particles at the interface. All images reprinted from Tan *et al.*[110] with permission from Elsevier.

sustainability, makes it a very attractive material for use in food. Cellulose has many applications in the food industry; it is commonly used as a fibre supplement, calorie reducer or thickener. A couple of recent studies have shown that cellulose could also be modified to act as Pickering particles.[37,111] In both publications, cellulose nanocrystals were shown to be a very efficient emulsifier (the presence of surfactant was not required to obtain stable emulsions). Kalashnikova *et al.*[111] showed that emulsions were stable against coalescence under quiescent conditions regardless of the particles concentration, but a minimum concentration was necessary when shear forces were applied, due to a better surface coverage at higher concentration. The long 'fibre' shape of the bacterial cellulose nanocrystals (length of ∼855 nm and width of ∼17 nm) was shown to bend and to overlap along the interface, providing an efficient steric barrier against droplet merging (Figure 10.13a). Using a different source of cellulose (ramie vs nata de coco), Zoppe *et al.*[37] observed that emulsion stability was obtained only when the nanocrystals were grafted with thermoresponsive poly(NIPAM) brushes. The production method of the cellulose used by Zoppe *et al.* is not described, and given that the celluloses used in both studies come from different sources, it is impossible to discuss the emulsion stability differences. Nonetheless, emulsions prepared with cellulose nanocrystals and poly(NIPAM) brushes exhibit a very interesting thermal behaviour, because the emulsions become unstable above 35 °C. Potential applications could be found in the food or pharmaceutical industry, as these emulsions could be used as a medium for drug or active delivery.

Chitin is a biopolymer that can be extracted from many different sources, *e.g.* fungi, crustaceans or insects (it is in fact the second most abundant polysaccharide on earth). Chitin nanocrystals form rod-like particles (240×20 nm) which were proved to stabilize corn oil-in-water emulsions for at least 30 days due to a strong interfacial adsorption (Figure 10.13b).[112]

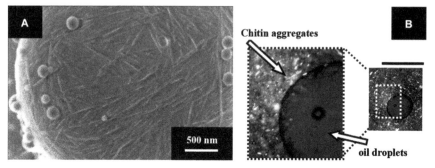

Figure 10.13 (A) SEM micrograph of polymerized styrene-in-water emulsions stabilized with bacterial cellulose nanocrystals (5 g L^{-1}) (reprinted with permission from Kalashnikova *et al.*,[111] © 2011 American Chemical Society). (B) Optical polarized micrograph of corn oil-in-water droplets stabilized with chitin nanocrystals (reprinted from Tzoumaki *et al.*[112] with permission from Elsevier).

Increasing the particle concentration above 0.1% has no effect on the droplet size, which stays constant at ~ 10 μm, but reduces the volume of serum released from the cream layer. This effect on creaming was attributed to the formation of an inner-droplet network formed by the particles present in excess. It was suggested that the emulsion microstructure was a gel-like network formed of flocculated droplets together with chitin nanoparticles non-adsorbed at the interface. This network, as well as chitin particles irreversibly adsorbed at the interface, also plays a role in the stability of lipid droplets during *in vitro* digestions by protecting them against lipolysis compounds.[113] Bile extract diffusion through the gel-like network is very slow, which retards the lipid digestion.

10.3.3.2   Oil-in-Water Emulsions Stabilized by Complexes

As aforementioned, polysaccharides can be associated with proteins by either covalent bonding or electrostatic interactions to form particulate complexes.[114] Given the wide range of proteins and polysaccharides, the number of combinations between these two compounds is almost infinite. This opens a broad new field for the design of particles suitable for use as food-grade Pickering stabilizers.[115–117]

In order to achieve emulsion stability, the two critical parameters are the protein/polysaccharide ratio and the pH of the solution. Neirynck *et al.*[118] studied the interactions between whey protein isolate (WPI) and pectin in the formation of soybean oil-in-water emulsions. At pH above the WPI isoelectric point, no complexes were formed and emulsion stability was mainly due to the adsorption of protein at the interface. Below the isoelectric point, WPI and pectin assemble to form complexes of which surface activity depends on the pectin concentration; around the electric equivalent point (low pectin concentration), insoluble complexes formed large emulsion droplets, while by adding more pectin, complex solubility in water increased, resulting in smaller droplets. Nonetheless, neither the interfacial tension nor the contact angle of insoluble particles was measured, and one would argue that the formation of bigger droplets could be due to the complex size rather than the lack of surface activity. Production of stable Pickering emulsions was achieved using both lactoferrin nanoparticles and their complexes with two polysaccharides (alginate and carrageenan).[31] Emulsion physical properties were very similar but the authors showed that the presence of polysaccharide improved emulsion stability to saliva and modulated the emulsion response to gastric conditions by preventing or delaying the proteolysis of lactoferrin. This could lead to very interesting food applications regarding topical delivery.

Emulsion stabilization using protein/polysaccharide interactions can also be achieved using another route. Rather than forming complexes before emulsification, protein and polysaccharide can be successively adsorbed at the interface to form a multilayer structure.[119] In this method, called the 'layer-by-layer' technique, a primary emulsion is formed with protein only, and then added to a polysaccharide solution to promote interfacial inter-actions between the two species. This emulsification methodology is very similar to the one used to produce stable water-in-oil emulsions that consists of primarily adsorbing liquid surfactant at the interface (typically GMO or PGPR) to promote the interfacial nucleation of TAG crystals. Nonetheless, despite the fact that the layer-by-layer technique is very promising for new food formulations,[120] polysaccharides, unlike TAG, do not crystallize at the interface to form a solid shell around oil droplets and cannot be considered, in this case, as a Pickering emulsifier. Moreover, Jourdain *et al.* demon-strated in a series of publications[121,122] that the mixed emulsions, *i.e.* emulsions stabilized with complexes, were more stable than multilayer emulsions. This was due to (i) the aggregation of primary emulsion droplets during layer-by-layer emulsification, and (ii) much stronger interfacial viscoelastic properties of the complexes.

10.3.3.3   Other Food Pickering Emulsifiers

Besides polysaccharide particles and polysaccharide/protein complexes, which have received significant interest over the last few decades, other types of food particles such as protein particulates,[27] flavonoids[123] or spores[124] have proved to be efficient Pickering emulsifiers. This section aims to review current knowledge regarding these particle-stabilized emulsions, but due to the need to develop new attractive and healthy food, it is very likely that the number of food-grade particles is going to dramatically increase in the next few years.

Proteins are one of the most commonly used emulsifiers in food formu-lation, but a large majority are water soluble, and despite the fact that they provide excellent emulsion stability through electrostatic repulsion and steric mechanisms, they are not considered as Pickering particles. Non-etheless, protein-rich systems often contain protein aggregates and pro-teinaceous colloidal particles, although these materials hardly contribute to emulsion stability.[125] A few recent studies demonstrated that protein par-ticulates could act as effective Pickering stabilizers.[25,27,31,126] Zein nano-particles (~ 100 nm), extracted from corn,[127] were successfully used to either encapsulate essential oils[126] or stabilize soybean-in-water emulsions.[27] Stable emulsions were formed due to the surface activity of the particles (contact angle $\sim 90°$), given that the pH of the aqueous phase was not close to the zein isoelectric point. Nonetheless, de Folter *et al.* observed that emulsion physical properties (*e.g.* droplet size or emulsion viscosity) could be tailored by adjusting the pH. The effect of pH on essential oil encapsu-lated nanospheres was also investigated by Parris *et al.*,[126] who showed that these particles have restricted digestibility in the stomach, which limits the release of oil.

Lipids are commonly used in the stabilization of water-in-oil emulsions as mentioned in Section 10.3.2, but due to their hydrophobic character, they are not considered as suitable stabilizers for oil-in-water emulsions.

Nonetheless, glyceryl stearyl citrate (GSC) particles, generated by quench-cooling of hot lipid dispersed in water, produced ~ 12 week-stable food-grade oil-in-water emulsions.[80] Contact angle measurements showed that these lipid nanoparticles could be wetted by both oil and water, which promoted their adsorption at the interface. Interestingly, emulsion stabilization was achieved despite a relatively low surface coverage, as TEM pictures revealed the 'bald' zones on the oil droplets (Figure 10.14c). Droplet coalescence was prevented by electrostatic repulsions due to the presence of negative charges at the particle surface. After 12 weeks, emulsion destabilization was observed and the authors suggested this was due to either particle Ostwald ripening or GSC solubilization in the oil phase.

Other examples of food-grade solid particles able to stabilize oil–water interfaces can be found in the literature such as sodium stearoyl lactylate (SSL),[128] flavonoids[123] or spores,[124] but these have been only little studied. Figure 10.14 shows interfacial structures formed by these particles. SSL, for

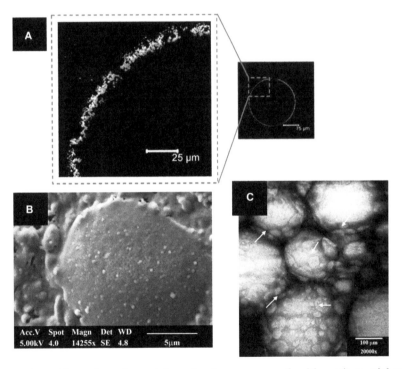

Figure 10.14 Examples of emulsion structures prepared with various Pickering particles. (A) CLSM images of fine n-tetradecanein-water emulsions prepared with tiliroside (reprinted with permission from Luo *et al.*,[123] © 2011 American Chemical Society). (B) SEM micrograph of sunflower oil-in-water emulsions stabilized with sodium stearoyl lactylate (reprinted from Kurukji *et al.*[128] with permission from Elsevier. (C) TEM micrographs of canola oil-in-water stabilized with solid liquid nanocrystals (reproduced from Gupta and Rousseau[80]).

example, is an anionic surfactant typically used as an emulsifier and/or rheology modifier in food products (*e.g.* bread and cake shortenings). When dispersed in water, SSL forms ordered structures (*e.g.* crystal aggregates). Kurukji *et al.*[128] showed that oil-in-water emulsion microstructure could be altered by modifying the SSL content; at high concentration, SSL forms crystal particles that adsorb at the interface (Figure 10.14b), providing long-term stability, while at low concentration, SSL acts as a more conventional low molecular weight surfactant, forming a liquid structure at the interface.

10.3.4   Multiple Emulsions

Multiple emulsions, also called double emulsions, are complex systems consisting of droplets within droplets dispersed in a continuous medium. The most common types of multiple emulsions are water-in-oil-in-water (W/O/W) and oil-in-water-in-oil (O/W/O).[129] These emulsions are usually produced using a two-step mechanism; a primary emulsion is formed using high-energy emulsification and then dispersed within a continuous medium under low shear. For instance, in the case of W/O/W emulsions, water droplets are first formed in the oil phase, and then the oil droplets containing internal water droplets are produced in an aqueous phase. One of the main challenges in the production of multiple emulsions is to keep intact internal droplets, *i.e.* to prevent their coalescence during secondary emulsification or even under post-production quiescent conditions or to prevent the transport of internal droplets to the continuous phase. This can be achieved by using appropriate processing conditions; for example membrane emulsification as a secondary process was found to be very efficient compared to a high-shear mixer.[130] Emulsion microstructure is also very important because designing strong interfaces may prevent inner droplet coalescence and outer droplet leakage.

In this context, Pickering stabilization seems particularly appropriate due to the irreversible adsorption of solid particles at the interface, which prevent coalescence by forming a steric barrier around the droplets. Evidence can be found in the literature that surfactant-free multiple emulsions can be stabilized using solid particles as sole emulsifiers.[131–134] Zhang *et al.*[133] reported the production of W/O/W emulsions using only one type of particle. Silica particles were produced and surface-modified during primary water-in-oil emulsification to stabilize the water droplets. After addition of (external) water, oil droplets were formed under shear and stabilized by modified silica particles contained in the oil phase. Nonetheless, particles used in these studies were not suitable for use in food. Food-grade particles have been used for the formulation of multiple emulsions, but surprisingly, only to stabilize either the inner or outer droplets.

Frasch-Melnik *et al.*[135] reported the use of a mixture of saturated monoglycerides and triglycerides to stabilize the inner water droplets of W/O/W emulsions. After primary emulsification and under cooling conditions, the lipid mixture solidifies at the water droplet ($\sim$ 3.5 μm) surface to form a solid crystalline shell. Outer oil droplets (~ 30 μm) formed in water under high-shear mixing were stabilized using sodium caseinate. Interestingly, the authors observed that the presence of shell preserved the integrity of the inner water droplets, even when subjected to high shear forces. Conversely, quinoa starch was proved to be a good emulsifier for outer oil droplets (~ 30 μm), inner water droplets (~ 1 μm) being stabilized with PGPR.[136]

Complexes made of proteins and polysaccharides were successfully used to stabilize oil/water interfaces in multiple emulsions. These complexes are actually the Pickering particles that have received the greatest interest over the last few decades. Their use in multiple emulsions was reviewed in detail by Dickinson.[137] Interestingly, Garti and coworkers[138,139] showed that the complexes, for example whey protein isolate (WPI) and xanthan gum, could stabilize oil droplets used as either inner or outer droplets to produce stable O/W/O or W/O/W emulsions, respectively. pH was found to play an important role in the emulsion stabilization, particularly in the case of W/O/W emulsions. Below the complex isoelectric point, protein and polysaccharide are strongly bound through electrostatic interactions and the stabilization is mainly due to the steric barrier formed by xanthan gum; above the isoelectric point, the system is mainly stabilized by a depletion mechanism and an electrostatic repulsion. The specific response of protein/polysaccharide complexes to pH may also be an answer to the creaming issue often occurring in double emulsions due to the large size of outer droplets. In particular, interactions between milk protein/pectin complexes may result in the formation of a gel-like structure under certain pH conditions. Murillo-Martínez *et al.*[140] showed that outer oil droplets stabilized with WPI/pectin complexes could be reduced to ~ 2.5 μm due to the highly viscous continuous phase.

10.3.5 Applications

Most of the studies regarding the use of food particles for stabilizing Pickering emulsions were carried out to get a better understanding of the stabilization mechanisms involved in such systems. Nonetheless, regardless of the emulsion type, some researchers have tried to develop Pickering emulsions that could have real applications in the food industry.

Lipid oxidation often occurs in food products that contain unsaturated fatty acids, oleic, linoleic, and linolenic compounds. This is undesirable as it leads to instability and loss of product appearance, taste, texture and nutritional profile. Norton and coworkers[141,142] investigated the effect of interfacial structure in order to reduce lipid oxidation of oil-in-water emulsion droplets. The authors initially showed that food-grade hydrophilic silica particles were very efficient at inhibiting lipid oxidation, compared to two classic food emulsifiers, Tween 20 and sodium caseinate. It was suggested that the presence of solid particles at the interface either separate the metal ions from the droplet surface or reduce the contact area between the oil droplets and the water. Microcrystalline cellulose (MCC) and modified starch (MS) particles were also found to (i) provide long-term stability (against coalescence and creaming) to oil-in-water emulsions, and (ii) enhance the emulsion oxidation stability. Nonetheless, the level of oxidation was higher with MS particles than MCC particles, probably due to the thicker interface formed by MCC and also MCC's ability to interact with the free radicals responsible for the oxidation through their negative surface charges.

Obesity in Western countries has dramatically increased over the last few decades. In the UK, for example, according to the OECD, the number of obese people has increased more than three times over the last 30 years, to affect about 24% of the population in 2009. Obesity causes chronic diseases such as diabetes, hypertension and strokes, and is considered a public health and policy problem because of its prevalence, costs and health effects. It is, therefore, desirable for the food industry to produce low-fat products. For example, Fryer, Norton and coworkers[101,143] have developed chocolate containing up to 60% water. Cocoa butter emulsion stability was attributable to the presence of a crystalline shell made of TAG sintered at the interface in the presence of PGPR around the water droplets. The polymorphic form of cocoa butter is very important in chocolate formulation as it determines the texture and the melting profile of the final product. Using a margarine line, the authors showed that it was possible to mimic tempering and to produce chocolate with a large amount of water and the most desirable polymorph, through a continuous process.

Controlling salt intake in everyday life is a major challenge for the food industry as a high consumption of salt may lead to cardiovascular diseases. In order to control salt release in food emulsion, Frasch-Melnik *et al.*[100,135] studied the effect of crystalline shell formed around water droplets. The shell, made of a mixture of monoglycerides and triglycerides, was very efficient because only 5% salt encapsulated in the water droplets of a water-in-oil emulsion was released after one month. This was attributed to the sintering of fat crystals at the interface, providing total surface coverage with no or very few defects. Salt release was enhanced by heating up the emulsion, which results in total or partial melting of the crystalline shell. This study was extended to double emulsions (W/O/W). Salt release was slightly higher compared to simple emulsions, due to the two-step process; the authors suggested that during secondary emulsification, the crystalline shell might have been cracked due to high shear forces. Salt encapsulation within inner droplets of W/O/W Pickering emulsions was also investigated by Lutz *et al.*[144] and compared to the encapsulation of sodium ascorbate. In this study, the inner water droplets were stabilized with PGPR and the outer oil droplets were stabilized with WPI/modified pectin complexes. Sodium ascorbate was released faster and to a greater extent than NaCl. The authors suggested that the release is due to the diffusion of the ionic species through the TAG oil droplets. Being more hydrated, NaCl diffuses more slowly, resulting in a slower release. More generally, the use of Pickering particles in the formulation of multiple emulsions offers great possibility in terms of encapsulation and release.[136,145]

10.4 Concluding Remarks

The interest in developing particle-stabilized food emulsions has grown significantly over the last few decades. This interest originated in the need to find alternatives for current high-calorie emulsion-based foods; the unique interfacial structure due to the presence of solid particles contrasts with the liquid interfaces formed by classic low molecular weight surfactants and may offer solutions to the current issues that the food industry is facing, in particular the increasing obesity in Western countries. Nonetheless, the field of food particle-stabilized emulsions is in its infancy and a lot of work is still to be done.

Triglycerides are a major component of many food products. Some TAG can play the role of Pickering stabilizers by either forming a shell around water droplets or a solid network within the bulk phase that prevents coalescence, or even both simultaneously. This has been known for a long time in the case of butter, for example, but interfacial phenomena, specifically the interactions between fat crystals and liquid surfactants, are only now being understood. Contrary to Pickering water-in-oil emulsions mainly stabilized by fat crystals, a few different particles have been investigated for the stabilization of oil-in-water emulsions. The focus has been mainly on polysaccharide particles, *e.g.* starch, because of their low cost and their insolubility in water, as well as the natural ability to form particles and protein/polysaccharide complexes. Other particles such as spores, flavonoids or protein particulates can also act as efficient emulsifiers, but there is little information available in the literature. Many food-grade particles, *e.g.* hydrophobins or some complexes, have not yet been used in emulsion formulations; it is very likely that the number of publications on food Pickering emulsions will grow significantly over the next few years.

Formulation of particle-stabilized emulsions for use in food is currently facing some limitations. The stability of such emulsions implies that particles should be an order of magnitude smaller than the droplets. However, the size range of many food-grade particulates is within 10–100 μm, which is problematic for both emulsion stability and food product properties. This is a major issue for which food researchers must find an answer in order to use these particles as food emulsifiers. Tailoring particle hydrophobicity represents another challenge for food scientists; particle surface activity must be carefully designed to optimize emulsion stability. However, surface hydrophobicity alteration often involves chemical substances that are not suitable for use in food, or only at very low content (*e.g.* OSA). It is very likely that in the near future, research will not only focus on the development of new types of particles but also on the use of food additives able to change the surface properties of existing particles such as starches or lipids.

References

1. D. J. McClements (ed.), *Food Emulsions – Principles, Practices, and Techniques*, 2nd edn, CRC Press, Boca Raton, FL, 2005.

2. M. J. Rosen (ed.), *Surfactants and Interfacial Phenomena*, 2nd edn, John Wiley & Sons, Inc., New York, 1989.
3. B. P. Binks, *Curr. Opin. Colloid Interface Sci.*, 2002, **7**, 21.
4. S. U. Pickering, *J. Chem. Soc.*, 1907, **91**, 2001.
5. P. Binks and O. Lumsdon, *Phys. Chem. Chem. Phys.*, 1999, **1**, 3007.
6. S. Stiller, H. Gers-Barlag, M. Lergenmueller, F. Pflücker, J. Schulz, K. P. Wittern and R. Daniels, *Colloids Surf., A*, 2004, **232**, 261.
7. S. Guillot, F. Bergaya, C. de Azevedo, F. Warmont and J. F. Tranchant, *J. Colloid Interface Sci.*, 2009, **333**, 563.
8. X. He, X. Ge, H. Liu, H. Zhou and Z. Zhang, *Colloids Surf., A*, 2007, **301**, 80.
9. H. E. Bergna and W. O. Roberts (eds), *Colloidal Silica: Fundamentals and Applications*, CRC Press, Boca Raton, FL, 2006.
10. E. F. Vansant, P. Van Der Voort and K. C. Vrancken (eds), *Characterization and Chemical Modification of the Silica Surface*, Elsevier, New York, 1995.
11. R. K. Iler (ed.), *The Chemistry of Silica: Solubility, Polymerization, Colloid and Surface Properties and Biochemistry of Silica*, Wiley-Interscience, Hoboken, NJ, 1979.
12. W. Stöber, A. Fink and E. Bohn, *J. Colloid Interface Sci.*, 1968, **26**, 62.
13. T. Matsoukas and E. Gulari, *J. Colloid Interface Sci.*, 1989, **132**, 13.
14. G. H. Bogush and C. F. Zukoski IV, *J. Colloid Interface Sci.*, 1991, **142**, 19.
15. T. Günther, J. Jupesta, F. Weigler, W. Hintz and J. Tomas, *PARTEC*, Germany, 2007.
16. I. Blute, R. J. Pugh, J. van de Pas and I. Callaghan, *J. Colloid Interface Sci.*, 2007, **313**, 645.
17. L. T. Zhuravlev, *Colloids Surf., A*, 2000, **173**, 1.
18. F. D. Osterholtz and E. R. Pohl, *J. Adhes. Sci. Technol.*, 1992, **6**, 127.
19. B. P. Binks and S. O. Lumsdon, *Langmuir*, 2000, **16**, 8622.
20. R. Pichot, F. Spyropoulos and I. T. Norton, *J. Colloid Interface Sci.*, 2012, **377**, 396.
21. E. Vignati, R. Piazza and T. P. Lockhart, *Langmuir*, 2003, **19**, 6650.
22. O. G. Jones and D. J. McClements, *Compr. Rev. Food Sci. F*, 2010, **9**, 374.
23. E. Dickinson, *Colloids Surf., B*, 2001, **20**, 197.
24. D. G. Dalgleish, P. A. Spagnuolo and H. Douglas Goff, *Int. Dairy J.*, 2004, **14**, 1025.
25. V. N. Paunov, O. J. Cayre, P. F. Noble, S. D. Stoyanov, K. P. Velikov and M. Golding, *J. Colloid Interface Sci.*, 2007, **312**, 381.
26. T. David-Birman, A. Mackie and U. Lesmes, *Food Hydrocolloids*, 2013, **31**, 33.
27. J. W. J. de Folter, M. W. M. van Ruijven and K. P. Velikov, *Soft Matter*, 2012, **8**, 6807.
28. H. Fessi, F. Puisieux, J. P. Devissaguet, N. Ammoury and S. Benita, *Int. J. Pharm.*, 1989, **55**, R1.
29. R. Santipanichwong, M. Suphantharika, J. Weiss and D. J. McClements, *J. Food Sci.*, 2008, **73**, N23.

30. I. Peinado, U. Lesmes, A. Andrés and J. D. McClements, *Langmuir*, 2010, **26**, 9827.
31. G. Shimoni, C. Shani Levi, S. Levi Tal and U. Lesmes, *Food Hydrocolloids*, 2013, **33**, 264.
32. M. B. Linder, *Curr. Opin. Colloid Interface Sci.*, 2009, **14**, 356.
33. A. R. Cox, F. Cagnol, A. B. Russell and M. J. Izzard, *Langmuir*, 2007, **23**, 7995.
34. J. Araki, M. Wada and S. Kuga, *Langmuir*, 2000, **17**, 21.
35. M. Rayner, A. Timgren, M. Sjöö and P. Dejmek, *J. Sci. Food Agric.*, 2012, **92**, 1841.
36. X. Song, G. He, H. Ruan and Q. Chen, *Starch - Stärke*, 2006, **58**, 109.
37. J. O. Zoppe, R. A. Venditti and O. J. Rojas, *J. Colloid Interface Sci.*, 2012, **369**, 202.
38. S. Elazzouzi-Hafraoui, Y. Nishiyama, J. L. Putaux, L. Heux, F. Dubreuil and C. Rochas, *Biomacromolecules*, 2007, **9**, 57.
39. A. Timgren, M. Rayner, M. Sjöö and P. Dejmek, *Procedia Food Sci.*, 2011, **1**, 95.
40. C. Li, Y. Li, P. Sun and C. Yang, *Colloids Surf., A*, 2013, **431**, 142.
41. A. Yusoff and B. S. Murray, *Food Hydrocolloids*, 2011, **25**, 42.
42. R. L. Whistler, Solubility of polysaccharides and their behaviour in solution, in *Carbohydrates in Solution*, ed. H. S. Isbell, American Chemical Society, Washington, USA, 1973, pp. 242–255.
43. E. Dickinson, *J. Sci. Food Agric.*, 2013, **93**, 710.
44. G. O. Phillips and P. A. Williams, in *Handbook of Hydrocolloids*, 2nd edn, G. O. Phillips and P. A. Williams, Woodhead Publishing, Cambridge, UK, 2009.
45. R. N. Tharanathan, *Crit. Rev. Food Sci. Nutr.*, 2005, **45**, 371.
46. M. Gudmundsson and A.-C. Eliasson, in *Carbohydrates in Food*, 2nd edn, ed. A.-C. Eliasson, CRC Press, Boca Raton, FL 2006, pp. 391–469.
47. C. G. Caldwell and O. G. Wurzburg, Polysaccharide Derivatives of Substituted Dicarboxylic Acids, 1953, USA, US2661349.
48. FDA Code of Federal Regulation, Title 21, *Chapt. 1 Part 172 Food starch-Modified*, section 172.982, 1981.
49. M. Rayner, M. Sjöö, A. Timgren and P. Dejmek, *Faraday Discuss.*, 2012, **158**, 139.
50. Y. Habibi and A. Dufresne, *Biomacromolecules*, 2008, **9**, 1974.
51. A. C. Lee and Y. H. Hong, *Food Res. Int.*, 2009, **42**, 733.
52. C. G. de Kruif and R. Tuinier, *Food Hydrocolloids*, 2001, **15**, 555.
53. J. L. Doublier, C. Garnier, D. Renard and C. Sanchez, *Curr. Opin. Colloid Interface Sci.*, 2000, **5**, 202.
54. D. J. McClements, *Biotechnol. Adv.*, 2006, **24**, 621.
55. D. J. McClements, E. A. Decker, Y. Park and J. Weiss, *Crit. Rev. Food Sci. Nutr.*, 2009, **49**, 577.
56. A. Ye, *Int. J. Food Sci. Technol.*, 2008, **43**, 406.
57. F. Weinbreck, R. de Vries, P. Schrooyen and C. G. de Kruif, *Biomacromolecules*, 2003, **4**, 293.

58. R. Tuinier, C. Rolin and C. G. de Kruif, *Biomacromolecules*, 2002, **3**, 632.
59. F. Weinbreck, M. Minor and C. G. de Kruif, *J. Microencapsulation*, 2004, **21**, 667.
60. Y. Yeo, E. Bellas, W. Firestone, R. Langer and D. S. Kohane, *J. Agric. Food Chem.*, 2005, **53**, 7518.
61. K. Sato and S. Ueno, *Curr. Opin. Colloid Interface Sci.*, 2011, **16**, 384.
62. K. Sato, *Eur. J. Lipid Sci. Technol.*, 1999, **101**, 467.
63. W. Skoda and M. van den Tempel, *J. Colloid Sci.*, 1963, **18**, 568.
64. M. L. Herrera, M. de León Gatti and R. W. Hartel, *Food Res. Int.*, 1999, **32**, 289.
65. A. Bell, M. H. Gordon, W. Jirasubkunakorn and K. W. Smith, *Food Chem.*, 2007, **101**, 799.
66. K. Sato, *Chem. Eng. Sci.*, 2001, **56**, 2255.
67. K. Sato and T. Kuroda, *J. Am. Oil Chem. Soc.*, 1987, **64**, 124.
68. S. Sonwai and M. R. Mackley, *J. Am. Oil Chem. Soc.*, 2006, **83**, 583.
69. D. Johansson and B. Bergenståhl, *J. Am. Oil Chem. Soc.*, 1992, **69**, 728.
70. D. Johansson and B. Bergenståhl, *J. Am. Oil Chem. Soc.*, 1992, **69**, 705.
71. D. Johansson and B. Bergenståhl, *J. Am. Oil Chem. Soc.*, 1992, **69**, 718.
72. D. Johansson and B. Bergenståhl, *J. Am. Oil Chem. Soc.*, 1995, **72**, 911.
73. D. Johansson, B. Bergenståhl and E. Lundgren, *J. Am. Oil Chem. Soc.*, 1995, **72**, 939.
74. D. Johansson, B. Bergenståhl and E. Lundgren, *J. Am. Oil Chem. Soc.*, 1995, **72**, 921.
75. D. Johansson and B. Bergenståhl, *J. Am. Oil Chem. Soc.*, 1995, **72**, 205.
76. D. Johansson and B. Bergenståhl, *J. Am. Oil Chem. Soc.*, 1995, **72**, 933.
77. R. H. Müller, K. Mäder and S. Gohla, *Eur. J. Pharm. Biopharm.*, 2000, **50**, 161.
78. H. Bunjes, *Curr. Opin. Colloid Interface Sci.*, 2011, **16**, 405.
79. E. Dickinson, F. J. Kruizenga, M. J. W. Povey and M. van der Molen, *Colloids Surf., A*, 1993, **81**, 273.
80. R. Gupta and D. Rousseau, *Food & Function*, 2012, **3**, 302.
81. J. Weiss, E. Decker, D. J. McClements, K. Kristbergsson, T. Helgason and T. Awad, *Food Biophys.*, 2008, **3**, 146.
82. T. Awad, T. Helgason, K. Kristbergsson, E. Decker, J. Weiss and D. J. McClements, *Food Biophys.*, 2008, **3**, 155.
83. W. Ramsden, *Proc. R. Soc. London, Ser. A*, 1903, **72**, 156.
84. R. Aveyard, B. P. Binks and J. H. Clint, *Adv. Colloid Interface Sci.*, 2003, **100-102**, 503.
85. Y. Chevalier and M. A. Bolzinger, *Colloids Surf., A*, 2013, **439**, 23.
86. E. Dickinson, *Curr. Opin. Colloid Interface Sci*, 2010, **15**, 40.
87. F. Leal-Calderon and V. Schmitt, *Curr. Opin. Colloid Interface Sci*, 2008, **13**, 217.
88. D. E. Tambe and M. M. Sharma, *J. Colloid Interface Sci.*, 1993, **157**, 244.
89. A. B. Pawar, M. Caggioni, R. Ergun, R. W. Hartel and P. T. Spicer, *Soft Matter*, 2011, **7**, 7710.
90. B. R. Midmore, *J. Colloid Interface Sci.*, 1999, **213**, 352.

91. D. E. Tambe and M. M. Sharma, *J. Colloid Interface Sci.*, 1994, **162**, 1.
92. S. Ghosh, T. Tran and D. Rousseau, *Langmuir*, 2011, **27**, 6589.
93. S. Ghosh and D. Rousseau, *Curr. Opin. Colloid Interface Sci.*, 2011, **16**, 421.
94. E. H. Lucassen-Reynders and M. van den Tempel, *J. Phys. Chem.*, 1963, **67**, 731.
95. N. Garti, H. Binyamin and A. Aserin, *J. Am. Oil Chem. Soc.*, 1998, **75**, 1825.
96. D. Rousseau, *Curr. Opin. Colloid Interface Sci.*, 2013, **18**, 283.
97. E. Davies, E. Dickinson and R. D. Bee, *Int. Dairy J.*, 2001, **11**, 827.
98. W. Skoda and M. van den Tempel, *J. Cryst. Growth*, 1967, **1**, 207.
99. S. Ghosh and D. Rousseau, *Cryst. Growth Des.*, 2012, **12**, 4944.
100. S. Frasch-Melnik, I. T. Norton and F. Spyropoulos, *J. Food Eng.*, 2010, **98**, 437.
101. J. E. Norton and P. J. Fryer, *J. Food Eng.*, 2012, **113**, 329.
102. B. S. Ghotra, S. D. Dyal and S. S. Narine, *Food Res. Int.*, 2002, **35**, 1015.
103. D. Rousseau, S. Ghosh and H. Park, *J. Food Sci.*, 2009, **74**, E1.
104. S. M. Hodge and D. Rousseau, *Food Res. Int.*, 2003, **36**, 695.
105. S. M. Hodge and D. Rousseau, *J. Am. Oil Chem. Soc.*, 2005, **82**, 159.
106. D. Rousseau and S. M. Hodge, *Colloids Surf., A*, 2005, **260**, 229.
107. R. Pichot, F. Spyropoulos and I. T. Norton, *J. Colloid Interface Sci.*, 2009, **329**, 284.
108. R. Pichot, F. Spyropoulos and I. T. Norton, *J. Colloid Interface Sci.*, 2010, **352**, 128.
109. D. Marku, M. Wahlgren, M. Rayner, M. Sjöö and A. Timgren, *Int. J. Pharm.*, 2012, **428**, 1.
110. Y. Tan, K. Xu, C. Liu, Y. Li, C. Lu and P. Wang, *Carbohydr. Polym.*, 2012, **88**, 1358.
111. I. Kalashnikova, H. Bizot, B. Cathala and I. Capron, *Langmuir*, 2011, **27**, 7471.
112. M. V. Tzoumaki, T. Moschakis, V. Kiosseoglou and C. G. Biliaderis, *Food Hydrocolloids*, 2011, **25**, 1521.
113. M. V. Tzoumaki, T. Moschakis, E. Scholten and C. G. Biliaderis, *Food & Function*, 2013, **4**, 121.
114. E. Dickinson, *Food Hydrocolloids*, 2009, **23**, 1473.
115. E. Dickinson, *Soft Matter*, 2008, **4**, 932.
116. C. Schmitt and S. L. Turgeon, *Adv. Colloid Interface Sci.*, 2011, **167**, 63.
117. E. Bouyer, G. Mekhloufi, V. Rosilio, J. L. Grossiord and F. Agnely, *Int. J. Pharm.*, 2012, **436**, 359.
118. N. Neirynck, P. Van der Meeren, M. Lukaszewicz-Lausecker, J. Cocquyt, D. Verbeken and K. Dewettinck, *Colloids Surf., A*, 2007, **298**, 99.
119. D. Guzey and D. J. McClements, *Adv. Colloid Interface Sci.*, 2006, **128-130**, 227.
120. T. Aoki, E. A. Decker and D. J. McClements, *Food Hydrocolloids*, 2005, **19**, 209.

121. L. Jourdain, M. E. Leser, C. Schmitt, M. Michel and E. Dickinson, *Food Hydrocolloids*, 2008, **22**, 647.
122. L. S. Jourdain, C. Schmitt, M. E. Leser, B. S. Murray and E. Dickinson, *Langmuir*, 2009, **25**, 10026.
123. Z. Luo, B. S. Murray, A. Yusoff, M. R. A. Morgan, M. J. W. Povey and A. J. Day, *J. Agric. Food Chem.*, 2011, **59**, 2636.
124. B. P. Binks, A. N. Boa, M. A. Kibble, G. Mackenzie and A. Rocher, *Soft Matter*, 2011, **7**, 4017.
125. E. Dickinson, *Trends Food Sci. Technol.*, 2012, **24**, 4.
126. N. Parris, P. H. Cooke and K. B. Hicks, *J. Agric. Food Chem.*, 2005, **53**, 4788.
127. R. Shukla and M. Cheryan, *Ind. Crops Prod.*, 2001, **13**, 171.
128. D. Kurukji, R. Pichot, F. Spyropoulos and I. T. Norton, *J. Colloid Interface Sci.*, 2013, **409**, 88.
129. G. Muschiolik, *Curr. Opin. Colloid Interface Sci.*, 2007, **12**, 213.
130. A. K. Pawlik and I. T. Norton, *J. Membr. Sci.*, 2012, **415-416**, 459.
131. H. Maeda, M. Okada, S. Fujii, Y. Nakamura and T. Furuzono, *Langmuir*, 2010, **26**, 13727.
132. Y. Nonomura, N. Kobayashi and N. Nakagawa, *Langmuir*, 2011, **27**, 4557.
133. J. Zhang, X. Ge, M. Wang, J. Yang, Q. Wu, M. Wu, N. Liu and Z. Jin, *Chem. Commun.*, 2010, **46**, 4318.
134. S. Zou, C. Wang, Q. Gao and Z. Tong, *J. Dispersion Sci. Technol.*, 2012, **34**, 173.
135. S. Frasch-Melnik, F. Spyropoulos and I. T. Norton, *J. Colloid Interface Sci.*, 2010, **350**, 178.
136. M. Matos, A. Timgren, M. Sjöö, P. Dejmek and M. Rayner, *Colloids Surf., A*, 2013, **423**, 147.
137. E. Dickinson, *Food Biophys.*, 2011, **6**, 1.
138. A. Benichou, A. Aserin and N. Garti, *Colloids Surf., A*, 2007, **294**, 20.
139. A. Benichou, A. Aserin and N. Garti, *Colloids Surf., A*, 2007, **297**, 211.
140. M. M. Murillo-Martínez, R. Pedroza-Islas, C. Lobato-Calleros, A. Martínez-Ferez and E. J. Vernon-Carter, *Food Hydrocolloids*, 2011, **25**, 577.
141. M. Kargar, F. Spyropoulos and I. Norton, *J. Colloid Interface Sci.*, 2011, **357**, 527.
142. M. Kargar, K. Fayazmanesh, M. Alavi, F. Spyropoulos and I. T. Norton, *J. Colloid Interface Sci.*, 2012, **366**, 209.
143. J. E. Norton, P. J. Fryer, J. Parkinson and P. W. Cox, *J. Food Eng.*, 2009, **95**, 172.
144. R. Lutz, A. Aserin, L. Wicker and N. Garti, *Colloids Surf., B*, 2009, **74**, 178.
145. R. Jiménez-Alvarado, C. I. Beristain, L. Medina-Torres, A. Román-Guerrero and E. J. Vernon-Carter, *Food Hydrocolloids*, 2009, **23**, 2425.

CHAPTER 11

Particle-Stabilized Emulsions in Heavy Oil Processing

DAVID HARBOTTLE, CHEN LIANG, NAYEF EL-THAHER,
QINGXIA LIU, JACOB MASLIYAH AND ZHENGHE XU*

Department of Chemical and Materials Engineering, University of Alberta,
Canada
*Email: Zhenghe.xu@ualberta.ca

11.1   Introduction

The landscape of oil production is shifting, with non-conventional oil resources seen as a credible option to meet the ever-growing demand for oil. The 2013 ExxonMobil report *The Outlook for Energy: A View to 2040* projected a rise in the total liquid fossil fuel demand to 113 million of oil-equivalent barrels per day in 2040.[1] A vast proportion of this demand is expected to be addressed through developments in oil sands, tight oil and deepwater oil. A common challenge in oil production from various oil resources is the formation of stable emulsions. The formed emulsions can be: (i) water-in-oil (W/O), (ii) oil-in-water (O/W) and (iii) multiple (W/O/W, O/W/O) emulsions, creating numerous operational challenges. Oil–water separation can often become the 'bottleneck' in production without ability to effectively de-stabilize emulsions. The challenge in oil–water separation is magnified in non-conventional oil production, where the high oil density and the entrainment of clays (fine particles) favour the production of stable emulsions that are increasingly difficult to destabilize. The stable emulsions of intermediate density accumulate between oil and water phases to form so-called 'dense-packed layers' or 'rag layers' consisting of complex emulsions (W/O,

RSC Soft Matter No. 3
Particle-Stabilized Emulsions and Colloids: Formation and Applications
Edited by To Ngai and Stefan A. F. Bon
Published by the Royal Society of Chemistry, www.rsc.org

O/W, W/O/W, O/W/O). These emulsions are known to be stabilized by sur-face active species that partition at the oil–water interface. The rag layer represents a pseudo-filter, hindering water dropout (W/O) or oil droplet creaming (O/W) and therefore reducing oil–water separation efficiency and oil production. The carry-over of micron-sized water droplets often leads to long-term corrosion issues and catalyst poisoning in downstream crude oil processing and refining. Chemical demulsifiers are frequently added to prevent rag layer formation and hence enable efficient oil–water separation. Common commercial demulsifiers are polymers based on ethylene oxide/propylene oxide (EO/PO) chemistry with the EO fraction more water-soluble and PO fraction more oil-soluble. Their surface activity enables interfacial partitioning to disrupt the protective interfacial film and promote fruitful coalescence upon direct contact of neighbouring droplets. The objective of this chapter is to address the key surface properties of solid particles that promote the formation of solid-stabilized petroleum emulsions, assess the properties of interfacial materials, and finally illustrate the mechanisms by which chemicals successfully disrupt interfaces and promote demulsification.

11.2   Theory of Particle-Stabilized Emulsions

For over a century,[2,3] solids-stabilized emulsions have been the subject of considerable scientific interest, with recent advancements focused on micro-scopic and theoretical understanding. The term 'Pickering emulsion' has been widely used to describe solids-stabilized emulsions after the inaugural work by Pickering, who observed in 1907 the interfacial partitioning of water-wetted particles and formation of stable oil-in-water emulsions.[3] The critical role of particle wettability on emulsion type and stability was further investigated by Schulman and Leja.[4] By controlling the contact angle of barium sulfate particles these authors observed formation of stable oil-in-water emulsions by particles with a contact angle $< 90°$ and water-in-oil emulsions by particles with a contact angle $> 90°$, as schematically shown in Figure 11.1. The stability of emulsions was observed to decrease with increasing particle hydrophilicity or hydrophobicity away from $90°$, demonstrating that the particles with a contact angle close to $90°$ would form the most stable emulsions.

To understand the relationship between particle wettability and emulsion stability we can consider a simple case of two droplets coalescing to form one larger droplet of reduced total surface area. The total Gibbs free energy change due to a coalescence event is given by:[5]

$$\Delta G = \gamma_{o/w} \times \Delta A - W_d \times \Delta_n \tag{11.1}$$

where $\gamma_{o/w}$ is the oil–water interfacial tension, ΔA is the change in interfacial area, Δ_n is the number of particles detached from the interface, and W_d given by Equation 11.2 is the work needed to detach a particle from a planar oil–water interface.[5,6]

$$W_d = \pi r^2 \gamma_{o/w}(1 \mp \cos\theta)^2 \tag{11.2}$$

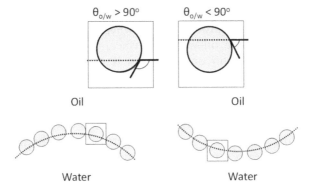

Figure 11.1 Preferential particle location at the oil–water interface governed by particle wettability.
Modified figure from Aveyard *et al.*[7] © 2003 Elsevier.

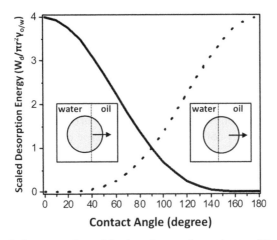

Figure 11.2 Scaled energy of particle detachment from the equilibrium position at the oil–water interface into the oil phase (solid line) or water phase (dashed line) as a function of the particle contact angle.

In Equation 11.2, θ represents the contact angle of particles, and $\mp$ sign describes particle detachment into the oil phase (positive) or water phase (negative). The particle radius, r is considered to be sub-micron such that the gravity contribution is negligible. Figure 11.2 shows the scaled energy of particle detachment into the oil phase (solid line) as a function of particle contact angle from $0°$ and $180°$. For an extremely low contact angle $(\theta \rightarrow 0°)$, the energy required to detach a strongly hydrophilic particle into a hydrophobic solvent approaches the maximum. With an increasing affinity for the particle to reside in the oil phase (reduced particle surface area in the water phase), the energy of particle detachment from the interface to the oil phase decreases to eventually reach zero when $\theta = 180°$. A second curve (broken line) considers the energy of particle detachment from the interface into water as a function of particle contact angle. With a uniform wettability the two curves are symmetrical around $\theta = 90°$ where $W_d/\pi r^2 \gamma_{o/w} = 1$. Hence for a particle partitioned at the oil–water interface, the maximum detachment energy into either the water or oil phases is observed at $\theta = 90°$. At this condition, solids-stabilized emulsions exhibit the greatest stability, which results from steric hindrance to compensate for the work of detachment.

Whilst assuming uniform particle wettability is often sufficient to describe the stability of many particle-stabilized emulsion systems, particles encountered during petroleum processing are often inhomogeneous in chemistry and hence in surface wettability. Contamination of clays and heavy minerals by organic matter results in a patchy particle of hydrophilic and hydrophobic segments.[8] These particles can be considered Janus-like with two distinct surface properties. Binks and Fletcher[9] provided a theoretical assessment of the detachment energy of Janus particles from an oil-water interface to a bulk oil or water phase by considering a particle geometry as shown in Figure 11.3. With polar and apolar patches the particle displays some similarity to a surfactant molecule and can be considered a colloidal surfactant.[10] Similar to the hydrophilic-lipophilic balance (HLB) of surfactants, the Janus balance (J) describes the energy needed to detach the particle from its equilibrium position into the oil phase, normalized by the energy needed to transfer the particle from its equilibrium into the water phase.[11]

The particle contact angle is dependent on the contact angle of the polar (θ_P) and apolar (θ_A) patches, and the relative areas occupied by these patches. The overall (or average) contact angle of this type of particles is given by:

$$\theta_{ave} = \frac{\theta_A(1 + \cos\alpha) + \theta_P(1 - \cos\alpha)}{2} \tag{11.3}$$

with the particle amphiphilicity characterized by:

$$\Delta\theta = \frac{\theta_P - \theta_A}{2} \tag{11.4}$$

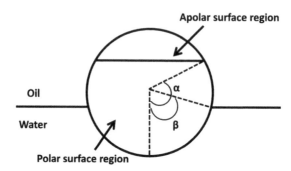

Figure 11.3   Geometry of a Janus particle partitioning at the oil–water interface. The parameters α and β describe the position of the surface boundary between the apolar and polar patches, and the immersion depth of the particle in the oil–water interface, respectively.
Reprinted with permission from Binks and Fletcher.[9] © 2001 American Chemical Society.

For a homogeneous particle $\Delta\theta = 0°$, and $\Delta\theta = 90°$ represents the maximum possible particle amphiphilicity. Similar to the example shown in Figure 11.2, the total surface free energy (E) for a Janus particle located at the planar oil–water interface is dependent on the immersion depth, β, and can be expressed as:

For $\beta \leq \alpha$:

$$E(\beta) = 2\pi r^2 \left[\gamma(\text{AO})(1 + \cos\alpha) + \gamma(\text{PO})(\cos\beta - \cos\alpha) \right.$$

$$\left. + \gamma(\text{PW})(1 - \cos\beta) - \frac{1}{2}\gamma(\text{OW})(\sin^2\beta) \right]$$

(11.5)

For $\beta \geq \alpha$:

$$E(\beta) = 2\pi r^2 \left[\gamma(\text{AO})(1 + \cos\beta) + \gamma(\text{AW})(\cos\alpha - \cos\beta) \right.$$

$$\left. + \gamma(\text{PW})(1 - \cos\alpha) - \frac{1}{2}\gamma(\text{OW})(\sin^2\beta) \right]$$

(11.6)

where γ is the interfacial tension between phases with A = apolar, P = polar, O = oil and W = water. Calculating the Janus particle detachment energy as a function of the particle amphiphilicity (Figure 11.4), Binks and Fletcher demonstrated a three-fold increase in detachment energy as $\Delta\theta$ increased from 0° to 90°.[9] The high detachment energy exhibited by Janus particles was

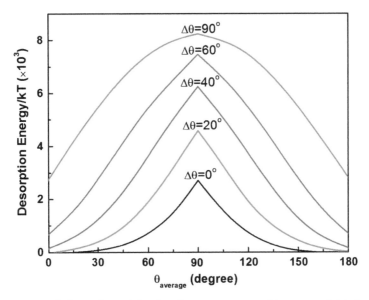

Figure 11.4   Variation of particle detachment energy with area-weighted average contact angle for particles of radius 10 nm and $\alpha = 90°$.
Unmodified figure printed with permission from Binks and Fletcher.[9]
© 2001 American Chemical Society.

calculated over a wide range of θ_{ave}, dissimilar to the narrow range of stability exhibited by homogeneous particles. Readers who are interested in more details on the detachment energy of Janus particles from oil–water interface into bulk oil or water phase are referred to finding more details elsewhere.[9,11]

11.3 Asphaltenes

Asphaltenes, as the heavy components of crude oil and bitumen are often blamed for causing emulsion stability in petroleum emulsions. Asphaltenes are commonly defined as a solubility class: soluble in aromatic solvents (toluene) and insoluble in n-alkanes (heptane, hexane, pentane). With hundreds of thousands of chemically distinct molecules acting as the building blocks of asphaltenes, this generalized definition of asphaltenes has often limited our understanding of asphaltene systems. Over the last decade or more, a renaissance in asphaltene research has seen advancement in the understanding of asphaltene molecular and colloidal structures. While there is no clear analogue for asphaltenes, their behaviour has been shown to provide similarities to both polymer and particle systems, although such science does not completely describe asphaltenes. Depending on the physical and chemical properties, asphaltenes are shown to exhibit different states of molecules from monomer (single unit) to nanoaggregates (several units) and colloidal clusters (multiple nanoaggregates), forming nano-particles in the size range of nanometres to tens of nanometres. Their partial polarity renders them interfacially active, readily partitioning at the liquid–liquid interface. Accumulation of asphaltene particles and their potential to associate and reorganize at the oil–water interface lead to the formation of an interfacial film. The presence of such films hinders the rapid and efficient separation of water-in-oil emulsions which are often unavoidably formed during crude oil production and processing.

11.3.1 Molecular Structure of Asphaltenes

Elemental composition analysis indicates that asphaltenes consist of carbon, hydrogen, nitrogen, sulfur and oxygen, with trace amounts of vanadium and nickel. The aromaticity of asphaltenes is typically on the order of 0.5 with the remaining carbon being aliphatic.[12] The molecular structure of asphaltenes has been the subject of strong debate in recent years, with different molecular structures proposed to account for the observed varying physicochemical properties of ashpaltenets. The first molecular structure of asphaltenes is described by the Yen-Mullins model[13,14] (modified Yen model) and is an extension of the Yen model proposed in 1967.[15] The basic feature of the Yen–Mullins model indicates a most probable island-like structure with a single polyaromatic core consisting of seven condensed aromatic rings and several alkyl and naphthenic side chains, as shown in Figure 11.5a. This structure is consistent with an asphaltene molecular weight around 750 Da.[14] With increasing asphaltene concentration the monomer units interact to form nanoaggregates at the

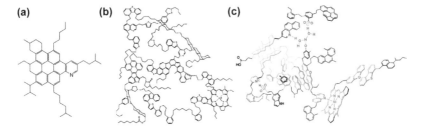

Figure 11.5 Proposed molecular architecture of asphaltenes: (a) island-like model, (b) archipelago model and (c) supramolecular model.
Reprinted with permission from Mullins,[14] © 2010 American Chemical Society; Strausz *et al.*,[16] © 1992 Elsevier; and Gray *et al.*;[17] © 2011 American Chemical Society.

critical nanoaggregate concentration (CNAC) of ~ 100 mg L^{-1} and larger clusters beyond the critical cluster concentration (CCC) of ~ 3 g L^{-1}.[13]

Based on the Yen–Mullins model, nanoaggregates are formed by π–π stacking of aromatic cores with the aggregate growth limited by steric repulsion from aliphatic side chains. The disordered structure of the central backbone allows for entrained solvent and limits the aggregation number to less than 10, with a nanoaggregate size of approximately ~ 2 nm. Subsequent aggregation of nanoaggregates under favourable conditions produces ~ 5 nm asphaltene clusters, with the structure entropy limiting the aggregation number to less than 10.[14] An alternative molecular structure was proposed by Strausz *et al.*,[16] who presented the archipelago model describing several different polyaromatic ring structures connected via aliphatic chains, as shown in Figure 11.5b. The inclusion of multiple active sites (functional groups) is supported by the recent three-dimensional aggregation model of asphaltenes, arising from supramolecular assembly.[17] The approach seeks to address inconsistencies that cannot be satisfied by an architecture dominated by a single aromatic backbone.

The supramolecular assembly model, on the other hand, broadens the basis for intermolecular interactions to include: (i) acid–base interactions, (ii) hydrogen bonding, (iii) metal coordination complexes, (iv) van der Waals interactions between apolar, cycloalkyl and alkyl groups, and (v) aromatic π–π stacking. Characteristics of host molecules, guest molecules and solvent(s) are identified, with the host molecules viewed as having two or more active sites, guest molecules consisting of a range of small and large alkyl aromatics, and solvents (low molecular weight species) to act as assembly terminators. By increasing the number of interaction mechanisms, the size and shape of aggregates can be highly variable, exhibiting polydispersity in the population (Figure 11.5c).[17] The polydispersed nature of asphaltene aggregates has been confirmed by the round-robin study on the size distribution of self-associated asphaltenes.[18] Analysing the same sample of asphaltenes with different analytical techniques (supported by co-contributors), 90% of the asphaltenes were identified as self-associating, forming loose, open and disc-like nanoaggregates spanning one order of magnitude in physical dimensions (2–20 nm).

11.3.2　Interfacial Activity

The mechanism of asphaltene transport from the bulk to the interface (liquid–liquid, liquid–solid) and subsequent association at the interface (interfacial aging) has been studied using a number of experimental techniques, including tensiometry,[19] light absorption,[20] UV spectrometry,[21] interferometry,[22] quartz crystal microbalance,[23] ellipsometry[24] and surface force apparatus.[25] Despite continued effort there remains uncertainty or areas related to the physical and mechanical properties of the adsorbed layer for discussion. Figure 11.6a shows the dynamic interfacial tension curves for different mass fractions of asphaltenes dispersed in a model oil containing 15 wt% toluene and 85 wt% aliphatic base oil.[19] For all mass fractions of asphaltenes studied, the interfacial tension is observed to decrease rapidly at short times, followed by a progressive reduction of the decay rate. A short

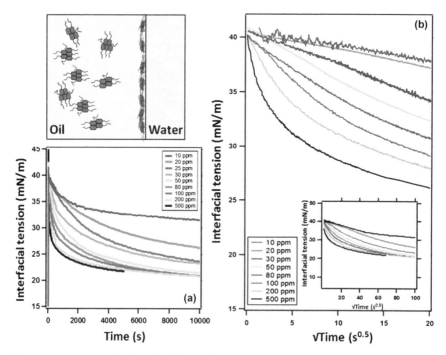

Figure 11.6　Dynamic interfacial tension for asphaltene mass fractions 10 to 500 ppm in model oil, (a) function of time, (b) function of square root time (inset – complete experimental time). Upper-left schematic shows the experimental configuration and illustration of the adsorption process of asphaltenes from the oil phase to the water-oil interface. Modified and reprinted with permission from Rane et al.,[19] © 2012 American Chemical Society.

time approximation for dynamic interfacial tension is given by a coupled Gibbs–Duhem and diffusion equation as:

$$\gamma(t) = \gamma_0 - 2RTC\sqrt{\frac{Dt}{\pi}} \tag{11.7}$$

where γ is the dynamic interfacial tension; γ_0, the pure solvent–water interfacial tension; R, the universal gas constant; T, the temperature; C, the bulk asphaltene concentration; D, the diffusion coefficient; and t, the time. For diffusion-limited adsorption, Equation 11.7 states that the change in the interfacial tension scales linearly with $\sqrt{t}$. As shown in Figure 11.6b, the interfacial tension decreases linearly with $\sqrt{t}$ for all mass fractions considered in the study, especially over a short adsorption time. The slope increases with increasing mass fraction of asphaltenes in solution. While diffusion is accepted as rate-determining of asphaltene adsorption at short time intervals, the adsorption kinetics over an extended period has been viewed differently. The deviation of asphaltene adsorption from diffusion-controlled linearity at high asphaltene concentration has been discussed in terms of: (i) energy barrier-controlled kinetics as a result of hindrance due to already adsorbed species, which causes a delay to the rate of adsorption by an energy barrier;[19] and (ii) slow evolution with time of relaxation and reorganization of the interfacial network constructed of asphaltene particles, as supported by pressure relaxation and interfacial shear rheology studies.[26,27]

11.3.3 Asphaltene 'Skins'

The adsorption of asphaltenes and/or their nano-aggregates with subsequent formation of an interfacial film contributes significantly to the stability of water-in-crude oil emulsions. Direct observation of these films and their inherent stability was demonstrated using the micro-pipette technique.[28,29] The technique involves generating a micron-sized water droplet in heavy crude (diluted bitumen) and aging of the droplet (adsorption of surface active species) before controlled water withdrawal and reduction in droplet size (and hence surface area). Figure 11.7a$_1$ shows a 15–20 μm spherical water droplet formed in 0.1 wt% bitumen diluted in heptol 1:1 (heptane to toluene at equivalent volumes). As the droplet is deflated and the interfacial area compressed, the surface of the droplet crumpled (Figure 11.7a$_2$), revealing a rigid interfacial film that is resistant to in-plane shear and hence providing a steric barrier to droplet–droplet coalescence.[28] The nature of the interfacial 'skin' is highly dependent on petroleum chemistry, with saturate, aromatic, resin and asphaltene (SARA) analysis providing a means to characterize chemistry of the oil through its constituent fractions. The resin and asphaltene fractions are often discussed as the problematic components influencing droplet stability, with their interplay both in the bulk and at the interface impacting the physical properties of the interfacial skin. Wu[30] designed an elegant experiment to

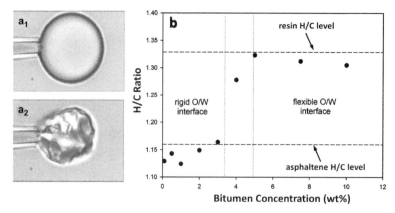

Figure 11.7   (a₁) Water droplet-in-diluted bitumen (0.1 wt% bitumen in 1:1 heptol), (a₂) interfacial crumpling observed during controlled droplet volume reduction, (b) H/C atomic ratio of the interfacial material extracted from emulsions of water-in-diluted bitumen.
Reprinted with permission from Yeung *et al.*,[28] © 2000 Elsevier; Wu,[30] © 2003 American Chemical Society; and Czarnecki and Moran[31], © 2003 American Chemical Society.

isolate and characterize accumulated material residing at the interface of a water-in-diluted bitumen emulsion with increasing bitumen concentrations from 0.1 to 10 wt% in 1:1 heptol. At low bitumen concentration (0.1 wt%) the micro-pipette study confirmed the formation of rigid interfacial film (as shown in Figure 11.7a₂), while at high bitumen concentration (10 wt%) the micron-sized droplet was observed to remain spherical as the interfacial area was compressed, leading to spontaneous emulsification of the water droplet as its size was further decreased. To study the properties of interfacial materials, a 3 wt% heavy water-in-diluted bitumen emulsion was prepared. The 'clean' interfacial material on stable heavy water droplets was isolated after washing by allowing the emulsified heavy water droplets passing through a clean oil phase and a planar oil–water interface, and then settling through clean water phase. After drying of the settled heavy water droplets, the H/C atomic ratio of the accumulated material was determined by elemental analysis. The data in Figure 11.7b indicates a step-change in H/C ratio for the recovered material from the heavy water emulsion droplets prepared in 3–5 wt% bitumen-in 1:1 heptol solutions. At lower bitumen concentrations an H/C ratio of ~1.15 suggests an interfacial material of asphaltene character (H/C ~1.16). At higher bitumen concentrations an H/C ratio of ~1.32 indicates an interfacial material containing other crude oil components, perhaps naphthenates, mixed in with a proportional amount of asphaltenes in the original crude.

Gao *et al.*[32] examined the crumpling ratio ($CR = \dfrac{A_f}{A_i}$, where A_i is the initial projected area of the droplet and A_f is the projected area of the droplet at the onset of crumpling) of micron-sized water droplets prepared in heptol solutions containing: (i) asphaltene (10^{-6} to 10^0 wt%) and (ii) asphaltene

$(10^{-6}$ to 10^{0} wt%) $+ 0.1$ wt% sodium naphthenate. Naphthenic acids are a complex mixture of surface active cyclic carboxylic acids that can compete for the interfacial area. In both cases, at low asphaltene concentrations $(10^{-6}$ to 10^{-5} wt% asphaltenes) droplet crumpling was not observed due to very little adsorption of asphaltene nanoaggregates. For asphaltenes at higher asphaltene concentrations $(10^{-4}$ to 10^{1} wt% asphaltenes) the crumpling ratio 'jumped' to ~ 0.6 with consecutive increases to ~ 0.8, confirming the formation of a rigid interfacial layer. With co-addition of sodium naphthenates the crumpling was significantly depressed, reducing the maximum crumpling ratio to ~ 0.3.[32] The mechanism by which rigid and non-rigid films form has been postulated by Czarnecki and Moran.[31] The model considers the competitive adsorption between a sub-fraction of asphaltenes and low molecular weight surfactants, with the surface competition dependent on the adsorption rate and reversibility, rather than the adsorption energy. The model assumes that: (i) asphaltenes adsorb slowly and irreversibly; (ii) low molecular weight surfactants adsorb quickly through a reversible process to reach an equilibrium state; and (iii) interfacial affinity of asphaltenes is lower than surfactants. To allow the slow diffusing asphaltene particles adsorb at the interface, the surfactant concentration should be sufficiently low so that the interface is not crowded by surfactant molecules. Under this scenario the asphaltene nanoaggregates can attach to the free interface and begin to interact with each other to form larger aggregates. As the asphaltene network begins to occupy large fractions of the interfacial area, the surfactant molecules which obey equilibrium are pushed out of the interface. With sufficient interfacial aging, the irreversible adsorption of asphaltene aggregates can span the droplet interface to form a thick, rigid film. At high concentrations of surface active species, rapid partition of the low molecular weight surfactants exhausts any available interfacial area for asphaltene aggregates to adsorb. Displacement of surfactant molecules by asphaltene particles cannot happen due to differences in their adsorption energies. As a result, the interface remains completely dominated by the surfactant molecules, which provide no shear resistance under interfacial area compression, and hence flexible.

11.3.4 Interfacial Film Rheology

Interfacial rheology is a technique that has been used to characterize the evolution of asphaltene 'skins'. Dilatational and shear are two common approaches to measure film rheology, with the latter receiving substantial attention in recent years.[27,33,34] Dilatational rheology considers the visco-elastic properties of the interfacial layer through harmonic oscillation of the interfacial area, while shear rheology describes the material deformation by shear at constant interfacial area.[35] Rheology measured by dilatation illustrates the rapid formation (typically < 15 min, depending on asphaltene concentration and solvent type) of an elastic dominant interfacial film $(G'$ (elastic) $> G''$ (viscous)) that remains almost insensitive to film aging (see Figure 11.8). In general, the link between dilatational rheology and emulsion

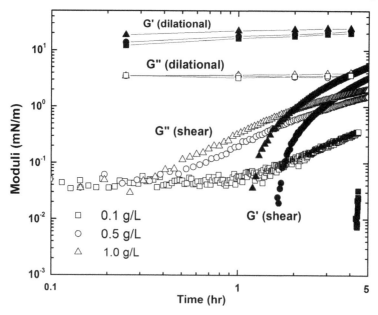

stability is at best qualitative, with a reasonable agreement between increasing dilatational elasticity and emulsion stability being observed. In recent years we studied the relationship between interfacial shear rheology and droplet–droplet coalescence using the double wall ring (DWR) geometry (shear rheology) and in-house built integrated thin film drainage apparatus (ITFDA) (coalescence), respectively. Readers are referred to other references for detailed discussion on instrument techniques.[34,36,37] The film properties as assessed by dilatational rheology are in contrast to the time evolution of interfacial shear properties. Under shear the viscoelasticity (G' and G'') exhibits greater sensitivity to film growth, with elastic and viscous properties showing substantial development with film aging (see Figure 11.8). At short aging time, the interfacial asphaltene film is shown to be dominant by viscous character. At long aging time, the elastic contribution develops and eventually exceeds the viscous contribution, resulting in an elastic dominated film (Figure 11.9a). For an equivalent system (0.4 g L^{-1} asphaltene in toluene), the coalescence times (square symbols in Figure 11.9a) for two water droplets aged between 0.5 and 4.0 hours were measured from the bimorph trace using the ITFDA (Figure 11.9b).

With the two techniques providing nearly equivalent surface area to volume ratios and interfacial aging governed by diffusion-controlled adsorption in the absence of advection diffusion, it is reasonable to relate both properties (shear rheology and droplet coalescence) to infer important physical parameters that govern droplet stability. With a viscous dominant film

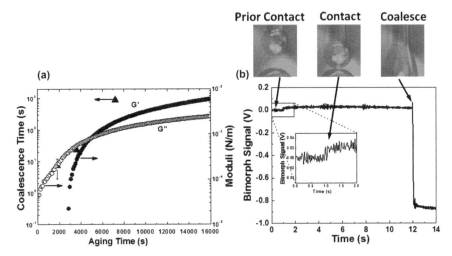

Figure 11.9   (a) Shear viscoelastic properties of an aging asphaltene film (initial bulk concentration 0.4 g L^{-1} asphaltene in 1:1 heptol); and direct comparison with droplet–droplet coalescence time as measured by the bimorph trace of ITFDA (b).
Reprinted with permission from Harbottle *et al.*[34] © (2014) American Chemical Society.

$(G'' > G' = 0)$, droplet coalescence is of the order of a few seconds. As the film develops and the elastic contribution impacts the film $(G'' > G')$, the droplet coalescence time increases to tens of seconds. Finally, as the elasticity dominates the film $(G' > G'')$ the droplets remain stable for 15 min and are considered to be non-coalescing. The characteristic behaviour has been observed for different asphaltene concentrations and solvent aromaticity.[34] It is interesting to note that the development of shear properties appears to be more sensitive when tracking changes in droplet stability as compared with dilatational properties. The transition to the condition $G' > G''$ could provide an insight into the physical mechanism that resists film rupture and droplet–droplet coalescence.

11.3.5   Thin Liquid Film Drainage

As two droplets approach one another an approximately flat thin oil film is formed between them. Drainage of the thin oil film to a critical film thickness and possible coalescence are governed by fluid and interfacial forces. The thin liquid film technique has been used to study drainage of water droplets in crude oil to elucidate the mechanism of droplet stabilization. A schematics of the thin liquid film apparatus is shown in Figure 11.10. The technique is based on a micro-interferometric method using a Scheludko-Exerowa measuring cell. The film holder is made from a porous glass plate with a 0.8 mm diameter hole drilled in the centre. The plate is soaked in the oil of interest and then immersed in the water-filled measuring cell. The thin

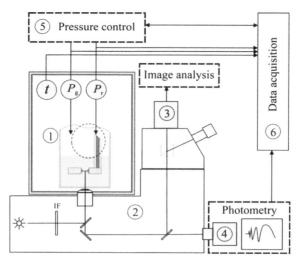

Figure 11.10 Thin liquid film apparatus: (1) thermostatted chamber, (2) inverted microscope, (3) CCD camera, (4) photodiode, (5) pressure control unit, (6) data acquisition.
Reprinted with permission from Tchoukov *et al.*[38] © 2010 Elsevier.

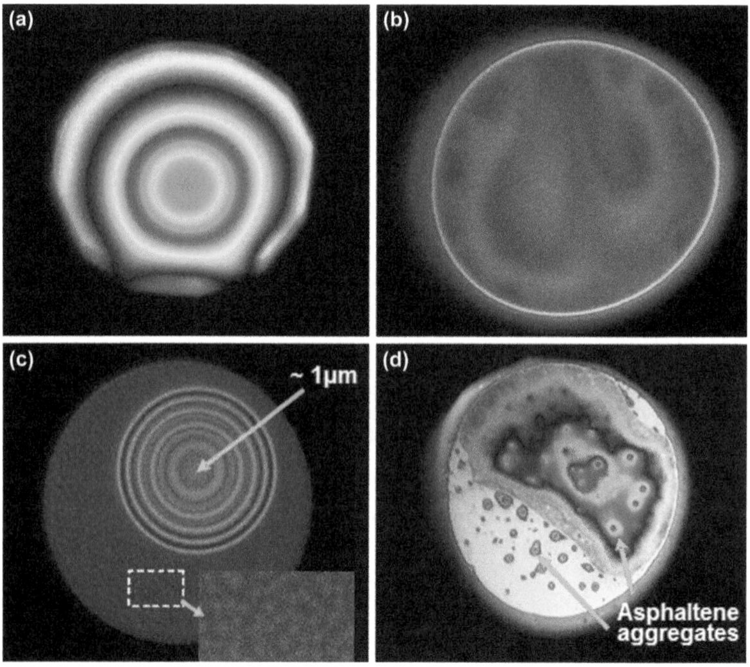

Figure 11.11 Thin film drainage: (a, b) 50 wt% bitumen in 80:20 heptol at time intervals 20 s (a) and 7 min (b); (c, d) 10 wt% bitumen in 80:20 heptol at time intervals 15 min (c) and 2.5 hours (d).
Reprinted with permission from Tchoukov *et al.*[38] © 2010 Elsevier.

liquid oil film is formed when liquid is withdrawn through the capillary by decreasing the pressure (P_r) in the capillary or by increasing the pressure (P_g) in the measuring cell. Drained films are observed in reflected light with the intensity used to calculate the film thickness.

The work of Czarnecki and co-contributors[38–40] elucidated characteristics of petroleum films that feature flexible and rigid interfacial 'skins' as shown in Figure 11.11a, b and Figure 11.11c, d, respectively. For flexible films (50 wt% bitumen in 80:20 vol/vol heptol), shortly after film formation a thick dimple develops and rapidly drains to the meniscus in one or several points. After several minutes the film reaches a critical film thickness ($\sim$30 nm for the highlighted example). The kinetics of the film thinning can be suitably modelled using Reynolds equation,[41] indicating similar characteristics of emulsion films stabilized by surfactants. At lower bitumen concentration (10 wt% in 80:20 heptol) rigid films formed affect both the drainage kinetics and film thickness. Figure 11.11c, d shows snapshots of the film after 15 min and 2.5 hours of interfacial aging, respectively. Similar to a flexible film, a thick dimple is initially formed. However, unlike a flexible film, the dimple in this case does not drain into the meniscus.

Examining the darker film area around the dimple, a large number of small lighter spots (inset Figure 11.11c) were visible, which the authors attributed to the formation of small asphaltene aggregates in the film. On further aging the film thickness became more heterogeneous with several lenses (thicker areas in the film) observed, indicating the formation of larger asphaltene aggregates. These aggregates are estimated to be $\sim$300 nm. The thin liquid film research identified the formation of multilayer structures that can resist coalescence. The authors introduced for the first time the critical importance of film yield stress that should be exceeded to destabilize the film.[39,40]

To determine the strength of asphaltene films, a continuous oscillation stress ramp can be applied to measure the critical stress for film rupture and fragmentation. Using the DWR technique, apparent (conversion by length) yield stresses up to 10^4 N m^{-2} are readily measured for aged films that satisfy the condition $G' > G''$.[34] These high shear yield stress values indicate that interaction stresses upon droplet–droplet collision in dynamic environments are not sufficient to break the films to attain coalescence. Alternative approaches for film 'softening' are highly desired and need to be applied.

11.4 Particles

11.4.1 Asphaltene Nanoaggregates as Emulsion Stabilizers

As indicated above, asphaltenes are considered as a major component in stabilizing water-in-crude oil emulsions. Asphaltenes defined by a molecular class soluble in toluene but insoluble in heptane or pentane represent the heaviest class of molecules in crude oil. The content of asphaltenes increases as the crude becomes heavier, typically much less than 1 wt% for conventional oil and could be as high as 20 wt% for heavy bitumen. Asphaltenes

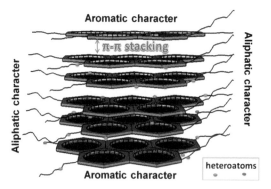

Figure 11.12 Schematic view of asphaltene nanoaggregates exhibiting anisotropic surfaces with aromatic planes preferring aqueous phase and aliphatic side surfaces preferring oil phase or lateral linkage/association. Courtesy of Dr. Robel Teklebrhan of University of Alberta, Edmonton, Canada.

are known to aggregate even at very low concentrations in a highly aromatic good solvent such as toluene.[42] The aggregation of asphaltenes is enhanced at the oil–water interface, forming a three-dimensional network of asphaltene nanoaggregates. Extensive molecular dynamics simulations confirmed molecular aggregation through π–π stacking.[43] These nanosized molecular aggregates, as shown schematically in Figure 11.12, have been blamed for stabilizing W/O petroleum emulsions.[44] With variable amounts of heteroatoms distributed within asphaltene molecules and hence their nanoaggregates, these molecular aggregates can be considered as Janus-like particles with their heteroatoms and fused aromatic rings creating hydrophilic patches in contact with the aqueous phase and the hydrophobic aliphatic side chains protruding into the oil phase to stabilize emulsions. The only difference between conventional Pickering and asphaltene-stabilized emulsions would be the special colloidal interactions between asphaltene nanoaggregates at the oil–water interface that lead to the formation of a three-dimensional protective interfacial film through inter-molecular bridging or interdigitation.[44,45] These interfacial layers make the emulsion extremely stable and difficult to break.

11.4.2 Mineral Particles as Emulsion Stabilizers

In petroleum production various types of inorganic solids such as sand, heavy metal minerals (pyrite, siderite, rutile, zircon, *etc.*) and clays are re-covered with the oil. Depending on the particle surface chemistry these particles can act as emulsion stabilizers, creating substantial above-ground processing challenges. Typically, the heavy metals and clays are naturally hydrophilic but become contaminated by indigenous compounds of crude, making them biwettable and hence increasing their potency to stabilize W/O and O/W emulsions.

Tetrahedron SiO₂ basal plane

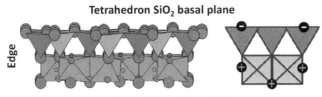

Octahedron Al(O,OH) basal plane

Figure 11.13 Structure of kaolinite clays exhibiting three different surface characteristics: a tetrahedron silica basal plane of pH-independent negative
surface charges, an octahedron aluminium oxy-hydroxyl basal plane of
weak-pH dependent surface charges, and two sides of edge surfaces of
strong pH-dependent surface charge characteristics. Each surface
exhibits different responses to the adsorption of a given solution
species such as surfactant, rendering them to different surface hydrophobicities and hence Janus type of particle behaviour.

Clay particles with distinct basal planes and edges are more surface active
due to their small size (< 2 μm) and surface anisotropy. The surface
anisotropy leads to a variety of possible interactions between clays and
indigenous components of crudes. A typical kaolinite clay particle is shown
schematically in Figure 11.13, exhibiting three types of surfaces: a tetrahedron silica (T-) basal plane, an octahedron aluminum oxyhydroxyl (O-)
basal plane and an aluminum-silicates edge (E-) surface. In an aqueous
solution of neutral pH, the T-basal plane carries a net negative surface
charge, while the O-basal plane and E-surface carry a net positive surface
charge. The magnitude of the charge on each basal plane depends on the
degree of isomorphic substitution of clay lattice ions.

Clay particles are known to interact with crude oil due to their anisotropic
nature of basal planes and edge surfaces as shown in Figure 11.13. This
characteristics allows clay minerals to become biwettable or even Janus-like
due to their interactions with other natural compounds in crude oil. For
example, the highly ionic nature of clay edge surfaces would interact with
natural surfactant more favourably than the basal planes. The edge surfaces
at neutral pH often carry positive charges which attract anionic surface active molecules, while the basal plane featuring permanent deficiency of
positive charges would attract cationic surfactants. Depending on the
abundance of cationic and anionic surfactants in crude oil and the type of
clays, the clay surfaces could be preferentially modified by natural surfactant
to exhibit Janus-like properties.

The chemical nature of toluene-insoluble organic carbon (TIOC), which
contaminates clay surfaces, remains poorly understood. In an attempt to
determine the chemistry of the organic contaminants that modify clay surfaces, Fu *et al.*[46] extracted the solids from crude with toluene, separated
residual organic matter under supercritical conditions with methanol as the
solvent, characterized the methanol extracts by various methods and compared the results with the SARA fractions. The properties of the methanol
extracts closely resembled those of the resins fraction. The organic matter

extracted by methanol under supercritical conditions was also found to be soluble in toluene after extraction. These findings indicate the challenge of breaking emulsions due to the strong affinity of contaminated hetero-geneous clay particles to the oil–water interfaces. However, Osacky *et al.*[47] proposed that the TIOC consists mainly of humic matter and asphaltene-like compounds rather than resins. Their study showed that the presence of ultra-thin illite correlated quite well with the amount of residual organic matter extracted by methanol at supercritical conditions. Pyridines were also found in the residual organic matter in high concentrations. The association of pyridines with illite is well known to alter the wettability of clays. Konan *et al.*[48] studied pyridine adsorption on illite and kaolinite minerals. Their results showed that pyridine mainly adsorbs on Lewis acid sites of clays, which accept the lone pair electrons of the pyridine nitrogen. The authors quantified Lewis acid sites on clay surfaces as a function of pretreatment temperature. Kaolinite was shown to feature very few Lewis acid sites whereas illite had significantly more sites, which decreased at higher pre-treatment conditions due to the removal of surface water. Lewis acid sites on clay increase the affinity of polar hydrocarbons.[49] The removal of surface water by thermal treatment of illite at high temperatures was demonstrated to reduce the affinity of alkenes and aromatics on illite, but not on kaolinite. On the other hand, the affinity of alkanes to illite and kaolinite was found to be unaffected by the thermal treatment of increasing temperature.[50] The presence of water appears to enhance the adsorption of resins and polar organic compounds on illite, thereby enhancing their biwettability.

Humic matter containing both hydrophilic and hydrophobic moieties can adsorb on clay minerals and make them biwettable.[51] Humic matter consists of humic and fulvic acids, and humin. Humic acid is soluble only at high pH, fulvic acid is soluble at high and low pH, whereas humin is completely insoluble in the aqueous phase. Fulvic acids are small, polar molecules with simple sugar and amino acid-related structures.[52] These characteristics make fulvic acids soluble in water and are therefore readily displaced by the larger, less polar humic acids on the surfaces of minerals.[53] Because of the chemistry of fulvic acids and humin, biwettability of clay minerals is therefore not expected to be greatly affected by their presence.

Humic acids are organic macromolecular polyelectrolytes. Soluble in both acid and base solutions, humic acids have the highest affinity to mineral surfaces and their interactions with clay have the greatest contribution to the biwettability of minerals. Because of their highly variable molecular struc-tures, the adsorption of humic acids is difficult to understand. The ad-sorption isotherms of humic acids on clays depend on the history of clay surfaces and surface-area-to-volume ratio of clays.[54] Individual humic acid molecules of 2–5 nm radius can aggregate to form hydrogen bonded supramolecules of a few hundred nm in diameter. Humic acids are also known to form complexes with metal cations. Combined with hydrogen bonding, humic acids show strong adsorption on various clay minerals.[55]

Collectively there are six mechanisms which explain humic acid adsorption on clay minerals: (i) cation bridges; (ii) anion/cation exchange; (iii) hydrogen bonding; (iv) van der Waals interactions; (v) hydrophobic interactions; and (vi) ligand exchange on a charged clay–water interface. The type of clay influences the adsorption mechanism. Ligand exchange and hydrogen bonding are the two most dominant driving forces for humic acid adsorption on clay edge surfaces which contain hydrolysed ferric, aluminium, calcium and magnesium cations. The presence of multivalent cations promotes adsorption of humic acids through cation bridges. Polysaccharides tend to adsorb strongly on kaolinite whereas aromatic compounds bind more favourably to smectite. Adsorption of humic acids on a clay surface is also enhanced with increased ionic strength or decreased pH due to poorer solvation of humic acid molecules by water. The molecular weight of humic acids also plays a role, with smectite minerals adsorbing lower molecular weight compounds than kaolinite.[56] Vermeer *et al.*[57] noted charge compensation and specific interactions as the driving force for humic acid adsorption, while the lateral electrostatic repulsion and loss of entropy hinder the adsorption. Positive charges on clay surfaces attract anionic moieties on the humic acid molecules. As a result, humic acid adsorption increases with increasing positive surface charges by multivalent cation activation. As more anionic sites from the humic acids form complexes with cations in the solution, lateral electrostatic repulsion is decreased, leading to an increase in humic acid adsorption on clay surfaces.

11.5 Rag Layers

These particle-stabilized emulsions exhibit a density intermediate to the two immiscible liquids. As a result, the emulsion partitions the oil and water phases by forming a layer that hinders creaming of oil droplets and settling of water droplets. The build-up of dispersed droplets leads to the formation of a complex multiphase layer, commonly referred to as a rag layer, composed of water and oil droplets stabilized by fine solids, asphaltenes and surfactants. Over time its presence eventually impacts separation performance. Despite the major operational challenge posed by rag layers, the nature and formation mechanism of rag layers remains poorly understood.[58] A recent study on an industrial sample revealed that rag layers are very stable complex oil-continuous emulsions of typically high solids content.[8] Rag layer formation is commonly encountered in oil sands processing and the extraction of bitumen where asphaltene precipitation coupled with the presence of biwettable fine solids prevents droplet coalescence and phase separation. Although low-quality oil sands contain a high fines fraction which is frequently shown to promote the formation of stable liquid–liquid interfaces, the role of fines in rag layer formation is less clear. There are at least two possible mechanisms postulated to describe the formation of rag layers: (i) slow coalescence rate of water droplets compared with the

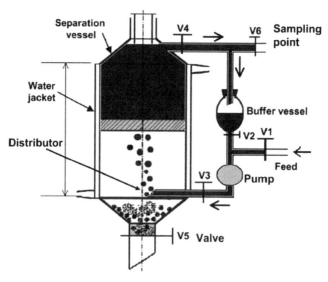

Figure 11.14 A schematic diagram of the vessel for generation of petroleum rag
layer materials.
Reprinted with permission from Gu *et al.*[60] © 2007 American
Chemical Society.

accumulation rate; and (ii) the formation of a solid barrier to settling materials due to the accumulation of biwettable fines at the oil–water
interface.[59]

To elucidate the mechanism of rag layer formation, a novel rag layer
generator (Figure 11.14) was designed to form well-controlled laboratory
samples through washing of solids-free naphtha-diluted bitumen froth.[60]
Despite the absence of inorganic solids, an ample amount of rag layer was
formed from naphtha-diluted bitumen froth. After fractionation of rag layer
materials into chloroform-insoluble, heptane-insoluble and heptane-soluble
fractions, it was determined by detailed characterization using elemental
analysis, FTIR characterization and TGA that the rag layer consists of 27 wt%
dry rag materials, 29.4 wt% naphtha and 43 wt% water. Among the dry rag
materials, chloroform-insoluble accounts for 45–49 wt%, while the asphaltene content ranges from 23 to 56 wt% of the chloroform-soluble fractions,
depending on the naphtha to bitumen ratio used for rag layer production.
The heteroatom concentration of asphaltenes isolated from the rag layer was
found to be twice as high as that from crude. The chloroform-soluble
components and asphaltenes from the rag layer exhibited the greatest
interfacial stability.

With repeated centrifugation of water-in-diluted bitumen froth emulsions,
Czarnecki *et al.*[58] showed that the rag layer is a mixture of clay particle-
mediated multiple emulsions. Saadatmand *et al.*[59] prepared rag layer
samples by diluting bitumen froth (with mixtures of heptane and toluene)
obtained from oil sands extraction using the Denver flotation cell. The

diluted froth was then centrifuged to assess possible mechanisms for rag layer stability. The type of solvent and its effect on asphaltene solubility were shown to govern rag layer formation. During the centrifugation, a mechanical barrier was speculated to accumulate at the oil–water interface due to the presence of oil-wet materials. The small size water droplets are known to be stabilized by oil-wet fine solids and asphaltene nanoaggregates. Slow coalescence of these emulsified water droplets results in their accumulate at the oil–water interface, forming a barrier that prevents water and fine solid particles passing through the interface and hence promotes the formation of rag layers. Such mechanical barrier was absent in a 50:50 heptane to toluene mixture, in which asphaltenes are highly soluble.

Addition of poor solvent is known to increase asphaltene precipitation, which promotes rag layer formation. A similar observation was made in a recent study by Kiran *et al.*[61] In the presence of water, poor solvents such as 80:20 heptol allowed asphaltenes to partition at the liquid–liquid interface, thereby increasing emulsion stability. In contrast, the addition of good solvents such as 50:50 heptol reduced rag layer formation. Good solvents are known to lower the affinity of asphaltenes to partition at the oil–water interface, exhibiting less potency for rag layer formation. The study by Saadatmand *et al.*[59] concluded that crude containing a greater amount of fine clays leads to the formation of significantly thicker rag layers. Accumulation of fine solids at the oil–water interface, which stabilizes water-in-crude emulsions, was found to be responsible for the greater rag layer formation. Jiang *et al.*[62] revealed the formation of a persistent rag layer when clays are present. Although the natural clays are considered hydrophilic (water-wettable), the clays in contact with crude oil contain toluene-soluble organic content, making them biwettable and hence promoting rag layer formation through the stabilization of multiple emulsions. Sodium silicate addition was found to make clay solids in a crude system more water-wettable, decreasing the volume of rag layers during a given crude–water separation.[62]

In a recent study, focus was placed on characterizing solids from a rag layer sample taken from the secondary cyclone overflow of a naphthenic bitumen froth treatment plant.[8] The received rag layer was found to be fragile and easily destroyed by handling. However a thick rag layer as much as 40 vol% of the total sample could be reformed even after vigorous mechanical agitation and then left undisturbed at room temperature for 14 days. The rag layer had a density of 960–980 kg m^{-3}, intermediate to the density of the oil phase and the water phase of 860 kg m^{-3} and 1000 kg m^{-3}, respectively. The rag layer contained 23–29 wt% water and 15 wt% solids. Solids recovered from the rag layer were found to be more hydrophobic than the solids recovered from the water layer, indicating the critical importance of particle wettability in stabilizing emulsions and hence inducing rag layer formation. The solids from the water layer consisted mostly of kaolinite and illite. In contrast, an elevated level of siderite and pyrite was measured in the solids isolated from the rag layer. The solids recovered from the rag layer appeared to be more severely contaminated with organic matter, exhibiting a stronger hydrophobicity. Chemical binding of carboxylic acid groups in heavy oil with metallic sites on clays and iron-containing siderite appeared to be the cause of severe contamination and significant wettability alteration. These biwettable fine particles attached strongly to the oil–water interface. Collectively with asphaltene nanoaggregates, the presence of hydrophobic particles contributed to rag layer formation and stabilization. Understanding the formation mechanism and role of particle wettability in promoting rag layer formation provides directions on mitigating rag layer formation.

To summarize, the contact of clay particles with petroleum leads to adsorption of indigenous components of petroleum such as asphaltenes, resins and humic acids. Subsequent contamination renders the clay particles highly biwettable. The particle-stabilized emulsions in petroleum are most likely a result of collective actions from contaminated biwettable clay particles and asphaltene nanoaggregates in the presence of resins. The emulsions formed through such complex mechanisms lead to great difficulties in petroleum processing and handling, as a result of the formation of so-called rag layers as described below.

11.6   Demulsification

The accumulation of asphaltene nanoaggregates and biwettable fine particles at the oil–water interface generates emulsions of extreme stability. These emulsions will remain stable for prolonged periods (longer than separator residence times) with little or no separation. To enhance the driving force for separation it is desirable to increase the droplet size either by: (i) droplet–droplet flocculation; and/or (ii) droplet–droplet coalescence. With a squared dependence on droplet diameter, the free-settling Stokes' equation indicates that the settling velocity increases four-fold as the droplet diameter doubles. To promote fruitful droplet interactions, the armoured films surrounding droplets have to be disrupted by molecules that are interfacially active and can compete for available liquid–liquid interfacial area. These molecules are referred to as demulsifiers and are frequently added into the process to minimize the formation and break-up of problematic emulsions and rag layers.

11.6.1   PO/EO Demulsifiers

The commonly used demulsifiers in petroleum processing include co-polymers of PO (propylene oxide) and EO (ethylene oxide). Xu et al.[63] studied this category of demulsifier by grafting PO/EO copolymers onto DETA (diethylenetriamine) as shown in Figure 11.15. The copolymers with dendrimer structures of an intermediate molecular weight (*i.e.* 7500–15,000 Da) were found effective at breaking W/O emulsions. These authors investigated demulsification performance of the copolymer by changing the EO:PO ratio. For PO ≫ EO, demulsification performance was shown to be poor at low

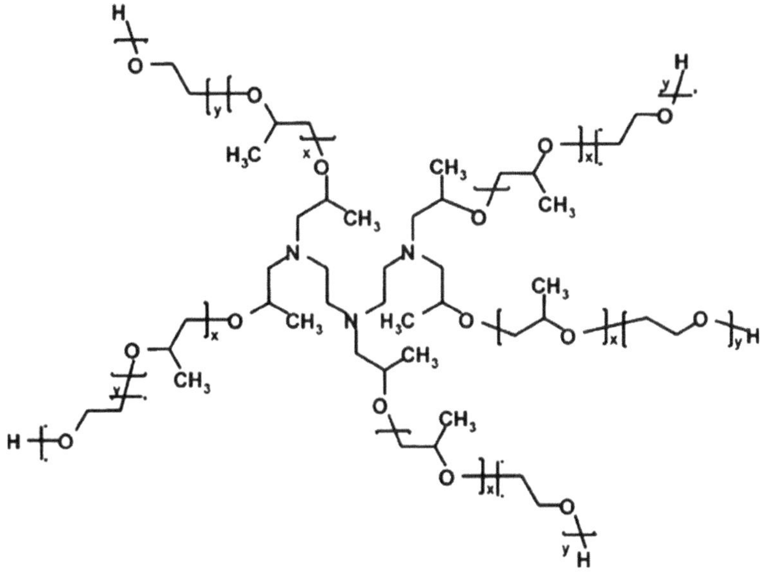

Figure 11.15 Chemical structure of DETA-based block copolymer with EO/PO units. Reprinted with permission from Xu *et al.*[63] © 2005 American Chemical Society.

demulsifier concentrations and effective at high demulsifier concentrations. For PO ≪ EO, the demulsifier was shown to be effective at low concentrations but overdosed at high concentrations to produce a stable middle-phase emulsion. For PO ≈ EO, effective emulsion separation was observed at low demulsifier concentrations with no concerns of overdoing and no formation of a stable middle-phase emulsion at high demulsifier concentrations. The authors concluded that the demulsifiers of balanced PO/EO number provided optimum demulsification performance.

The hydrophilic and hydrophobic nature of the EO/PO copolymer promotes solubility in both liquid phases and hence a capability to partition at the liquid–liquid interface. The hydrophilic–lipophilic balance (HLB) describes the potential of the surfactant to control the emulsion type. For demulsification, the HLB value of the surfactant should be in the region of 7–9. Due to the definition of the HLB concept, it is not easy to measure the HLB value of demulsifier molecules due to lack of exact structure of the molecules. Alternatively, relative solubility numbers (RSNs) can be used to represent the hydrophobicity of large molecules such as PO/EO demulsifiers. The RSN values of the demulsifiers are measured using ethylene glycol dimethyl ether and toluene as titration solvents. In this method, 1 g of demulsifier is dissolved in 30 mL of solvent consisting of toluene and ethylene glycol dimethyl ether. The resultant solution is titrated with deionized water until the solution becomes turbid. The volume of water (mL) titrated is recorded as the RSN.

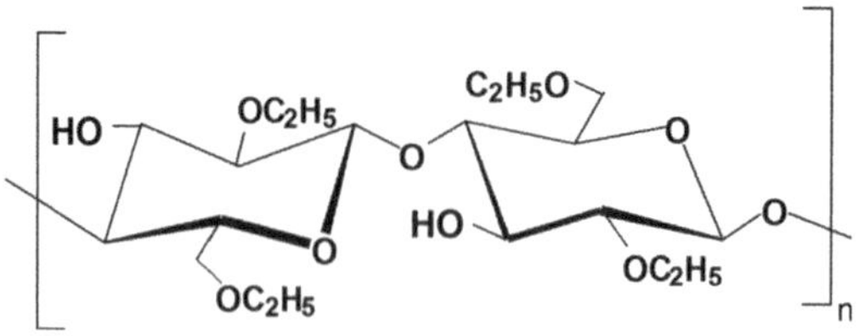

Figure 11.16   Structure of EC with a degree of substitution of 2.
Reprinted with permission from Feng *et al.*[64] © 2009 American
Chemical Society.

For an equivalent PO number in an EO/PO copolymer of increasing EO
content, the interfacial activity of the copolymer increases, further decreasing the water–oil interfacial tension with increasing RSN. Xu *et al.*[63]
reported that the demulsification performance of EO/PO copolymers increased with a decrease in the bitumen–water interfacial tension. Hence a
good demulsifier should reduce the interfacial tension by partitioning at the
oil–water interface. However, it is important to note that the magnitude by
which the interfacial tension decreases is not the sole guide for demulsification performance. Other parameters such as interfacial rheology, film
pressure and film structure also need to be considered.

11.6.2   Demulsification Mechanism

In several recent studies, ethylcellulose (EC) has been identified as a suitable
biodegradable demulsifier for bitumen froth.[64,65] EC is a polymeric material
produced by treating alkali cellulose with chloroethanol, resulting in a cellulose ester. In a typical EC molecule, the hydroxyl and oxyethyl side chains
are either equatorial or axial, located on both sides of the ring as shown in
Figure 11.16. The structure of EC makes it distinguishable from conventional linear or star-like EO/PO copolymer demulsifiers. Various grades
of EC have been synthesized with different molecular weights and degrees of
substitution. By controlling the cellulose/sodium hydroxide/iodoethane
molar ratio in the etherification reaction, EC polymers of varying molecular
weight and hydroxyl content can be synthesized.[66]

EC is miscible with organic solvent and is interfacially active. The interfacial activity of EC is attributed to the amphiphilic nature of its molecular
structure of hydrophilic cellulose backbone and the side chain of hydrophobic ethyl substituents. On the basis of the monomer unit of EC and
Davies formula,[67] the HLB value of EC was estimated to be around 8, which
is within the range for W/O demulsifiers. Feng *et al.*[64] studied the interfacial
tension between naphtha and water in the presence and absence of 60 wt%
bitumen in naphtha solution as a function of EC concentration, as shown in
Figure 11.17a. Clearly, EC is more efficient at reducing the naphtha–water
interfacial tension than reducing naphtha-diluted bitumen (naphtha/
bitumen mass ratio = 0.6)-water interfacial tension, indicating that EC

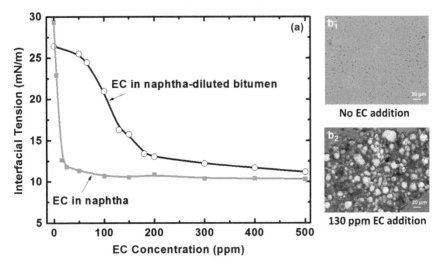

Figure 11.17 (a) Interfacial tension between naphtha and water in the presence and absence of 60 wt% bitumen as a function of EC concentration at 20 °C. (b) Micrographs of emulsions without demulsifier addition and 130 ppm EC addition. Samples were taken at 6.5 cm from the emulsion surface (bottom of the test tube).
Reprinted with permission from Feng *et al.*[64] © 2009 American Chemical Society.

molecules have to compete with the indigenous surface active species in bitumen to lower the naphtha-diluted bitumen-water interfacial tension. At an appropriate demulsifier dosage EC is able to coalesce water droplets (Figure 11.17b$_2$) causing a 5- to 10-fold increase in the droplet diameter than the case without EC addition (Figure 11.17b$_1$), and thus promote a rapid decrease in water content.

To understand the role of molecular structure on demulsification performance, Feng *et al.*[66] synthesized six different EC demulsifiers of different molecular weights and degrees of hydroxyl substitution (DHS). The demulsification performance of EC was shown to depend on both the hydroxyl content and molecular weight. For an equivalent molecular weight (EC3, EC4, EC5 = 143,000 Da), greater water removal was measured with increasing degrees of hydroxyl substitution (EC5, DHS = 1.33; EC4, DHS = 1.70; EC3, DHS = 2.14). The dependence on hydroxyl substitution was evaluated at high demulsifier concentrations (290 ppm). The observed dewatering performance is reasonable because higher DHS increases the interfacial activities of EC due to the decreased hydroxyl content. However, excessive hydroxyl substitution (EC1 = 45,000 Da, DHS = 2.79) increases the solubility of the EC molecules in the oil phase and reduces the effectiveness of the demulsifier to partition the oil–water interface. Such drop-off in performance is highlighted by comparing the performance of EC1 and EC2 (EC2 = 45,000 Da, DHS = 2.40) demulsifiers (Figure 11.18). Comparing

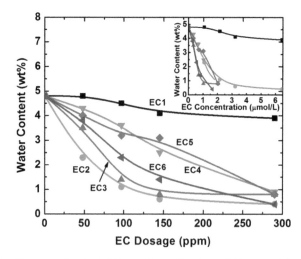

Figure 11.18   Water content at 2.5 cm from the top of the emulsion surface as a function of EC dosage for six EC demulsifiers.
Reprinted with permission from Feng *et al.*[66] © 2011 American Chemical Society.

performance on the basis of molar concentration (data shown in inset of Figure 11.18), EC demulsifiers of similar hydroxyl content and higher molecular weight (EC3 = 143,000 Da and EC6 = 177,000 Da) were shown to provide the best demulsification performance. EC1 with the lowest hydroxyl content and lowest molecular weight was shown to be the worst demulsifier.

To better understand the demulsification performance based on molecular weight and hydroxyl substitution, the authors measured the water-diluted bitumen interfacial tension as a function of EC concentration (data not shown). On a molar basis, in general, the demulsifiers of high molecular weight were shown to be more effective in reducing interfacial tensions. However, no clear correlation could be established between interfacial tension and dewatering performance. The lack of correlation was attributed to the hydroxyl content that governs the molecular conformation of the EC molecule and its ability to flocculate water droplets. Higher hydroxyl content would lead to stronger intra- and intermolecular interactions between cellulose molecules, causing the molecule to self-coil, thus decreasing its ability of bridging flocculation.

To further elucidate the demulsification mechanism by EC molecules, droplet–droplet interactions were visualized *in situ* using the micropipette technique.[64] Figure 11.19 shows sequential still images of two micron-sized water droplets in naphtha-diluted bitumen with and without EC addition. In the absence of EC, two micron-sized water droplets were brought in contact for a few minutes under a given applied compression force and did not coalesce. Upon release of the compression force the two droplets separated and returned to their original spherical shape with no visible droplet deformation under contact or retraction (Figure 11.19 A1 and A2). The

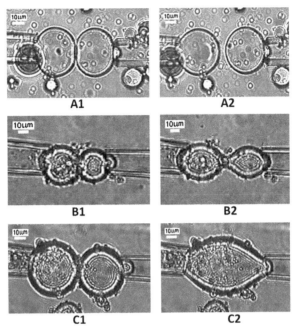

Figure 11.19 Interactions of water droplets visualized by the micropipette technique. Water droplets were present in 0.1 wt% naphtha-diluted bitumen emulsion. (A) no demulsifier addition, (B) 35 ppm EC addition and (C) 130 ppm EC addition.
Reprinted with permission from Feng *et al.*[64] © 2009 American Chemical Society.

characteristic nature by which the droplets reside in contact and remain spherical upon retraction indicates that there was no attractive force (coagulation) between the two droplets. The lack of interaction between the two droplets is a result of a steric barrier surrounding the water droplets. Wang *et al.*[68] reported strong steric repulsive forces between two asphaltene-coated surfaces in toluene and heptane/toluene mixtures, measured by atomic force microscopy (AFM). The repulsion was attributed to the asphaltene brush-like layers resisting adhesion unless a critical compressive force is exceeded.[25]

Figure 11.19 B1 and B2 shows the interaction between two water droplets in the presence of 35 ppm EC. Applying a compressional force the two droplets remained stable and did not coalesce (Figure 11.19 B1). However, when the compressional force was removed the water droplets adhered to each other and stretched substantially before final separation (Figure 11.19 B2). The strong adhesion confirms flocculation of the two droplets in the presence of EC without coalescence. By increasing the concentration of EC further to 130 ppm, the two water droplets flocculated immediately upon contact (Figure 11.19 C1) and coalesced shortly to one larger droplet (Figure 11.19 C2). The coalescence time occurred over several seconds, with the coalescence rate increasing with higher demulsifier concentration.

From the micropipette study it was hypothesized that the concentration dependence on flocculation and coalescence behaviour was closely related to the degree of interfacial material substitution by EC molecules. Wang *et al.*[69] considered the interaction between adsorbed asphaltenes and EC molecules in solution by quartz crystal microbalance with dissipation monitoring (QCM-D) and AFM. From their QCM-D studies the authors noted the positive interaction between EC molecules and the asphaltene layer coated on a silica substrate, gradually transforming the adsorbed layer over several hours. Washing of the newly formed film resulted in no mass loss, indicating that the new layer was irreversibly adsorbed. With limited structural information the authors supplemented their findings with AFM imaging of adsorbed asphaltene layers soaked in EC solutions for up to 7 hours. Figure 11.20 AE-0 ($t = 0$ hours) shows the topographical features of asphaltenes adsorbed on hydrophilic silica. The asphaltenes are randomly distributed in the form of close-packed colloid-like nanoaggregates with a mean square roughness of ~1 nm. With increasing soaking time (Figure 11.20 AE-6, $t = 6$ hours) in the EC solution the topography of the asphaltene film shows two distinct changes: some aggregates grew in size while the discrete flat areas expanded on the surface. Those discrete flat areas are considered as a result of adsorption of EC molecules on silica (data not shown). After 7 hours of soaking (Figure 11.20 AE-7, $t = 7$ hours) the changes in film properties become even more apparent, showing higher rough domains (larger aggregates) and much expanded flat open areas.

The process of EC displacing asphaltenes from a silica surface is illustrated in Figure 11.21. Initially, the more surface active EC molecules adsorb on the silica surface at the defects of the asphaltene layer. The distribution of polar groups on asphaltenes is not dense and nor uniform, hence the asphaltene layer cannot occupy all the binding sites on a substrate. As such, opportunity is provided for sequential adsorption of EC molecules. This initial EC adsorption step is rapid. Compared with asphaltenes, the polar binding groups on EC molecules are more uniformly distributed along the polymeric chains and of higher density. As a result, each EC molecule can

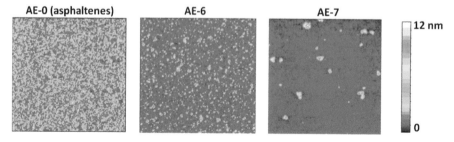

Figure 11.20   AFM topography images of asphaltene-coated sample surfaces soaked in EC-in-toluene solution for 0–7 hours.
Reprinted with permission from Wang *et al.*[69] © 2011 American Chemical Society.

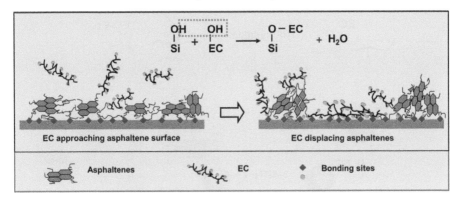

Figure 11.21 Schematics of asphaltene displacement by EC on a hydrophilic solid surface.
Reprinted with permission from Wang *et al.*[69] © 2011 American Chemical Society.

form multiple binding sites with the substrate. Because the EC molecules interact more strongly with the substrate, the EC molecules gradually expand, displacing the asphaltene layer and weakening the binding between asphaltenes–asphaltenes and asphaltenes–substrate. The driving force for EC to displace the adsorbed asphaltenes is attributed to its higher number of stronger hydrogen bonding sites toward the hydroxyl groups on the hydrophilic substrate. This process gradually squeezes the asphaltene aggregates into small areas of larger size domains of significantly greater thickness. Once these voids are created in the film there is opportunity for water to connect and hence promote droplet coalescence. Although the mechanism for asphaltene displacement by EC molecules has been demonstrated at the solid–liquid interface, the system is a good analogue for the liquid–liquid system. Asphaltene displacement and the formation of voids in the film has also been confirmed by imaging (AFM) Langmuir–Blodgett films transferred from the liquid–liquid interface after EC addition.[64,70]

11.6.3 Magnetic Responsive EC Demulsifier (M-EC)

One drawback from chemical demulsification is the continual supply of demulsifier to treat the petroleum emulsion. Recently, a novel approach to chemical demulsification has been demonstrated through the synthesis of a magnetically responsive demulsifier that can be manipulated under a magnetic field and hence recovered for reuse.[71,72] The concept of synthesizing magnetic-ethylcellulose (M-EC) nanoparticles is illustrated in Figure 11.22. Firstly, magnetite (Fe_3O_4) nanoparticles are coated with a thin layer of silica using a dense liquid silica coating method to protect the magnetic particle and make the surface amenable for further functionalization. The silica-coated magnetic nanoparticles are then modified by a silane coupling agent, 3-aminopropyltriethoxysilane (3-APTES), to render the

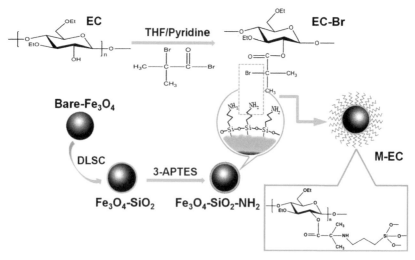

Figure 11.22   Schematic illustration of the synthesis of magnetic ethylcellulose (M-EC) nanoparticles (M-EC).
Reprinted with permission from Peng *et al.*[72] © 2012 Wiley.

surface of amine functionality. EC molecules are modified through an esterification reaction of acryl halide (O=C-Br) with hydroxyl (–OH) to replace some hydroxyl groups of EC with 2-bromoisobutyryl bromide, forming bromoesterified EC, or EC-Br. The modified EC interacts through the chemical reaction between –NH_2 groups on the Fe_3O_4-SiO_2-NH_2 surface and bromine on EC-Br, chemically anchoring onto the amine-functionalized Fe_3O_4 nanoparticle (Fe_3O_4-SiO_2-NH_2) surface to form a polymer EC layer, leading to the formation of M-EC. As prepared, M-EC nanoparticles possess a strong magnetic core with an interfacially active organic EC layer on the surface.[72]

Figure 11.23a compares demulsification performance of a 5 wt% water-in-naphtha-diluted bitumen emulsion (naphtha/bitumen mass ratio = 0.65) by M-EC and conventional chemical EC at 130 ppm. The control experiment (blank without demulsifiers) shows a negligible difference in the water content along the test tube after 1 hour of settling, remaining at around 4.7 wt%. This observation indicates a very stable W/O emulsion without noticeable water separation under gravity alone. In contrast, the addition of 1.5 wt% M-EC with magnetic separation reduced the water content of the emulsion to less than 0.4 wt% at a location of 6 cm below the emulsion surface. Thereafter, the water content increased sharply to well above 20 wt%. The results clearly show that more than 90% of the water in the original emulsion settled under magnetic forces to less than 14% of the original emulsion volume. In comparison to a water content of 18.5 wt% at the depth of 6 cm for the case with EC addition, the addition of M-EC clearly shows a better water separation into a much smaller volume of sludge and less oil loss because of the enhanced separation of M-EC tagged water

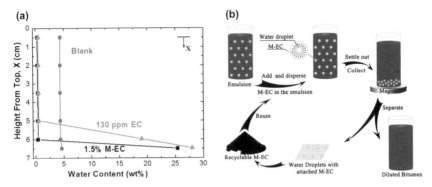

Figure 11.23 (a) Water content at different depths of diluted bitumen emulsions treated with 1.5 wt% M-EC or 130 ppm EC after settling on a hand magnet for 1 hour at 80 °C. (b) Schematic illustration of the demulsification process and recycling tests using M-EC as demulsifiers. Reprinted with permission from Peng *et al.*[71] © 2012 Wiley.

droplets by magnetic forces. The rate of M-EC water separation was of the order of seconds rather than minutes or hours. A further study showed the potential regeneration and reuse of M-EC (Figure 11.23b) by effective (~90%) water removal from a water-in-diluted bitumen emulsion over 10 consecutive cycles. With improved demulsification efficiency and kinetics, the reusable M-EC demulsifier is also economically attractive for continuous operations. Preliminary studies showed that the demulsification performance could be extended to industrial bitumen froths, with the original water content in the froth reduced to below 0.5 wt% in 5 minutes at room temperature.[71]

11.7 Conclusions

In petroleum production there are many operational challenges that have galvanized a research community over several decades. One of the greatest challenges is to achieve rapid and efficient separation of problematic emulsions that form due to the accumulation of surface active species (indigenous oil compounds and fine biwettable particles) which stabilize liquid–liquid interfaces and prevent fruitful droplet–droplet interactions. To enhance separation, chemical demulsifiers are added to compete for available interfacial sites and disrupt the rigid interfacial networks. It is important to note that this chapter is by no means exhaustive and is not meant to cover all aspects of the emulsification and demulsification problems facing the petroleum industry. Work has been selected to provide an overview of the key fundamental mechanisms that contribute to the formation and break-up of emulsions and provide scientific directions. It is our intention to clearly illustrate the importance of understanding basic science in emulsion stabilization before tackling the technology for effective demulsification.

References

1. ExxonMobil, *The Outlook for Energy: A View to 2040*, Exxon Mobil Corporation, Irving, TX, 2013.
2. W. Ramsden, *Proc. R. Chem. Soc.*, 1903, **72**, 156–164.
3. U. Pickering, *J. Chem. Soc. Trans.*, 1907, **91**, 2001–2021.
4. J. H. Schulman and J. Leja, *Trans. Faraday Soc.*, 1954, **50**, 598–605.
5. R. J. G. Lopetinsky, J. H. Masliyah and Z. Xu, in *Colloidal Particles at Liquid Interfaces*, ed. B. P. Binks, Cambridge University Press, New York, 2006.
6. D. E. Tambe and M. M. Sharma, *Adv. Colloid Interface Sci.*, 1994, **52**, 1–63.
7. R. Aveyard, B. P. Binks and J. H. Clint, *Adv. Colloid Interface Sci.*, 2003, **100-102**, 503–546.
8. M. M. Kupai, F. Yang, D. Harbottle, K. Moran, J. Masliyah and Z. Xu, *Can. J. Chem. Eng.*, 2013, **91**, 1395–1401.
9. B. P. Binks and P. D. I. Fletcher, *Langmuir*, 2001, **17**, 4708–4710.
10. A. Kumar, B. J. Park, F. Tu and D. Lee, *Soft Matter*, 2013, **9**, 6604–6617.
11. S. Jiang and S. Granick, *J. Chem. Phys.*, 2007, **127**, 161102.
12. R. I. Rueda-Velásquez, H. Freund, K. Qian, W. N. Olmstead and M. R. Gray, *Energy Fuels*, 2013, **27**, 1817–1829.
13. O. C. Mullins, H. Sabbah, J. Eyssautier, A. E. Pomerantz, L. Barré, A. B. Andrews, Y. Ruiz-Morales, F. Mostowfi, R. McFarlane, L. Goual, R. Lepkowicz, T. Cooper, J. Orbulescu, R. M. Leblanc, J. Edwards and R. N. Zare, *Energy Fuels*, 2012, **26**, 3986–4003.
14. O. C. Mullins, *Energy Fuels*, 2010, **24**, 2179–2207.
15. J. P. Dickie and T. F. Yen, *Anal. Chem.*, 1967, **39**, 1847–1852.
16. O. P. Strausz, T. W. Mojelsky and E. M. Lown, *Fuel*, 1992, **71**, 1355–1363.
17. M. R. Gray, R. R. Tykwinski, J. M. Stryker and X. Tan, *Energy Fuels*, 2011, **25**, 3125–3134.
18. H. W. Yarranton, D. P. Ortiz, D. M. Barrera, E. N. Baydak, L. Barre, D. Frot, J. Eyssautier, H. Zeng, Z. Xu, G. Dechaine, M. Becerra, J. M. Shaw, A. M. McKenna, M. M. Mapolelo, C. Bohne, Z. Yang and J. Oake, *Energy Fuels*, 2013, **27**, 5083–5106.
19. J. P. Rane, D. Harbottle, V. Pauchard, A. Couzis and S. Banerjee, *Langmuir*, 2012, **28**, 9986–9995.
20. S. Acevedo, M. A. Ranaudo, C. García, J. Castillo and A. Fernández, *Energy Fuels*, 2003, **17**, 257–261.
21. A. W. Marczewski and M. Szymula, *Colloids Surf. Physicochem. Eng. Asp.*, 2002, **208**, 259–266.
22. J. Castillo, M. A. Ranaudo, A. Fernández, V. Piscitelli, M. Maza and A. Navarro, *Colloids Surf. Physicochem. Eng. Asp.*, 2013, **427**, 41–46.
23. K. Xie and K. Karan, *Energy Fuels*, 2005, **19**, 1252–1260.
24. H. Labrador, Y. Fernández, J. Tovar, R. Muñoz and J. C. Pereira, *Energy Fuels*, 2007, **21**, 1226–1230.
25. A. Natarajan, J. Xie, S. Wang, J. Masliyah, H. Zeng and Z. Xu, *J. Phys. Chem. C*, 2011, **115**, 16043–16051.

26. E. M. Freer and C. J. Radke, *J. Adhes.*, 2004, **80**, 481–496.
27. P. M. Spiecker and P. K. Kilpatrick, *Langmuir*, 2004, **20**, 4022–4032.
28. A. Yeung, T. Dabros, J. Masliyah and J. Czarnecki, *Colloids Surf. Physicochem. Eng. Asp.*, 2000, **174**, 169–181.
29. A. Yeung, T. Dabros, J. Czarnecki and J. Masliyah, *Proc. R. Soc. Math. Phys. Eng. Sci.*, 1999, **455**, 3709–3723.
30. X. Wu, *Energy Fuels*, 2003, **17**, 179–190.
31. J. Czarnecki and K. Moran, *Energy Fuels*, 2005, **19**, 2074–2079.
32. S. Gao, K. Moran, Z. Xu and J. Masliyah, *J. Phys. Chem. B*, 2010, **114**, 7710–7718.
33. Y. Fan, S. Simon and J. Sjöblom, *Colloids Surf. Physicochem. Eng. Asp.*, 2010, **366**, 120–128.
34. D. Harbottle, K. Moorthy, L. Wang, Q. Chen, S. Xu, Q. Lu, J. Sjoblom and Z. Xu, *Langmuir*, 2014, **30**, 6730–6738.
35. P. Erni, *Soft Matter*, 2011, 7, 7586–7600.
36. L. Wang, D. Sharp, J. Masliyah and Z. Xu, *Langmuir*, 2013, **29**, 3594–3603.
37. L. Wang, Z. Xu and J. H. Masliyah, *J. Phys. Chem. C*, 2013, **117**, 8799–8805.
38. P. Tchoukov, J. Czarnecki and T. Dabros, *Colloids Surf. Physicochem. Eng. Asp.*, 2010, **372**, 15–21.
39. J. Czarnecki, P. Tchoukov, T. Dabros and Z. Xu, *Can. J. Chem. Eng.*, 2013, **91**, 1365–1371.
40. J. Czarnecki, P. Tchoukov and T. Dabros, *Energy Fuels*, 2012, **26**, 5782–5786.
41. I. B. Ivanov and D. T. Dimitrova, in *Thin Liquid Films Fundamentals and Applications*, eds I. B. Ivanov and D. T. Dimitrova, Marcel Dekker Inc., New York, 1988.
42. A. M. McKenna, L. J. Donald, J. E. Fitzsimmons, P. Juyal, V. Spicer, K. G. Standing, A. G. Marshall and R. P. Rodgers, *Energy Fuels*, 2013, **27**, 1246–1256.
43. R. B. Teklebrhan, L. Ge, S. Bhattacharjee, Z. Xu and J. Sjöblom, *J. Phys. Chem. B*, 2014, **118**, 1040–1051.
44. P. Tchoukov, F. Yang, Z. Xu, T. Dabros, J. Czarnecki and J. Sjöblom, *Langmuir*, 2014, **30**, 3024–3033.
45. A. Natarajan, N. Kuznicki, D. Harbottle, J. Masliyah, H. Zeng, and Z. Xu, *Langmuir*, 2014, **30**, 9370–9377.
46. D. Fu, J. R. Woods, J. Kung, D. M. Kingston, L. S. Kotlyar, B. D. Sparks, P. H. J. Mercier, T. McCracken and S. Ng, *Energy Fuels*, 2010, **24**, 2249–2256.
47. M. Osacky, M. Geramian, D. G. Ivey, Q. Liu and T. H. Etsell, *Fuel*, 2013, **113**, 148–157.
48. K. L. Konan, C. Peyratout, J.-P. Bonnet, A. Smith, A. Jacquet, P. Magnoux and P. Ayrault, *J. Colloid Interface Sci.*, 2007, **307**, 101–108.
49. T. J. Bandosz, J. Jagiełło, B. Andersen and J. A. Schwarz, *Clays Clay Miner.*, 1992, **40**, 306–310.
50. N. El-Thaher and P. Choi, *Ind. Eng. Chem. Res.*, 2012, **51**, 7022–7027.

51. A. M. Shaker, Z. R. Komi, S. E. M. Heggi and M. E. A. El-Sayed, *J. Phys. Chem. A*, 2012, **116**, 10889–10896.

52. R. Baigorri, M. Fuentes, G. González-Gaitano, J. M. García-Mina, G. Almendros and F. J. González-Vila, *J. Agric. Food Chem.*, 2009, **57**, 3266–3272.

53. A. W. P. Vermeer and L. K. Koopal, *Langmuir*, 1998, **14**, 4210–4216.

54. Q. Zhou, P. A. Maurice and S. E. Cabaniss, *Geochim. Cosmochim. Acta*, 2001, **65**, 803–812.

55. Z. R. Komi, A. M. Shaker, S. E. M. Heggi and M. E. A. El-Sayed, *Chemosphere*, 2014, **99**, 117–124.

56. X. Feng, A. J. Simpson and M. J. Simpson, *Org. Geochem.*, 2005, **36**, 1553–1566.

57. A. W. P. Vermeer, W. H. van Riemsdijk and L. K. Koopal, *Langmuir*, 1998, **14**, 2810–2819.

58. J. Czarnecki, K. Moran and X. Yang, *Can. J. Chem. Eng.*, 2007, **85**, 748–755.

59. M. Saadatmand, H. W. Yarranton and K. Moran, *Ind. Eng. Chem. Res.*, 2008, **47**, 8828–8839.

60. G. Gu, L. Zhang, Z. Xu and J. Masliyah, *Energy Fuels*, 2007, **21**, 3462–3468.

61. S. K. Kiran, E. J. Acosta and K. Moran, *Energy Fuels*, 2009, **23**, 3139–3149.

62. T. Jiang, G. J. Hirasaki, C. A. Miller and S. Ng, *Energy Fuels*, 2011, **25**, 545–554.

63. Y. Xu, J. Wu, T. Dabros, H. Hamza and J. Venter, *Energy Fuels*, 2005, **19**, 916–921.

64. X. Feng, P. Mussone, S. Gao, S. Wang, S.-Y. Wu, J. H. Masliyah and Z. Xu, *Langmuir*, 2010, **26**, 3050–3057.

65. X. Feng, Z. Xu and J. Masliyah, *Energy Fuels*, 2009, **23**, 451–456.

66. X. Feng, S. Wang, J. Hou, L. Wang, C. Cepuch, J. Masliyah and Z. Xu, *Ind. Eng. Chem. Res.*, 2011, **50**, 6347–6354.

67. J. T. Davies, A quantitative kinetic theory of emulsion type. 1. Physical chemistry of the emulsifying agent. In: *Gas/Liquid and Liquid/Liquid Interfaces*, Proceedings of 2nd International Congress Surface Activity, Butterworths, London, 1957, vol. 1, pp. 426–438.

68. S. Wang, J. Liu, L. Zhang, J. Masliyah and Z. Xu, *Langmuir*, 2010, **26**, 183–190.

69. S. Wang, N. Segin, K. Wang, J. H. Masliyah and Z. Xu, *J. Phys. Chem. C*, 2011, **115**, 10576–10587.

70. J. Hou, X. Feng, J. Masliyah and Z. Xu, *Energy Fuels*, 2012, **26**, 1740–1745.

71. J. Peng, Q. Liu, Z. Xu and J. Masliyah, *Energy Fuels*, 2012, **26**, 2705–2710.

72. J. Peng, Q. Liu, Z. Xu and J. Masliyah, *Adv. Funct. Mater.*, 2012, **22**, 1732–1740.

Subject Index